Richard Owen, And Others

Report of a geological Reconnoissance of Indiana

Made during the Years 1859 and 1860, under the Direction of the late David Dale

Owen

Richard Owen, And Others

Report of a geological Reconnoissance of Indiana
Made during the Years 1859 and 1860, under the Direction of the late David Dale Owen

ISBN/EAN: 9783337192471

Printed in Europe, USA, Canada, Australia, Japan

Cover: Foto ©ninafisch / pixelio.de

More available books at **www.hansebooks.com**

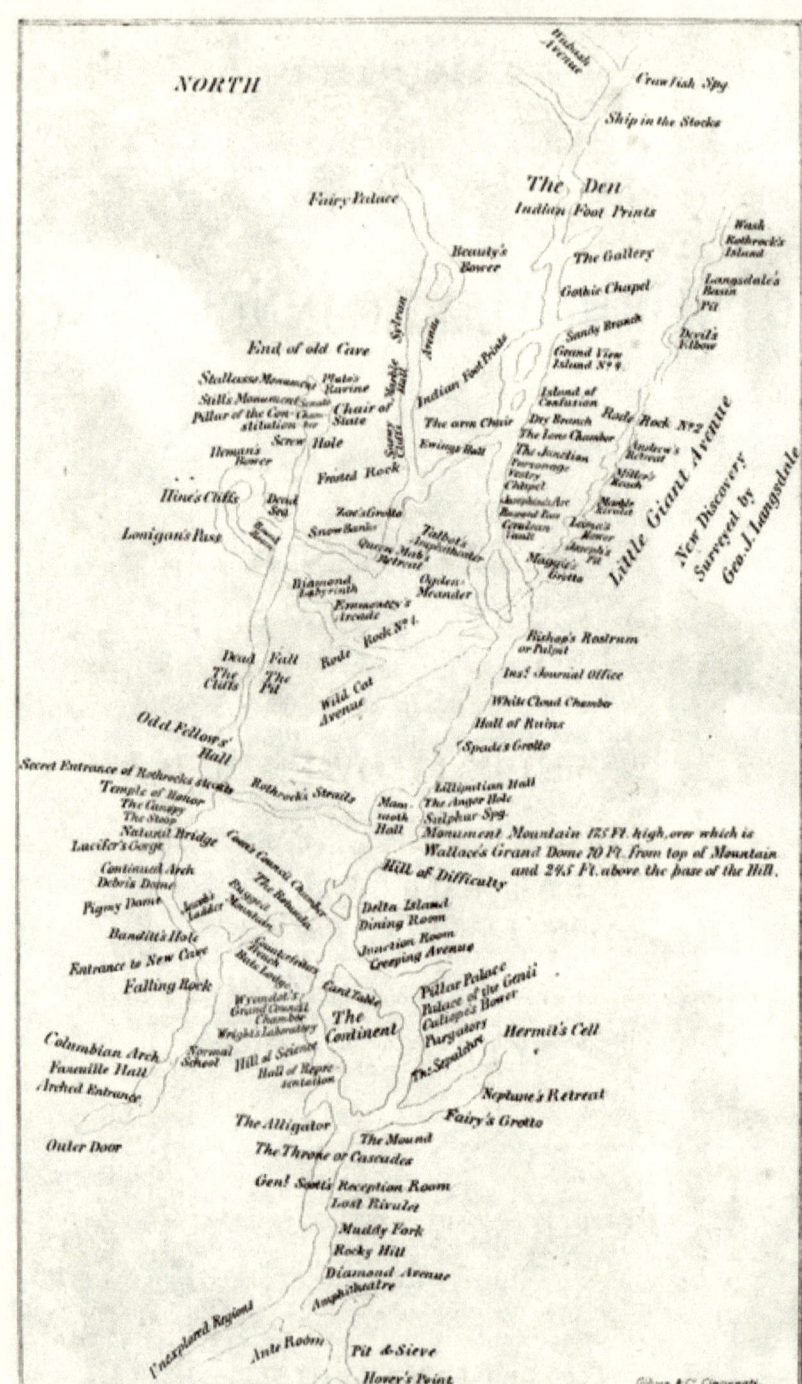

MAP OF WYANDOT CAVE.

REPORT

OF A

GEOLOGICAL RECONNOISSANCE

OF

INDIANA,

MADE DURING THE YEARS 1859 AND 1860, UNDER THE DIRECTION OF THE LATE

DAVID DALE OWEN, M. D.,

STATE GEOLOGIST,

BY RICHARD OWEN, M. D.,
PRINCIPAL ASSISTANT, NOW STATE GEOLOGIST.

ALSO, REPORTS ON THE ANALYSIS OF THE SOILS, BY R. PETER, M D., CHEMIST;
SURVEY OF THE COAL FIELDS, BY LEO LESQUEREUX, FOSSIL BOTANIST,
AND TOPOGRAPHICAL WORK, BY JOSEPH LESLEY,
TOPOGRAPHICAL GEOLOGIST.

PUBLISHED BY AUTHORITY OF THE INDIANA LEGISLATURE.

INDIANAPLOIS:
H. H. DODD & CO., BOOK PRINTERS.
1862.

TABLE OF CONTENTS.

	PAGE.
Prefatory Letter.	vii
Introduction—Objects of the Survey	ix

REPORT OF RICHARD OWEN.

CHAPTER I.

PRELIMINARY OBSERVATIONS	3
Laws governing geological investigations	4
Palæontology our chief guide in determining the relation of strata	7
Comparative anatomy the basis of Palæontology	8
On the question, whether superposed strata exhibit a succession of fossilized beings, apparently more highly organized in recent than in older deposits.	9
Tabular view of the Aqueous Rocks	14
Conspectus of the Igneous Rocks	16
Geographical distribution of formations over the world.	18
Geographical distribution of formations in Indiana	23

CHAPTER II.

DETAILS OF COUNTIES	27
SEC. 1.—Counties in the Lower Silurian Formation	30
Sub-Section 1.—General description of the formation	30
Sub-Section 2.—The resulting soil; its analysis, adaptation, &c.	31
Sub-Section 3.—Rock quarries	33
Sub-Section 4.—Metallic ores, &c.	35
Sub-Section 5.—Growth of timber and other predominant vegetation	36
Sub-Section 6.—Mineral springs, Artesian wells, &c.	37
Sub-Section 7.—Miscellaneous facts, as the prevalence of milk-sickness, potato rot, &c.	38
Sub-Section 8.—Characteristic fossils	39
A. Radiates	39
B. Mollusks	39
C. Articulates	40
D. Vertebrates	40
Sub-Section 9.—A more detailed description of each county in this formation	40
Wayne county	40
Union county	44

	PAGE.
Fayette county	45
Franklin county	49
Dearborn county	50
Ripley county	52
Ohio county	53
Switzerland county	55
Sec. 2.—Counties in the Upper Silurian Formation	57
Sub-Section 1.—General description	57
Sub-Section 2.—Soil, &c	58
Sub-Section 3.—Rock quarries	59
Sub-Section 4.—Metallic ores	60
Sub-Section 5.—Timber and predominant vegetation	60
Sub-Section 6.—Springs	61
Sub-Section 7.—Miscellaneous facts	61
Sub-Section 8.—Characteristic fossils	62
A. Radiates	62
B. Mollusks	62
C. Articulates	63
D. Vertebrates	63
Sub-Section 9.—A more detailed description of each county	63
Adams and Wells counties	63
Huntington county	66
Wabash county	67
Miami county	72
Jay and Blackford counties	74
Grant and Howard counties	75
Delaware county	76
Madison county	77
Randolph county	80
Henry and Hancock counties	82
Rush county	85
Decatur county	86
Jennings county	88
Jefferson county	88
Sec. 3.—Counties in the Devonian System	92
Sub-Section 1.—General description	92
Sub-Section 2.—Soil, &c	93
Sub-Section 3.—Rock quarries	93
Sub-Section 4.—Metallic ores	93
Sub-Section 5.—Timber and predominant vegetation	93
Sub-Section 6.—Springs	93
Sub-Section 7.—Miscellaneous facts	94
Sub-Section 8.—Characteristic fossils	94
A. Radiates	94
B. Mollusks	95
C. Articulates	95
D. Vertebrates	95

	PAGE.
Sub-Section 9.—A more detailed description of each county	95
Cass county	95
Carroll county	97
Tipton and Hamilton counties	101
Shelby county	103
Bartholomew county	104
Jackson and Scott counties	104
Clarke county	106
SEC. 4¹.—Counties in the Sub-Carboniferous Sandstone formation	108
Sub-Sections 1 to 9	108
Tippecanoe county	111
Clinton and Boone	113
Marion county	114
Hendricks and Johnson counties	116
Morgan and Brown counties	117
Washington and Floyd counties	120
SEC. 4².—Counties in the Sub-Carboniferous limestone	124
Sub-Sections 1 to 9	124
Montgomery county	132
Putnam county	134
Monroe county	135
Lawrence county	137
Orange and Harrison counties	140, 146
Crawford county	149
SEC. 4³.—Counties in the Coal Measures	160
Sub-Sections 1 to 9	160
Warren county	163
Fountain and Parke counties	165
Vermillion and Clay counties	167
Vigo and Owen counties	170
Greene and Sullivan counties	171
Martin and Daviess counties	173
Knox and Dubois counties	178
Pike and Gibson counties	180
Perry and Spencer counties	181
Warrick and Vanderburgh counties	189
Posey county	190
SEC. 5.—Counties in the Drift or Quaternary formation	192
Sub-Sections 1 to 9	192
Steuben, LaGrange and Elkhart counties	198
St. Joseph and LaPorte counties	199
Porter and Lake counties	204
DeKalb and Noble	207
Kosciusko, Marshall and Starke counties	208
Jasper and Newton	211
Allen and Whitley counties	214
Fulton and Pulaski counties	217
White and Benton counties	218

CHAPTER III.

PHYSICAL GEOGRAPHY.—

	PAGE
SEC. 1.—Water sheds and plateaus	223
SEC. 2.—Valleys and hydrographic basins, including prairies, sand plains, lakes, &c	228
Remarks on miscellaneous subjects connected with the survey	240

DR. PETER'S REPORT.

Remarks on Agricultural Chemistry	245
Chemical analysis of Indiana soils	249
SEC. 1.—Soils from the Lower Silurian formation	249
SEC. 2.—Soils from the Upper Silurian formation	252
SEC. 3.—Soils from the Devonian formation	255
SEC. 4.—Soils from the Sub-Carboniferous formation	258
SEC. 5.—Soils from the Coal Measures group	259
SEC. 6.—Soils from the Quaternary formation	263
Tabular views of soils analyzed	266

REPORT OF PROF. LESQUEREUX.

Introductory remarks	273
Directions for searching for coal	275
Quality of the coal and its value	279
Geological horizon of the coal strata of Indiana	291
Connected section of Coal Measures	299
Conclusions	341
Report of Mr. J. Lesley, Topographical Geologist	343
Appendix—Tabular views, &c., useful for reference	347

PREFATORY LETTER.

To the Members of the State Board of Agriculture,
Indianapolis, Indiana:

GENTLEMEN:—The Legislature of Indiana having, on your recommendation, passed an act for a Geological Reconnoissance of the State, approved March 5, 1859, which should "prepare the way for a more full and systematic system herafter," and having appropriated $5,000 "for the present purpose of making the geological reconnoissance, collections and analysis of specimens of minerals, ores, earths and stones," placed the whole under your control. Your Board, in accordance with the above act, secured the services of my lamented brother, late State Geologist of Kentucky and Arkansas, as well as of this State, to superintend the work. He accepted with hesitation, as you will remember, because he was still engaged in the field explorations of Arkansas, and had also his last report to make to Kentucky; but he finally consented with the understanding that, until those duties were completed, I would perform most of the field work and report on the same. In the condensed report submitted to you, the promise was made by him to prepare those details of our field work by general observations from himself on agricultural chemistry and milk sickness, particularly the connection of the latter with peculiar geological formations.

But his untimely death arrested the labors continued with unflagging perseverance until within three days of his decease, and to which he had so entirely devoted his life, as not to permit himself the relaxation necessary for health; and the world was thus too soon deprived of his valuable services. His own State, more especially, will feel and deplore his loss; even the two articles above alluded to, as expected from his pen, might have greatly promoted the health of our population, and increased the wealth derivable from our soil, through the useful practical suggestions designed to be conveyed.

The report, thus necessarily deprived of a leading feature, must lose part of its interest; and even those portions, for which I am alone responsible, are rendered less complete and more subject to inadvertancies than they would have been under circumstances less trying. When to these we add the fact that the administration of my late brother's affairs require a settlement with three States, and a general superintendence of the work so nearly completed by his exertions, perhaps those to whom you present this report may find some other excuse for apparent meagreness or omissions, than the deficiency of our State in objects of mineral and agricultural interest and wealth.

The result of my labors, orginally addressed to my late brother, I have now the honor to present to you.

The other accompanying reports from the distinguished gentlemen who consented to undertake the departments in which they are preeminent, could receive no additional lustre from my commendations; but, through you, I would tender to them my late brother's unfeigned admiration and warm feeling of obligation.

Permit me, gentlemen of the State Board, in closing these prefatory remarks, to express my sense of indebtedness to you for prompt and efficient assistance, as well as for personal generous hospitality.

Throughout the entire survey the courtesy with which we were also aided by those to whom we have been referred, sometimes even by entire strangers, was highly gratifying, as well as the disinterestedness with which collegiate and private collections were thrown open for inspection. For all these obligations, I avail myself, with pleasure, of this opportunity to return, officially and individually, my warmest acknowledgments.

I am, gentlemen, very respectfully,
Your obedient servant,
RICHARD OWEN,
Principal Assistant.*

---

*Since appointed by the State Board of Agriculture State Geologist of Indiana.

INTRODUCTION.

In accordance with the directions given, after your consultation with the late State Geologist, the corps under my charge proceeded to examine, first in the fall of 1859, the counties not traversed by railroads, along the Ohio river; afterwards, in succession, each agricultural district,* chiefly by railroad travel, commencing in the north, making as prolonged a stay in each district as the time intervening before the probable setting in of winter would permit. That time being necessarily very limited, our chief endeavor was to ascertain where objects of importance and interest were to be found for the Spring Survey, rather than to make a critical and detailed examination of any one locality. It was arranged, as you will remember, to meet each member of the State Board in the District over which he presided, and thus obtain interesting general facts, as well as to collect for examination and analysis in the laboratory during the winter, a supply of minerals and soils.

---

*For the information of those not acquainted with the manner in which the various counties of our State were grouped by you into Agricultural Districts, the following statement is subjoined: The Legislature provides that each county, having a regularly organized Agricultural Society, may send a delegate to the annual meetings held in Indianapolis. These Delegates elect a State Board of Agriculture consisting, according to the present organization, of sixteen members, (eight elective annually,) each one of whom presides over the interests of the particular district to which his election assigns him, besides enacting regulations for the general farming interests of the State, and development of agricultural knowledge and prosperity in Indiana. The Geological Survey of the State was recommended by you, who now constitute the above Board; and the Legislature, on the adoption of that recommendation, placed the direction of the Survey under your fostering care.

The basis upon which the State is districted for agricultural purposes is expressed in the resolutions, adopted by your Board at the meeting in January, 1859:

"1. *Resolved*, That the State be districted into sixteen agricultural districts, upon the following basis: A meridian line to be drawn from north to south, passing through the centre of Indianapolis, and seven parallel lines to be drawn from east to west, the location of these parallels to be decided by a committee immediately.

"2. *Resolved*, That the eight members of the State Board now to be elected, be taken from

From your remarks I considered that this survey was designed to subserve various useful purposes and interests:

1st. That of the *Farmer*, by analyzing the soils collected from the different geological formations, and showing which were best adapted for any given crop. Also, by comparing fields in long cultivation, or worn out, with the nearest virgin-soil, ascertaining what materials were deficient, or had been exhausted, and informing the agriculturalist whether sub-soiling would return the necessary ingredients to his land, or whether lime, plaster, barn yard or green manures, poudrette, guano or other fertilizers would most improve that agricultural region.

2d. The Practical *Miner* was to be aided by theoretical calculations, after due examination of the coal shales and other distinctive characteristics, in his search after cheap fuel, by pointing out to him the probable depth at which other seams might be reached, if he were dissatisfied with the one which showed itself at the surface level; giving also the comparative analytical results of different coals, some suitable for furnaces or blacksmithing, for generating steam or heating apartments, and others for the manufacture of coal oil, paraffine candles, &c.

3d. The Miner was further to be assisted in his search after *Iron* and the production of the best quality of that valuable metal. The furnishing of cheap and good coal and iron was deemed of interest to the whole State, and therefore, although the mining of coal in Indiana may be chiefly confined perhaps to twenty or twenty-two of the south-west-

---

alternate districts on the east and west side of the meridian line, and the other districts to be filled at the next annual meeting of the Board.

"3. *Resolved*, That those counties divided by this meridian shall fall to that side upon which shall lie the greater portion of its territory."

In accordance with the above, the sixteen districts are thus arranged:

1st. District—Posey, Vanderburgh, Gibson, Warrick and Spencer counties.
2d. District—Pike, Dubois, Martin, Daviess, Knox and Sullivan counties.
3d. District—Perry, Crawford, Harrison, Floyd and Washington counties.
4th. District—Orange, Lawrence, Jackson, Greene, Monroe, Brown and Scott counties.
5th. District—Clark, Jefferson, Switzerland, Jennings, Ohio and Ripley counties.
6th. District—Dearborn, Franklin, Decatur, Bartholomew and Rush counties.
7th. District—Johnson, Shelby, Morgan and Marion counties.
8th. District—Owen, Clay, Vigo, Parke and Vermillion counties.
9th. District—Putnam, Hendricks, Montgomery and Boone counties.
10th. District—Fayette, Wayne, Union and Henry counties.
11th. District—Randolph, Delaware, Madison, Hamilton, Hancock, Tipton and Jay counties.
12th. District—Clinton, Tippecanoe, Warren, Fountain, Benton and White counties.
13th. District—Blackford, Grant, Huntington, Wells, Adams, Wabash and Howard counties.
14th. District—Carroll, Cass, Miami, Fulton, Pulaski, Jasper, Porter and Lake counties.
15th. District—Marshall, LaPorte, Starke, St. Joseph and Elkhart counties.
16th. District—Allen, LaGrange, Whitley, DeKalb, Noble, Steuben and Kosciusko counties.

ern counties, and the production of iron to the region of the low coals, or the swamp lands, throughout some ten or twelve of our northern counties, yet the whole State is likely to be benefitted by any information which would diminish the cost of production, or improve the quality of these two articles; for these reasons attention was to be first directed to the examination of the above staples.

4th. As being closely connected with the solution of these and similar practical questions, it was deemed important also to determine the exact limits of each *Geological Formation*, already approximately laid down in the reconnoissance made over twenty years since by the late State Geologist, but now more readily defined in detail in consequence of the opening of numerous coal banks, stone quarries, &c., the exposures in railroad cuts, as well as the minute description and limitation of characteristic fossil species. Bearing upon the above point would be the determination of areas over which certain fossil plants and animals extended, as well as the vertical range* they enjoyed in the pre-adamitic seas; just as we might now trace the prolongation of our New England, Maryland and Carolina oyster beds to their extreme southern limit, follow the warmth-loving coral to its highest northern latitude, or dredge the ocean to know at what number of fathoms in depth any given species of its inhabitants ceased to be found.

5th. The instructions were held in view to ascertain any facts bearing on the mysterious disease variously termed milk-sickness, slows, tires and trembles, particularly such as might point to some connection with geological peculiarities affecting soil, water, vegetation and the like;

---

*To verify the topographical heights in this connection and for similar purposes, I carried constantly the Aneroid Barometer. I found it worked well and was to be relied upon whenever the weather was somewhat settled, and the instrument was guarded from exposure to the sun. To enable me to work out, after my return home, the necessary corrections for changes in the weight of the atmosphere, irrespective of level, Dr. A. Clapp, kindly undertook to make, during the continuance of the fall reconnoissance, tri-daily observations referable to low water in the Ohio river, at New Albany, the height of which above high tide in the Gulf of Mexico is known. Latterly these corrections were made from observations conducted under the direction of the late State Geologist, at New Harmony, during our absence in the field. Some heights were verified by a comparison with tables obtained from a work of Charles Ellet, Esq., on the Physical Geography of the Mississippi Valley, published in the "Smithsonian Contributions to Knowledge." Other hypsometrical facts were also courteously furnished from the railroad and canal surveys; still many points remain, the determination of whose exact topography and altitude would throw much light on geological investigations; indeed, as often remarked by the late State Geologist, it is highly important in a survey, that stratigraphical investigations should be aided by topography, rather than depend wholly on palæontology; and it was for the purpose of demonstrating this, that the valuable services of Mr. Lesley were secured and employed in a prominent coal region, Perry county.

but unfortunately the testimony given was often conflicting, sometimes vague and unsatisfactory; such items, however, as could be elicited will be found noted.

6th. It was considered important and useful to examine and report on the numerous quarries throughout the State from which *building materials* are, or could be, obtained. To render this investigation of the highest practical value, specimens should be submitted to repeated and severe trials for strength, durability, &c.; and then at least approximate decisions could be furnished from the laboratory, whereas, without them or similar tests, samples would otherwise demand years of natural exposure for the determination of their relative qualities. The same applies to materials for the construction of roads and various engineering purposes.

7th. The examination for other metals than iron, such as lead, zinc, copper, silver and gold. Particularly regarding the latter, information was desired by many in order to know whether they would be justified in continuing their washings or even in risking capital for the apparatus, implements, aqueducts, &c., sometimes necessary for the successful prosecution of this work.

8th. A report on the clays suitable for pottery, fire or other superior brick; on limestones having hydraulic properties, or even superior adaptation to the manufacture of lime; on marls suitable for making artificial rock, for fertilizing, &c.; on gypsum, salt boring, and a variety of similar items.

9th. Information regarding Artesian wells, and where they may be attempted with fair prospect of success; the analysis of medicinal waters and report on their applicability to the cure of various diseases. In some regions, particularly near summit levels, there is often an anxious enquiry as to the obtaining of better water than they possess, by deeper digging or boring: on these points the geologist, after due examinations regarding dip, &c., is qualified to advise with considerable precision.

10th. The *Natural History* of the country might well justify some examination and report regarding the quality of timber, and its relation to geological peculiarities of soil, the prevalence of certain plants, animals, &c.

11th. Some useful practical inferences might be drawn from an exhibit of the formation of our prairies and *swamp lands*, regarding the best mode of drainage, improvement and the like.

12th. Under the head of Miscellaneous Facts or Statistics many items might be mentioned strictly within the field of geological investigation, or at all events as useful and interesting for observation, if time and means permitted. Suffice it to cite, as samples, the record of facts bearing upon the geographical distribution of hog-cholera, potato-rot, the Canada thistle and other injurious weeds, grain destroying birds, the Hessian fly, grain moth, army worm, fruit tree or timber borers and similar objects, so as at least to know whether any were peculiar to certain geological districts, or were strictly bounded by lines of latitude and longitude. By the collection and diffusion of such information we might hope to aid the farmer in the selection of his crop for a given locality, or perhaps to induce him to modify to some extent the period of his thrashing grain or cutting timber, &c.

Sometimes, too, I may here remark, it is not so much that facts regarding for instance a coal bank, a deposit of iron ore, a rock quarry or the like, are new and unknown in their immediate neighborhood, as that such information has not been diffused; whereas it ought to be disseminated through the State for the benefit of our citizens, as well as published out of the State to induce actual settlers to improve our vacant lands, and attract capitalists to work up our raw materials; the facts and statements thus receiving, after critical investigation on the part of the geologist, disinterested verification and authentic publicity through the medium of an official report.

I am well aware that the present appropriation of $5,000 would only suffice to make a beginning in examinations so extensive as the above, especially when we consider that the accurate analysis of a set of soils, upon the very satisfactory and reliable methods pursued by Dr. Peter, of Lexington, for the surveys of Kentucky and Arkansas, can not be made, even at the moderate per diem charged by that distinguished chemist, at a less cost than from fifteen to twenty dollars for each set of three, virgin, surface and sub-soil. The minute quantitative analysis of a *single* soil, submitted separately to almost any good chemist, would cost fully double that amount. When we further reflect that even at our census of 1850 we had nearly 100,000 farms cultivated and yet more than half of our lands unimproved, with an area of 33,809 square miles, comprising ninety-two counties, we can readily see how much practicable and profitable work there is to be done. But holding the rank in population and resources which our State is shown to possess according to the partial census returns of 1860 already reported, it does not seem unreasonable to hope that Indiana may one day emulate her sister States.

Meantime, however, in order to obtain the greatest amount of useful practical information, with the means at our command, primary attention was paid, as above stated, to the development of those pre-eminently useful minerals, coal and iron; notes on the other subjects enumerated being taken incidentally when passing from one point to another. This may serve as some explanation or apology to those five or six counties which were not reached, as well as to some others which received but a passing or partial investigation. If any examinations were neglected which could have been prosecuted, all the circumstances considered, the neglect was certainly not intentional and it is hoped will therefore be excused.

The above being the construction put upon the duties to be performed, and the hope being entertained that circumstances might favor more varied and extended research the efforts of our corps were directed, during the spring and fall field-work of 1860, chiefly to the examination of the coal openings, beds, banks, shafts or pits in the twenty-two* counties from which that valuable combustible had been mined or raised, and to the defining of the coal-field limit, particularly so as to ascertain its greatest eastern workable extension, its western boundary reaching into Illinois. Mr. Lesquereux having been engaged, in consequence of his intimate acquaintance with Fossil Botany, and long experience in the eastern coal fields, to decide on the identity or the distinctive difference of the seams inspected while we were associated in our examinations, it will be unneccessary for me to report on the same, except, perhaps, regarding the analysis, economic value, and the like, of different coal beds. Our attention was also directed to such an examination of the chief iron localities, partly in the coal field, partly in swampy lands of some northern counties, as would enable a corps to trace them more readily in detail hereafter. Such samples as time permitted were submitted to chemical analysis.

The result of the above general exploration or rapid reconnoissance of 1859, and the facts ascertained in the detailed survey thus commenced in 1860, it has been the endeavor to embody in such language as it was hoped would be intelligible even to the general reader who can not usually be supposed to have devoted much previous attention to the science of geology.

To carry out this view, it was thought appropriate to define such tech-

---

*Two of these counties, affording chiefly sub-conglomerate coal, are not embraced in the description of the counties situated in the Coal Measures proper; they will be found under the head of sub-carboniferous limestone counties.

nicalities, as it was sometimes considered necessary to employ, when these seemed better calculated to express the ideas designed to be conveyed than could be done by language more vague or less purely geological.

The full execution of this design seemed even to justify the devoting of a few pages, under the head of "Preliminary Observations," to a brief exposition of the more prominent principles upon which that useful and attractive science, geology, is based.

If the appropriation for printing and illustrating the report had been considered as permitting the execution of all necessary maps, sections and illustrative sketches, together with the plates giving the most characteristics fossils of each period, such as, if means permit, should always accompany a geological report, the attempted explanations and descriptions might have been rendered much more intelligible, especially in connection with an appended glossary and tabular views, which could be consulted where unexplained terms occurred. This may to some seem a useless expense; but the miner, the agriculturalist, and the general reader can not be expected to provide themselves with expensive text-books* in science, or to be posted in technicalities other than those employed in their own department; yet these readers constitute

---

*For the benefit of those desiring to examine the principals and facts of Geology in a somewhat extensive course, it may be permitted to suggest the following works for perusal, beginning with the enumeration of the easiest first: *General Geology.* Some articles on Geology, written for the "Indiana Farmer," may perhaps prepare the way for others, such as Prof. St. John's Elements or Prof. Emmon's small work, Prof. Hitchcock's Geology, or Page's or Chamber's. Then should follow Sir Charles Lyell's "Elements," also his "Principles," or Ansted or De La Beche; the whole study being much aided by the use of Prof. Hall's "Chart of the Geological Formations," and a reference to the various State Geological reports.

Special Geology. Dr. Mantell's "Medals of Creation," for the study of palæontology in all formations, or Prof. Pictet's Palæontology for those who read French, as the work is not translated. Prof. Hall's volumes on the Palæontology of New York are admirable for the study of United States poleozoic fossils; Sedgwick and McCoy, or Murchison, or Verneuil for European organic remains of the Silurian strata; Phillips and Hugh Miller for the Devonian; DeKoninck's fossil animals of Belgium, also the Kentucky Reports are excellent for the Carboniferous system; Dr. Morton, Messrs. Conrad and Lea, for the American cretaceous fossils, Mantell for European cretaceous, Dr. Grateloup for the Miocene Tertiary of Europe, and the work of the late Prof. Tuomey and Prof. Holmes, of Alabama, on the Tertiary of the United States. To go yet more into detail, Barrande and Burmeister have fully examined the Trilobites, D'Orbigny, Austin, Raumer and Miller the Crinoids; Woodward's Manuel is admirable for recent and fossil shells, and Davidson for his Specialty, the Brachiopods; Brongniart, Lindley and Hutton, and Gainitz are standard authority on fossil plants; Dana or Edwards and Haime on corals; Agassiz and Gould's Principles of Zoology should be read as introductory to the study of palæontology, or Ruschenberger's Natural History, then Prof. Owen's small work on the skeleton and teeth; or for a more extended course of comparative Physiology, Carpenter's work, or the General Structure of the Animal Kingdom, by Thos. R. Jones, besides other valuable works too extensive to be here enumerated.

a large class, who, it is hoped, may derive profit and pleasure from the survey and its report. It is perhaps better to repeat somewhat similar language twice, or to give popular terms in addition to scientific, rather than be misunderstood; and the brief statement of principles can readily be passed over without perusal by those already posted in Geology.

The design being, as already remarked, to make such a statement of well ascertained facts as might benefit the miner or mechanic in developing the mineral wealth of our State, and also to throw light on the operations of the agriculturalist, by pointing out, after accurate analysis, the peculiarities of soil resulting from different geological formations and their consequent adaptation to different crops; by disclosing, through the same source, the ingredients lost by cultivation and the means of renovating worn soils, or of rendering others still more productive, it seemed not unreasonable to conceive that the statements made would inspire more confidence in the minds of those unacquainted with the minutiæ of Geology, if some account were previously given of the general principles and methods by which Geologists arrive at facts or data in their explorations, as well as conclusions in their reasoning deduced from those data, carefully separating any mere theories, or opinions advanced, from facts or truths well established by repeated observations, and acknowled as such by the great mass of geologists, perhaps by all.

After a brief discussion of principles and enumeration of the geological formations constituting the earth's crust, a few words seemed in place regarding the geographical distribution of those rocks over the globe, with a somewhat more extended notice of their predominating prevalence in different parts of the United States.

This prepares us for a more full understanding and appreciation of such formations as are found in Indiana generally; leading us gradually and intelligibly to the details of county geology.

Some general remarks suggested by a comparison of the collated facts obtained in the brief detailed examinations which time permitted, have been reserved for description after giving the data upon which they are based; such theoretical remarks, (designed, however, also as practical deductions,) admitting thus of more ready acceptance or rejection.

Pursuant to the instructions and plan exhibited above, I now proceed to offer the following Report:

REPORT

OF A

GEOLOGICAL RECONNOISSANCE

OF INDIANA.

BY RICHARD OWEN, M. D.

CHAPTER I.

PRELIMINARY OBSERVATIONS.

Let us suppose a Geologist to be transported to a mass of rocks, even in a distant and foreign country, where the ledges or layers exposed are so continuous as to show that they are in the position originally occupied, perhaps ages gone-by, technically termed *in situ*. That Geologist, particularly if well versed in palæontology, or a knowledge of ancient beings, will soon be enabled to say, by an inspection of these rocks, supposing them to contain, as most aqueous rocks do, some petrifactions or organic remains, whether these were deposited after or before the coal-bearing period, consequently whether or not, by digging there, there is a probability of finding valuable coal deposits. This and similar facts he can predict, not by divining rods, not by "exorcising spirits from the vasty deep," or any black arts, but by applying to practice some general geological principles or ascertained truths, such as these: That certain animals and plants occupied the earth's surface and afterwards became extinct, leaving their organic remains, or at least traces of their forms,* in the subsequently solidifying rocks; and that to these animals and plants succeeded another set, having such marked differences as to be classed under *species* distinct from the former, perhaps to constitute even different *genera*. The sedimentary or aqueous rocks, in which these organic remains are found, derive their name

---

* These organic remains are generally termed *fossils;* and for the space left exhibiting only the form, when the organic substance wholly disappears, as is sometimes the case, particularly in Magnesian limestones, the word *cast* is used. The true geological meaning of petrifaction is the same as fossil, although in common language the term is sometimes applied to a mere incrustation or deposition, usually of carbonate of lime, around the animal or vegetable. In a fossil the organic body, animal or vegetable, is replaced, to a greater or less extent, by inorganic or mineral matter, often silicious and calcareous in character, sometimes aluminous, or even metallic, through chemical decomposition, recomposition and infiltration. Animals and vegetables are called organic bodies or organisms, because they have special organs, such as those of nutrition, &c., to develop their growth or sustain life, whereas inorganic or mineral bodies only add to their growth by accretion or the deposition of additional similar particles around the original nucleus.

from the supposition that they were originally deposited as sediment from water, Latin *aqua;* their distinctive lithological or stony character they receive from the different particles, siliceous, aluminous, &c., thus carried down from off the high igneous mountains, (the formation of which is hereafter explained,) into the lower portion of the earth's crust, and there deposited until fitted for animal and vegetable life. At the close of certain periods, sometimes perhaps after thousands of years, the whole of these organisms, the flora and fauna, or plants and animals, as already alluded to, gradually, and apparently often quietly, (but sometimes suddenly and with evidence of terrestial convulsion,) died out and were imbedded in the clays, sands, mud, &c., of that epoch, to be covered afterwards by accumulations of somewhat differing materials. By the pressure of these incumbent masses, as well as by the cementing power of some of the infiltering materials, aided perhaps by heat and electricity, these lower strata were consolidated into stratified rocks, (those having regular layers or strata,) some furnishing no evidence of animal or vegetable existence, but the majority, at least of the somewhat later deposits, containing fossils or organic remains, and hence being termed fossiliferous rocks.

The great law of the difference in the organic remains of successive layers of strata, amounting, as already stated, usually to a specific difference, often to a generic and sometimes to a marked distinction of order, class and department, as hereafter also explained, is the science lately developed and now being studied under the name of Palæontology. A thorough knowledge of these minute distinctions is, however, the study of a lifetime, inasmuch as in the department of the Molluscous animals alone, (sometimes called shell-fish,) there are 15,000 fossil species and 12,000 recent. But its importance can scarcely be over-estimated, when we consider that the earth's crust has been found made up of very similar successions of aqueous strata, at parts the most remote from each other and characterized by a sequence of animal and vegetable remains having many points of resemblance in common. Nor has this regular succession (except locally over a small area evidently disturbed,) ever been found in an inverted order, such as would place, for instance, the Mastodon in an early deposit, with a large number of trilobites in strata above, evidently of more recent deposition. As this has never been found to be the case, although occasionally some of the intervening members or even a whole system may be absent, (either because circumstances did not favor the formation or because after deposition, perhaps before thorough consolidation, subsequent washings carried off some layers from considerable areas, leaving a valley of deundation,) we consider ourselves justified in expecting to find this law hold good universally. Here it seems necessary to digress long enough to offer some explanation of the principles upon which these specific, generic and other distinctions are based.

The classification generally adopted, in modern times, depends upon the internal structure of the animal or plant, which is found to be much more important than the difference or resemblance in external appearance. The basis, therefore, of zoological or animal classification, first into great divisions or departments, and afterwards into minor sub-divisions, depends upon the most important organs, or sets of organs, termed a system. Thus the nervous system, or particular plan on which the nerves are arranged, giving energy and vitality to the whole being, forms the ground-work for Cuvier's great division into four *departments*.

The sub-division of these departments into *classes* is dependent to a great extent, on difference in the circulatory or respiratory systems, in other words, on the manner in which the animal breathes and has its blood purified. Further sub-divisions into *orders* are often founded, especially among the more highly organized animals, on differences in the nutritive system or mode of receiving and converting food to nourish the animal economy. This usually involves a variation in the form of the teeth,* hence their palæontological importance. More minute sub-divisions into *genera*, (the plural of genus, a latin word signifying origin, stock, family or kind,) are sometimes dependent on variations in the locomotive or in the prehensile organs, designed to enable the animal to move about and to obtain food, such as feet, fins, wings, feelers, tentacles, and the like. At other times these generic groupings, particularly of lower animals, throw together species having a resem-

---

* The great Cuvier, in his work on Fossil Remains, has the following observations bearing on this point: "Every organized being forms an entire system of its own, all the parts of which mutually correspond and concur to produce a certain definite purpose by reciprocal action or by combining to the same end. Hence none of these separate parts can change their forms without a corresponding change in other parts of the same animal; and consequently each of these parts, taken separately, indicates all the other parts to which it has belonged. Thus, if the viscera of an animal are so organized as only to be fitted for the digestion of recent flesh, it is also requisite that the jaws should be so constructed as to fit them for devouring prey; the claws must be constructed for seizing it and tearing it to pieces; the teeth for cutting and dividing its flesh; the entire system of limbs or organs of motion for pursuing and overtaking it; and the organs of sense for discovering it at a distance. Nature must also have endowed the brain of the animal with instincts sufficient for concealing itself and for laying plans to catch its necessary victims."

blance in the form of shell or hinge teeth, of the crustaceous covering, of the horny or calcareous framework, &c., but these frequently indicate also similarity in anatomical structure and should rarely, if ever, be employed for classification, unless thus characterized. Animals that naturally breed together or have minute characteristics alike, which are not liable to change, are usually considered of the same *species*, (Latin for appearance, quality.) When changes arise from accidental causes and are not likely to be so permanent but that circumstances may again modify them, or when the intermediate gradations can be traced between the two organisms thus differing, then the term *variety* is used in the animal as well as in the vegetable world.

As a diversity of opinion exists regarding the origin of the specific, generic and other difference above alluded to, and as a considerable amount of controversy has latterly been elicited on the subject, it may not be out of place to state briefly the antagonistic views, the bearing of neither, however, affecting materially the utility, or diminishing the certainty, of the observed palæontological records, as tests of the relative ages of any given rocks, which present a sufficient amount of organic remains in a moderately well-preserved state.

Darwin, a naturalist long and favorably known in the scientific world, has published a work designed to show that all organic beings have a tendency to reproduce themselves in a geometrical ratio; but that, from various conflicting causes, only a small number in this "struggle for life," of those endowed with a structural or functional difference, usually somewhat superior, reaches maturity. This difference, imparted to the offspring, again influences the new variety, causing further improvement; which general fact or law, operating from the beginning, he considers sufficient to produce not only varieties, but the distinctions we assign to species, genera, orders, &c., and consequently he thinks that all organisms, living and extinct, animal and vegetable, have proceeded from the simplest primordial or original form of life.

The celebrated Agassiz, on the other hand, in extracts, given in the July number of Silliman's Journal for 1860, from the advance sheets of his "contributions to the Natural History of the United States," thus states his opposite views: "I have attempted to show that branches," (a division equivalent to departments formerly used by him, and to sub-kingdoms, used by Prof. R. Owen, of London,) "in the animal kingdom are founded upon different plans of structure and for that very reason have embraced from the beginning representatives between which there could be no community of orgin; that classes are founded upon different modes of execution of these plans, and therefore they also embrace representatives which could have no community of origin; that orders represent the different degrees of complication in the mode of execution of each class, and therefore embrace representatives which could not have a community of origin any more than the members of different classes and branches; that families are founded upon different patterns of form, and embrace representatives equally independent in their origin; that genera are founded upon peculiarities of structure, embracing representatives which, from the very nature of their peculiarities, could have no community of origin; and that finally, species are based upon relations and proportions that exclude, as much as all the preceding distinctions, the idea of a common descent."

Let us now return from this digression, regarding the best mode of classifying organic remains, to the re-enunciation of the great law.

Whatever may be the original cause of the differences observed, whether created thus distinctly by a periodical interposition of Divine fiat, or so modified by an eternal and immutable law of the same Omnipotence, as to produce, through physical changes in the inorganic elements of the earth, corresponding modifications of structural adaptation to the new external circumstances, the important fact still remains an unquestioned truth, that *a certain vertical range, or ascertained thickness of fossiliferous rock, is characterized by the organic remains of plants and animals, differing more or less from the plants and animals in the rocks above, as well as those in the rocks below the given layers or strata*; just as the hieroglyphics and coins of one nation, while having some characters in common, are found differing from those of another nation preceding or succeeding it.

To the vertical range of beds having thus something in common, (even if persistent as regards a given thickness only over a moderately extended horizontal area, thickening or perhaps thinning out beyond that limit,) geologists give the name periods, systems, formations and the like, prefixing some adjective explanatory either of the geographical prevalence somewhere in that sub-division, or of its lithological peculiarity. Thus the beds below the Coal Period, and deposited before it, are often called the "Devonian system," because developed in Devonshire, England, or sometimes "Old Red Sandstone," because, where first studied, its lithological character was that of a sandstone highly colored with peroxide of iron.

But in the same manner that all the coins of one nation might have something in common and yet those of each dynasty or successive reign, differ in certain particulars, so too the separate geological strata of a period, while presenting a general resemblance in animal and vegetable form of organic remains, may yet have minor distinctions, in the separate component deposits, justifying a sub-division into subordinate members, such as "upper, middle, lower," and the like. To carry the analogy still further, as there may be certain signs or words common to two of those nations, and even similar to those used at the present day, so there are certain fossils common to several systems or formations and even existing at the present day; the analogue being of the same genus, if not specifically identical with its ancient prototype; and as the study of these ancient relics requires an acquaintance with the key to the hieroglyphics, in like manner palæontology, to be useful, demands a knowledge of minute distinctive characters.

Having the above great fundamental law before us, we have next to examine whether there is any general law, (of progression or some other character,) which may aid us in distinguishing the organisms in all the older strata from those in the newer, before we proceed to study the more minute differences. Many geologists concede that evidence of such a law is observable throughout the entire series of aqueous deposits; others do not admit that we have yet sufficient data for such a generalization. A brief examination of these points, bearing both on vegetable and animal life, may lead to a more thorough comprehension of the subject: it is therefore subjoined.

Some animals are very simple in their structure, consisting only of a sack-shaped, gelatinous material, the opening to which is furnished with a few tentacles, or organs of feeling and motion, for the purpose of enabling them to procure food; sometimes this is connected with a strong framework or skeleton of calcareous matter. Such radiates, and some brachiopod mollusks, also comparatively simply-formed animals, with the nutritive sack more elongated, and the addition of a liver, the whole usually protected by a shell or shells; as also trilobites, crustaceans not high in the scale of organization, but resembling some of the earlier embryonic stages of our modern King-crab, have thus far been found relatively most abundant in the early geological formations, and are therefore supposed by most geologists to have predominated, or to have been comparatively more numerous, soon after the earth became adapted for animal existence, than the highly organized animals appear to have been. Thus, for instance, the corals, which constitute the skeleton or framework of the simple, sack-like animals just mentioned, as well as some species of the bivalve mollusks and trilobites, next alluded to, above, form whole rock masses in the earlier layers of the earth's crust, presenting very much the same appearance which a coral reef or one of our Atlantic oyster beds would have, if compressed by great weight and the aid of a natural cement into a compact, solid mass. In the later deposits, overlying the former, we find comparatively more of the complicated structures or animals, such as elephants, mastodons and the like, having relatively more fully developed the nutritive, circulatory, respiratory and nervous systems, upon a plan similar to that found in man.

On this subject, Hugh Miller, in his "Testimony of the Rocks," remarks in his usual powerful and appropriate language: "It is a marvellous fact, whose full meaning we can as yet but imperfectly comprehend, that myriads of ages ere there existed a human mind, well nigh the same principles of classification now developed by man's intellect in our better treatises of geology and botany, were developed on this earth by the successive geologic periods; and that the by-past productions of our planet, animals and vegetables, were chronologically arranged in its history according to the same laws of thought which impart regularity and order to the works of the later naturalists and phytologists." * * * "Commencing at the bottom of the scale, we find the thallogens or flowerless plants, which lack proper stems and leaves—a class which includes all the algæ. Next succeed the Acrogens or flowerless plants that possess both stems and leaves—such as the ferns and their allies. Next, omitting an inconspicuous class, represented by but a few parasitical plants incapable of preservation as fossils, come the endogens—monocotyledonous flowering plants that include the palms, the liliaceæ and several other families, all characterized by the parallel venation of their leaves. Next, omitting another inconspicuous tribe, there follows a very important class—the gymnogens—polycotyledonous trees represented by the coniferæ and cycadaceæ. And last of all came the dicotyledonous exogens, a class to which all our fruit, and what are known as our 'forest trees' belong, with a vastly preponderating majority of the herbs and flowers that impart fertility and beauty to our gardens and meadows. This last class, though but one, now occupies much greater space in the vegetable kingdom than all the others united."

"Such is the arrangement of Lindley, or rather an arrangement the slow growth of ages, to which this distinguished botanist has given the last finishing touches. And let us now mark how nearly it resembles the geologic arrangement, as developed in the successive stages of the earth's history."*

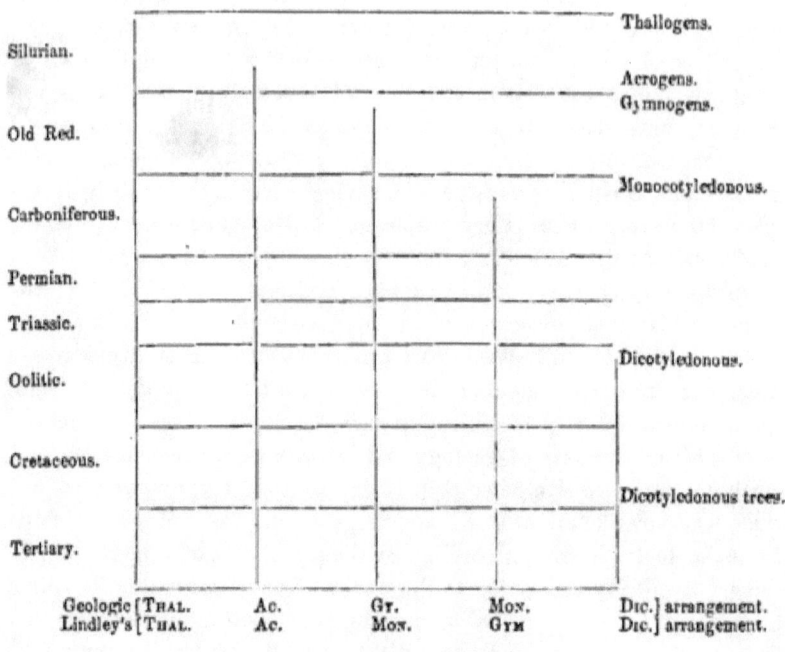

"*The Genealogy of Plants.*"

* * * "And such seems to be the order of classification in the vegetable kingdom, as developed in creation and determined by the geologic periods."

"The parallelism which exists between the course of creation, as exhibited in the animal kingdom, and the classification of the greatest zoologist of modern times, is perhaps still more remarkable. Cuvier

---

* "The horizontal lines in this diagram indicate the divisions of the various geologic systems; the vertical lines the sweep of the various classes or sub-classes of plants across the geologic scale, with, so far as has been yet ascertained, the place of their first appearance in creation; while the double line of type below shows in what degree the order of their occurrence agrees with the arrangement of the botanist. The single point of difference indicated by the diagram between the order of occurrence and that of arrangement, viz., the transposition of the gymnogenous and monocotyledonous classes, must be regarded as purely provisional. It is definitely ascertained that the Lower Old Red Sandstone has its coniferous wood, but not yet definitely ascertained that it has its true monocotyledonous plants, though indications are not wanting that the latter were introduced upon the scene at least as early as the pines or araucarirns; and the chance discovery of some fossil in a sufficiently good state of keeping to determine the point, may, of course, at once re-transpose the transposition, and bring into complete correspondence the geologic and botanic arrangements."

divides all animals into vertebrate and invertebrate; the invertebrate consisting, according to his arrangement, of three great divisions—mollusca, articulata and radiata; and the vertebrates of four great classes—the mammals, the birds, the reptiles and the fishes. From the lowest zone at which organic remains occur, up to the higher beds of the Lower Silurian System, all the animal remains yet found belong to the invertebrate divisions. The numerous tables of stone, which compose the leaves of this first and earliest of the geological volumes, correspond in their contents with that concluding volume of Cuvier's great work, in which he deals with the mollusca articulata, and radiata; with, however, this difference, that the three great divisions, instead of occurring in a continuous series, are ranged, like the terrestrial herbs and trees, in parallel columns. The chain of animal being on its first appearance is, if I may so express myself, a three-fold chain; a fact nicely correspondent with the further fact, that we cannot in the present creation range *serially*, as either higher or lower in the scale at least two of these divisions—the mollusca and articulata."

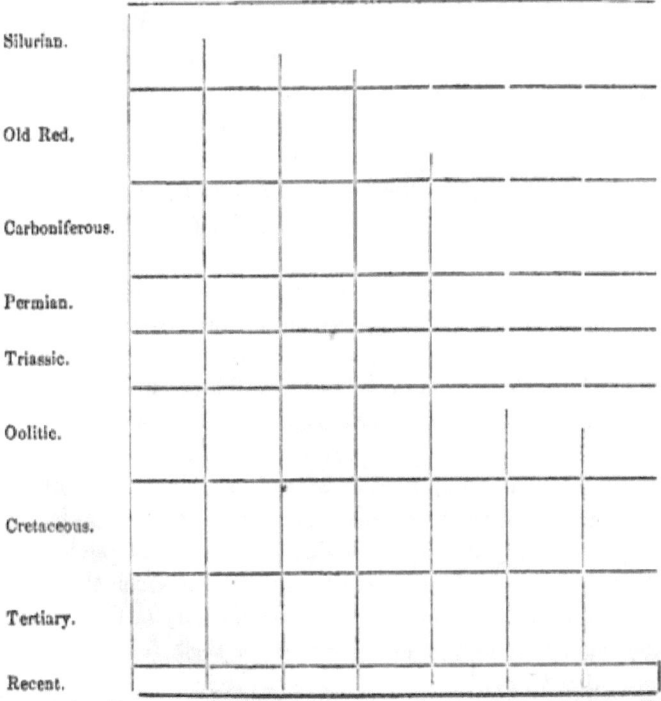

The Genealogy of Animals.

After these extracts from Hugh Miller's works, perhaps a few quotations from the great comparative anatomist, Dr. Wm. B. Carpenter, may not be out of place, as restrictive of this generalization, which is objected to by some geologists as being not yet fully proved, although they admit the accuracy and importance of palæontology. He remarks, in his "Principles of Comparative Physiology:" "The 'idea' of progress from the more general to the more special, which we have thus found to prevail alike in the completed structure of the existing types of vegetable and animal organization and in the developmental process by which they attain it, may also be traced in that long series of organic forms which have successively appeared and disappeared on the face of this globe, and have finally given place to those of our own epoch. The entombment of the remains of many of these, in the strata in progress of formation at the time of their existence, has enabled the Palæontologist to reconstruct, to a certain extent, the Fauna and Flora of each of those great epochs in the earth's history, which are distinctly marked out in geological time, both by extensive disturbances in the earth's crust, and by striking changes in the structure and distribution of the living beings which dwelt upon it. Each of these epochs was characterized by some peculiar forms or combinations of forms of animal and vegetable life, which existed in it alone; and the further we go back from the existing period, the wider are the diversities which we encounter, both in the general aspect of these kingdoms of nature, which depends upon the relative proportions of their different subordinate groups, and in the features and structure of the beings composing these groups. The attempt has been made to prove that these changes might be reduced to a law of progressive development." * * * "A more satisfactory account of the succession of organic life on the surface of the globe, may probably be found in the general *plan* which has been shown to prevail in the development of the existing forms of organic structure; namely, *the passage from the more general to the more special.* This seems to be manifested in two modes. In the first place we find a certain class of cases in which extinct animals, especially the earliest forms of any class that may be newly making its appearance, present indications of a closer conformity to 'archetypal generality' than is shown in the existing animals to which they bear the closest approximation; and hence their conformity to the latter is closer in the embryo condition of these than in their fully developed and more specialized state."

Having now examined the principles upon which the distinctions are based that have given rise to separate and distinct names for successive geological layers of sedimentary deposits, it will probably be useful to exhibit, in a tabular form, for more ready comparison of relative thickness and other distinctive characters, the arrangement or classification of the entire aqueous or sedimentary period, into subordinate ages, systems, formations, groups, members, or other subdivisions, to which unfortunately, as yet there are not data sufficient to affix very definite limits of time or vertical space.*

---

*For the Indiana Farmer a suggestive, approxmiate table was furnished by me, attempting therein to attach definite ideas to those terms in general use. This tabular view will be found in the appendix.

CHRONOLOGICAL SUCCESSION OF SEDIMENTARY OR AQUEOUS ROCKS.

(BEGINNING WITH THE MOST RECENT OR LAST DEPOSITED.)

Period, Formation or System.		Sub-divisions into groups or members.	Synonyms, or terms used by some as equivalent.	REMARKS.
Cainozoic.	Quaternary.	Newer.	Alluvium or modern or Historical periods.	Remains of man fossil. Mammals abundant.
		Older.	Post-Tertiary. Dilluvium, Bowlder, or erratic group.	Great Northern drift, by some classed in the tertiary as pleistocene.
	Tertiary.	Pliocene. Newer.		Containing about 95 per cent. of shells, identical with recent species.
		Pliocene. Older.		Containing from 35 to 50 per cent. of shells, identical with recent species.
		Miocene.		Containing about 17 per cent. of shells, identical with recent species.
		Eocene.		Containing about 3½ per cent. of shells, identical with recent species.
Mesozoic.	Carboniferous.	Chalk. Gault. Greensand. Wealden.		Echinoderms abundant. Only this member found in the United States. Iguanodons numerous.
	Oolite.	Upper. Lower.	Jurassic formation because prevalent in the Jura Mountains.	
	Lias.	Upper. Middle. Lower.		Huge reptiles abundant in this formation, also found in the previous.
	Trias.	Upper. Middle. Lower.	New red sandsone. Saliferous. Poikilitic.	Gigantic bird tracks in Connecticut.
Palæozoic.	Permian.		Magnesian limestone.	
	Carboniferous.	Coal measures.		Land plants abundant in coal period.
		Millstone grit.	Carboniferous or coal conglomerate.	Some thin coals found in this group.
		Sub-carboniferous limestones and sandstones.	Cavernous or mountain limestone, Knob sandstone.	
	Devonian.	Upper.	Old red sandstone of European writers; Catskill group of New York.	Remarkable fossil fishes during this period.
		Lower.	Chemung, Portage, Genesee, Tully, Hamilton and Marcellus.	Of New York Geologists
			Corniferous and Onondago limestone; Schoharie and Cauda galli grits; Oriskany sandstone groups.	Lyell extends the Upper Silurian to the Hamilton group, inclusive.
	Silurian.	Upper.	Upper pentamerus limestone. Delthyris shaly limestone. Pentamerus galeatus limestone. Onondaga salt group.	Lower Helderburg, Brachiopods common. Ludlow formation of English writers.
			Niagara group. Clinton group. Medina sandstone.	Wenlock or Dudley formation, of English writers.
			Hudson river group. Trenton limestone.	
		Lower.	Black River limestone. Bird's Eye limestone. Chazy limestone. Calciferous limestone.	Caradoc, of English writers. Llandeilo, of English writers.
			The Potsdam sandstone, of New York Geologists. This last sub-group, is by some English writers, considered as belonging to a separate system: The Cambrian.	Trilobites and corals numerous.

After examining this tabular view of aqueous rocks, one sub-division of which alone may be 10,000 feet thick, it is very natural for those, who are aware that the deepest mining operations do not lead us much over 2,000 feet below the earth's surface, to inquire how we became acquainted with these lower strata. The reply is that we could not study them if they occupied the same horizontal position in which they were originally deposited; but as they have been disturbed and their edges occasionally brought to the surface by upheaval or force acting from below upwards, it becomes necessary, towards a right understanding of our subject, for us to examine this phenomenon, and become familiar with the terms applied to its different parts and phases.

It is generally supposed that the first film, or inner portion of the earth's crust, consists of rocks which resulted from cooling after being in a molten condition, somewhat like the slack we see thrown from large furnaces, having either a solid crystalline structure or a porous spongy appearance. These hypogene (nether formed,) or igneous or crystalline rocks, although constituting probably the inner film, and thus deriving their origin deep in the earth, may, in consequence of internal commotion and expansion, either raise portions of the superincumbent aqueous rocks or even break through and pour over them. This may take place after deposition of the palæozoic (older) sedimentary rocks, also called primary fossiliferous, or of the middle aged (mesozoic) rocks, called secondary, or even during part of the cainozoic age, which embraces the Tertiary or modern epochs. Thus these eruptive rocks, such as granite and basalt, may be called primary-granite, secondary-granite, tertiary-basalt, &c., according to the period at which they burst through the earth's crust.

Of the various Igneous rocks, including some sedimentary deposits which it is supposed have been altered or metamorphosed by the action of fire, and hence called metamorphic, a tabular view, conspectus or synopsis, is now also given as an additional aid.

CONSPECTUS OF IGNEOUS ROCKS.

Division.	Sub-divisions.		COMPOSITION AND REMARKS.
IGNEOUS OR HYPOGENE ROCKS. (Sometimes also termed Hypogoic, Eruptive, Crystalline, Non-Sedimentary or Non-Fossiliferous.)	Metamorphic.	Quartzite.	A rock composed of quartz.
		Hypogene marble or limestone.	A carbonate of lime destitute of fossils, such as statuary marble.
		Clay slate.	Composed chiefly of clay, (Latin, alumina or argilla.)
		Chlorite slate.	Composed chiefly of the mineral chlorite, a silicate of alumina and iron.
		Talcous slate.	Composed chiefly of talc, a silicate of magnesia, with some potash and iron.
		Hornblende slate.	Composed chiefly of hornblende: silica, alumina, magnesia, lime and iron.
		Mica slate.	Composed chiefly of mica: a silicate of alumina, with some lime and iron.
		Gneiss.	Composed of quartz, felspar and mica, disposed in regular layers; it is, in fact, stratified granite.
	Volcanic. (Modern.)	Pumice.	A spongy form of trachyte, so light as to float on water.
		Pearlstone.	Resembling obsidian, but less glassy.
		Obsidian.	Vitreous lava, like melted glass, sometimes white, at others black.
		Tuff or Tufa.	Volcanic tufa is made up of fragments of scoriæ and pumice.
		Scoriæ.	Volcanic cinders.
		Trachytic lava.	A form of felspathic lava which feels rough to the touch.
		Lava.	Melted matter which has flowed from a volcanic crater.
		Recent Amygdaloid.	Volcanic material, permeated by gases, which leave almond-shaped cavities into which mineral matter frequently filters.
	Volcanic. (Ancient.)	Ancient Amygdaloid.	Formed in the same manner as the recent, often of trap containing zeolite or quartz minerals.
		Trachyte or Domite.	Chiefly glassy felspar
		Serpentine rock.	Serpentine, a magnesian mineral; with limestone disseminated, forms Verd antique marble.
		Diallage rock.	Diallage (a variety of hornblende,) and felspar.
		Hypersthene rock.	Hypersthene and Labrador felspar.
		Greenstone.	Hornblende and felspar. } Trap Rocks.
		Basalt.	Usually augite, felspar, iron and olivine.
		Earthy porphyries.	Regular crystals in a compact base, such as trachyte, clinkstone or claystone.
	Plutonic.	Crystalline porphyries.	Regular crystals in a crystalline base, such as granite or syenite.
		Schorl rock.	An aggregate of schorl (black tourmaline) and quartz.
		Protogine.	A talcose granite: felspar, quartz and talc.
		Syenite.	Quartz, felspar and hornblende.
		Pegmatite.	Quartz and felspar.
		Graphic granite.	Quartz so disposed among felspar as somewhat to resemble Hebrew writing.
		Granite.	Quartz, felspar and mica.

Allusion has already been made to the fact that these Igneous rocks have at various periods burst through the sedimentary rocks, in consequence of internal action, thereby disturbing considerable areas of these aqueous deposits, usually elevating, most, the portions nearest to the igneous upheaving source. Sometimes eruptive rocks, thus breaking through, may rise to form the highest mountains, and the higher they rise the greater the angle or inclination, or dip, they will give to the originally horizontal aqueous deposits through which they break. Sometimes the igneous rocks are elevated sufficiently to disturb these horizontal sedimentary strata, yet without breaking through the crust in such a manner as to be detected anywhere on the surface.

We shall not now stop to inquire whether this internal force is due to a disturbance of electrical equilibrium, and this again dependent on inequality of temperature in different portions of our planet's interior, or whether it arises from chemical action, because these discussions, however interesting, would lead us too far; suffice it that to the extended lines of such upheaval, geologists give the name of "strike," from the German "Streichen," to extend or have a certain direction. The strike, then, of hills or mountains, would be their ridge, (range being usually applied to several parallel ridges,)* or line of greatest extension. This may be compared to the ridge or comb of a roof on a building. The deviation from a true horizontal line, which the sedimentary rocks are made to assume by this upheaving source is technically called the "dip." This may be compared to the sides of the long mountain or to the slope of a roof, and necessarily lies square across, or forms a right angle with the line of strike; then, if we suppose the ridge of a house to occupy a north and south line, the slope of the roof will be east and west, one prolongation of each slope running upwards, the other downwards, beneath the horizon or general level of the earth's surface. When this is the case with a bed of rocks, geologists term their first appearance on the surface, the prolonged line of which would run into the sky, the "out-crop" of those rocks; whereas, when rocks disappear under the general level, they are said to "dip under." A line, usually on a ridge, from which rocks dip each way, is said to have an anticlinal axis, whereas a union of two lines converging towards the earth is called a synclinal axis, and is commonly found in a valley. Occasionally layers, when in a plastic condition, appear to have experienced side or lateral pressure, in which case the strata are said to be "folded" or "curved." At other times, part of a bed seems to have been detached from the other portion, with or without disturbing their horizontality, and then either elevated or depressed; in such case, the vertical line extending from the dislocated bed to the place of original junction, is called by miners a "fault." The marks, nearly horizontal or gently undulating, that indicate different materials to have been deposited on each other from water, before consolidating into rock, are termed lines of deposition, and others, which do not correspond with them, and yet show a facility in certain rocks to split indefinitely in planes parallel to each other, are called lines of slaty cleavage. To the great cracks, often at right angles to the bed, of which quarrymen avail themselves in getting out rock, geologists apply the name of "joints."

The *rate* of dip, or angle with the horizon, formed by a bed of rock

---

*Thus the Appalachian Range is composed of the Blue Ridge, the Allegheny Mountains or Ridge, and the Cumberland Ridge.

may be measured by any straight edge with a pendulum attachment. Geologists generally employ a clinometer compass, the magnetic needle, when afterwards allowed to vibrate, giving the particular *direction* of the dip.

The rate of say 2° westerly dip would, supposing rocks exposed along a lake or sheet of quiet water, so as to show their original true bedding, cause any seam or layer of those rocks, which was about a foot above the surface of the water at the east end of the lake, to disappear under its surface, when traced in a westerly direction the distance of only twenty-eight feet. Consequently, a seam of coal disappearing at any given point with a dip of 2° could only be reached say at the distance of half a mile west of that place, supposing the same rate of dip continuous, by sinking a shaft nearly a hundred feet deep, and then working in, somewhat horizontally. Or, taking the inverse of the proposition, a seam of coal which, at a given point, just emerges from the surface of a lake or pond, might, by tracing it in the direction of its out-crop, the contrary of the dip, be found half a mile east, in hills nearly a hundred feet above the level of said water.

The explanation here of this geological phenomenon may, in connection with what had previously been stated, serve to render intelligible the apparent paradox that rocks *first* deposited, and consequently *geologically the lowest*, may, by upheaval at some point, not only come to the surface, ("crop out,") but even be elevated to such a hight as to be geographically or *topographically much higher* than the more recently deposited strata, beds or layers.

After this definition of technicalities in connection with the short explanatory account of igneous rocks, as well as the previous description and tabular view of sedimentary deposits, it may be interesting to know, as these do not show themselves everywhere on the earth's crust, at which portions of the globe any given rocks are most prevalent, in other words, their geographical distribution. Tracing these variations on a large scale may prepare us better to follow the geological differences observable in Indiana.

The peaks of the Himalaya mountains, the highest range in Asia and indeed on the globe, are composed of gneiss (stratified granite,) and other *Igneous or Crystalline rocks*, flanked by late aqueous deposits, as the Cretaceous and Tertiary; the Alps, the highest European mountains, are of the same type, and Mount Atlas, in Africa, is also of Plutonic origin. The Grampian and other hills in the north of Scotland, some of the northern parts of Ireland, of the north and west of England, portions of Norway and of Spain are likewise hypogene, chiefly of the plutonic sub-division. Passing to the western Hemisphere we find igneous rocks prevalent in the highest parts of the Andes, metamororphic schists, with beds of limestones resting on the slopes, forming the Rocky Mountains, while the Slates visible in the Appalachian range are by some assigned to the same type, by others considered only a metamorphic condition of silurian strata. The northern regions of our continent have also mountains of hypozoic, crystalline rocks, which probably furnished in part the bowlders* so common in northern Indiana. Part of the north shore of Lake Superior furnishes splendid samples of the ancient volcanic rock, basalt, &c., such as are found also at the well-known localities: Fingal's Cave in Scotland, and the Giant's Causeway, in the north of Ireland. Evidences of extinct volcanoes of more recent origin can be found in Auvergne, about the centre of France; (where some of the old craters are in a high state of cultivation,) the same type prevailing in portions of the Pyrenees. Active volcanoes, furnishing lava and other rocks mentioned in the tabular view, are most abundant in equatorial or at least tropical regions, such as Sumatra, Java, the Moluccas, Phillipine Islands, Central India, Sandwich Islands, the Azores, Canary and Cape Verd Islands, Mexico, the West Indies, Central America, and part of the Andes. Active volcanoes are, however, abundant in some of the northern regions, as the Northern Rocky Mountains, Italy, Sicily, and Japan, also in the arctic and antarctic countries, as the Aleutian and Kurile Islands, Iceland and the antarctic continent discovered by the United States exploring expedition. The slates furnished by the metamorphic rocks are abundant in Norway, Sweden, Scotland, Switzerland and especially Wales.

The palæozoic rocks of *Silurian* date are found in the north and west part of Russia in Europe, in eastern Siberia, in Sweden, Asia, Africa and eastern Australia, in parts of Spitzbergen, Nova Zembla and Terra del Fuego. The system receives its name from prevailing in the west of England and adjoining parts of Wales, in which an ancient tribe of Britains lived, called the Silures. On our continent, the lower part of this system exists in the south of Canada and northwest part of Russian America, extends from the upper Mississippi to

---

* This word being derived from the verb to bowl, the orthography recommended by Webster is employed, as indicative of the derivation, instead of the more common mode of spelling the word with the letter "U."

the southern portions of Minnesota and Wisconsin; sweeps from the Gulf of Saint Lawrence through part of Vermont, a considerable region in New York State, somewhat in Pennsylvania, and is continued in a narrow strip along the eastern slope of the Appalachian range. This lower part of the Silurian system has also formed a plateau of considerable elevation and extent in Ohio, part of Kentucky and Indiana, as well as an upheaval of more limited dimensions in Tennessee; while the upper silurian formation is traceable in New Brunswick, Nova Scotia, Lower Canada, along the south of Lake Ontario, through Niagara, north of Lake Erie, the northern parts of Ohio, Indiana, Illinois and Iowa, besides occupying less extensive tracts in Missouri, Arkansas and Tennessee. The extensive and valuable lead deposits of the United States are chiefly situated in the lower part of the silurian system.

The *Devonian* system, as its name implies, prevails in Devonshire, England, as well as in Herefordshire, Shropshine, Worcestershire and South Wales; it furnished, in northern and middle Scotland, the remarkable fishes and other organisms so ably described in Hugh Miller's Old Red Sandstone; it is not much developed on the continent of Europe except in Russia. In our country it forms a great part of the Catskill Mountains, New York, besides showing itself usually in a more or less narrow belt around the numerous coal fields hereafter described.

The sub-carboniferous sandstones and limestones rest on these Devonian rocks, and in their turn sustain the various sandstones, limestones, clay beds and coal seams which constitute our true coal measures, the whole being embraced under the names of the carboniferous system. The coal deposits are found usually in the form of a basin, at least when not much disturbed by igneous action. Of the fields, in North America, one comprises a considerable portion of New Brunswick and Nova Scotia; another, of an elongated form, extends, under the name of the Appalachian coal field, from Pennsylvania through part of Ohio, Virginia, Kentucky, Tennessee and a portion of Georgia, into Alabama. A third occupies a great part of Illinois, about a quarter of Indiana, and a portion of western Kentucky. A fourth great coal field is situated chiefly in Missouri and Iowa; a fifth and smaller one in Michigan; a sixth in Arkansas, besides other coal deposits, (some perhaps of Tertiary date,) the limits of which have not yet been fully defined, such as those in Texas, Kansas, Van Couver's Island, &c.

In the old continent more than fifteen million tuns of coal are annually mined and consumed in Great Britain and Ireland; one million six hundred thousand tuns in Belgium; one million one hundred and fifty thousand tuns in France; one million tuns in Germany. Coal is also found in Bohemia, Hungary, along the Persian Gulf, in parts of India, China, Japan, Australia, Tasmania, Southern Chili, besides existing in other localities either less known or unimportant.

The *Permian* formation is well developed in the province of Perm, Russia, and in Thuringia, Germany; it also exists in parts of England and Scotland, and has latterly been discovered in Kansas, United States.

The *New Red Sandstone* has been most successfully studied in Connecticut and Massachusetts, but also extends into New Jersey, Pennsylvania and Virginia. It exists in Cheshire, England, furnishing large quantities of salt; also in Saxony, Germany, &c. It is characterized by the tracks of gigantic birds, of which, in Connecticut, President Hitchcock has already distinguished more than thirty different species. In the succeeding system, the *Oolitic*, which embraces, according to some authors, the Lias, Oolite and Wealden, huge reptiles appear to have abounded. Their remains are found chiefly in Germany and England, yet some of the groups of this system exist in Switzerland, Russia and India. Portions of Virginia which furnish coal are considered as Oolite. This formation we must carefully distinguish from an Oolitic limestone, which occurs in our own State and elsewhere among the sub-carboniferous limestones. According to some authors the Wealden forms part of the next division.

The *Cretaceous* system, next in the ascending series, is well developed in England and France, where it supports the short herbage of the high and dry South Down pastures, so well adapted for sheep raising. Some subdivisions of this system exist in Germany, Russia, Italy, northwestern Africa, India and South America; also in parts of the Carpathian mountains, and Pyrenees. One of the lower groups, the Greensand, extends in the United States from Alabama through parts of Mississippi, and Arkansas, besides being found in Nebraska, and locally in New Jersey, Virginia, the Carolinas, and Georgia.

It is in an upper member of this system that the chalk of commerce is found, often enclosing nodules of flint, as well as silicified organisms, around which these flinty particles were deposited. The cretaceous system closes the Secondary or Mesozoic age, and brings us to the Tertiary age, the newer part of which contains fossil shells, of which a large percentage is identical in species with those now living. The oldest formation of the *Tertiary* epoch, called the *Eocene*, is found forming the basins in which London and Paris are built, and in which Baron Cuvier and Prof. Richard Owen studied the gigantic mammalian remains so prevalent in those formations. Eocene Tertiary also exists at Mt. Bolca, in Northern Italy, at Mt. Lebanon, in Syria; also in Greece, Morocco, Algiers, Egypt, Persia and India. On our continent it is traced in the Patagonian Andes, and at various points near our Atlantic seabord, from Virginia to Mississippi.

The *Miocene Tertiary* prevails in the valley of the Adour, Southern France, constitutes the "molasse" of Switzerland, the basin of the Danube around Vienna, and that of the Rhine near Mayence, extends into Poland and Hungary, and is found in India and Siam. In the United States it is most prevalent in Delaware, Maryland, Virginia and North Carolina.

The *Older Pliocene* constitutes heavy deposits on both sides of the Apennines, forms the hills of Rome, and extends also into Greece and Asia Minor. In England it exists as the Suffolk crag.

The *Newer Pliocene* is found in Sicily and the Cyclopean Isles. The English and Australian cave breccias, as well as the Pampas plains of South America, are usually considered as of Newer Pliocene age.

Besides these regular deposits, we have materials that, from their rounded appearance, have evidently been transported a greater or less distance; some of these fragments are hence called Drift, and by some writers are included as part of the Tertiary. Other writers commence a new era with these transported materials, calling it Quaternary; and denominate the older materials as Drift or Diluvium, which they suppose has been conveyed by the action of ice, or water, or both, from the original situation, in which it constituted a rock mass. Under this head would also be included the deposits of marl, supposed to have constituted ancient lake beds. To the latter materials of transportation these geologists apply the term Alluvium, from the Latin word to wash, because these are chiefly the result of late washings, being found in the beds of rivers, particularly at their mouths; also in low, swampy places. Others designate these deposits as Recent, because belonging to the Historical period, and make them include, also, the limestone now forming in the West Indies, the calcareous tufas, depositing wherever water trickles slowly from a soil, or over rocks highly impregnated with lime, as well as the bog iron ores, forming where water impregnated with iron is arrested in its course.

The older drifted materials are abundant in Scandinavia, Siberia, and the northern part of our great Mississippi Valley. In Indiana, the larger bowlders extend as far as 39° south, and smaller materials are abundant where valleys have permitted their passage, as far, at least, as the Ohio river.

The whole subject, however, connected with the Drift phenomena, being very complicated, and at the same time interesting and useful, it has been thought best to devote part of another chapter in the report to an investigation of the facts collected, and of the inferences which it appears fair to deduce from the same.

After this general summary of the geographical distribution of the various geological formations, it may be well to recapitulate again briefly those designated as occurring in Indiana. Our own State has no crystalline rocks coming to the surface, although it is doubtless entirely underlaid by them; and in some cases these hypozoic or igneous rocks may be at no very great distance from the surface. The points nearest to our own State at which, as yet, they have been observed to reach the surface, are in the Cumberland ridge of the Appalachian range of mountains, some two hundred miles from our south-eastern boundary; and the upheaval of these hypogene rocks is considered by geologists as having taken place since the period when the sedimentary rocks visible in Indiana were deposited. This is deduced from the fact that, as a general rule, all our rocks in Indiana have an inclination or dip *away* from this upheaving source. As water finds its own level and deposits its materials usually horizontally, it seems reasonable to draw the above inference, inasmuch as the rocks, west of the Allegheny range, have a general slope or inclination below the horizon, in a westerly direction, (disregarding in this generalization the partial dips effected by the Silurian upheavals and other yet more local disturbances,) until they come within the influence of the Rocky Mountain range, west of the Mississippi river.

Thus, then, in Indiana, we have usually a gentle westerly dip, sometimes a little north of west, sometimes south of west, and occasionally west of south. The dip, or variation of the rocks from a true horizontal line, estimated by their disappearing under the surface of the water in descending the Wabash, the fall of the river being known, appears to be commonly only a few feet in a mile, although accasionally as high as 2;° while some rare local or partial dips are as high, in Indiana, as 45.°

The aqueous or sedimentary rocks observable in Indiana comprise the following systems:

1st. The *Lower Silurian*, chiefly in the south-eastern counties.

2d. The *Upper Silurian*, extending from these south-eastern counties over most of the north and north-west, although partly concealed by Drift.

3d. The *Devonian* or Old Red Sandstone,* having the same direction, but occupying a less extensive area somewhat more southerly than the Upper Silurian.

4th. The sub-carboniferous sandstones and limestones extend from Floyd and Harrison counties, in a belt thirty or forty miles wide, to Tippecanoe county, and thence under the drift probably to Lake Michigan.

5th. The Coal measures, embracing at least twenty south-western counties, besides portions of five others adjoining.

6th. The *Drift*, made up of rounded materials, which, from their form, as already stated, give evidence of having been detached from rocky masses, and transported a greater or less distance. On this point most geologists agree; but they are not yet of one opinion regarding the agency employed in this transportation. Suffice it here briefly to state that, while some attribute the agency to glaciers, many think icebergs carried the larger masses from the northern regions, (in which similar rocks are often found constituting high mountains,) into warmer seas, where they finally stranded on the shallow shores, or sometimes sank to the bottom at deeper places, not however as yet freed from ice. The icebergs or ice masses, acted on by winds and waves, still carried these hard rock-massses, causing them to drift along the rocky bottom, and thus to wear off their own edges and corners, while grooving the flat surfaces of the underlying, usually softer aqueous, rocks, into striæ, or channeled furrows, which can yet be distinctly seen at various places, after the removal of drift from a quarry rock. These particulars will, however, be more fully detailed in a subsequent part of this Report.

The northern portion of Indiana, especially, has had its rocks, and originally deposited materials, covered up in most places, sometimes to a considerable depth, by this drift, or *Quaternary* deposit, as it is termed by some writers, who think a new and marked era was thus in-

---

* The name "Red Sandstone" is not appropriate in the United States, because the fossils characterizing that period, instead of being found, as in Europe, in a sandstone highly colored red by oxide of iron, are, in this country, usually found in limestone.

augurated, after the Tertiary; hence the name,* as before remarked, designed to denote a fourth great epoch in the geological history of the earth.

At the first mention of a section of country thus covered up by drifted and rounded fragments, almost destitute of fossils, we might expect such a region to be comparatively devoid of interest for the Geologist; but it proves far otherwise, when we trace out facts bearing on these drift phenomena, on the formation of prairies, and incidentally, perhaps, on the deposition of materials in basins, now termed coal fields or coal basins, as it will be attempted to show in a later part of this Report. These northern regions of Indiana are, however, besides, by no means devoid of mineral wealth, having tracts of swampy muck, rich in organic materials, often underlaid by deposits of bog iron ore, of marl and of clays suitable for the manufacture of earthen ware, and of the so-called Milwaukie brick; as well as being rich in agricultural wealth, in consequence of the excellent soils formed by the disintegration of the varied quaternary materials. In places too, as will be shown hereafter, this Drift, especially towards the center of our State, contains considerable deposits of gold-dust, brought with particles of its quartz matrix from the original mountain site.

It will appear evident, from the above facts, that, as we travel from the middle-eastern border of our State westward, we go *with* the dip.

Starting from the highest levels in the State, whence our largest streams take their origin, and passing gradually from these geologically low formations (the Lower and then upper Silurian) to the Devonian regions, topographically lower, although geologically higher, and thence to the sub-carboniferous limestones and sandstones, which disappear under the true coal measures, we thus reach finally our valuable coal deposits. This coal-bearing formation is the uppermost and last true geological deposit in Indiana, (if we consider the Drift, as some authors do, too partial and erratic to be classed as such,) but topographically

---

* To render this quite intelligible to the general reader, he is reminded that the early geologists applied the term primary (from the Latin *primus*, first,) to the Igneous rocks, supposing them to be first formed; as, however, these may come to the surface at various periods of the earth's history, later geologists use the word *primary* for the earlier fossiliferous rocks from the Lower Silurian to the Permian inclusive, to denote the first existence or great era of animal and vegetable life; *secondary* (from the Latin *secundus*, second,) embraces the formations from the New Red Sandstone to the Cretaceous, both inclusive; the *Tertiary* (Latin *tertius*, third,) includes, as may be seen in the Table, Eocene, Miocene, and Pliocene; while the Quaternary (from the Latin numeral *quatuor*, four, or the ordinal *quartus*, fourth,) includes, besides the Drift, all the deposits and modifications up to the present day, and such as may continue to be made until some other *great* change takes place.

the lowest, as indicated by the convergence of the Ohio and the Wabash, until the latter empties into the former in the extreme southwest corner of our State.

Each geological formation has its marked differences of soil, forest growth, and adaptation for peculiar agricultural products, as well as its varying materials for the construction of works of art, buildings, bridges, roads, pottery, &c., and also its differing water as a beverage, and to some extent atmospheric variations, hygrometric, miasmatic, &c., producing consequent varieties in the diseases to which vegetable as well as animal life are exposed.

The definite limits of these formations will be found more accurately described, and other details more exactly pointed out, in the subsequent portions of this report, where each county* is taken up in succession, or where several counties, varying but little in character, are embraced under one head, as a District.

---

* Eighty-five out of our ninety-two counties were examined by me personally to some extent; but I do not consider any one of the counties to have been yet nearly as thoroughly explored as it merited, or as the mineral and agricultural interests of the State demand. Time and means did not permit more to be done; indeed, had it not been for the earnest desire if possible to visit at least each county seat, and explain the course adopted, so as to avoid any appearance of partiality, it would have been, perhaps, better to have taken fewer counties, and to have completed the work as we went.

CHAPTER II.

DETAILS OF COUNTIES.

In order to save time and space, as well as less to interrupt the continuity of the subject, it has been thought best to include, under this head, not only the preparatory information obtained, regarding any county or counties, during the fall reconnoissance of 1859, but also the more detailed facts observed or communicated in the spring and autumn surveys of 1860, as well as the data kindly furnished by the different members of the State Board of Agriculture in reply to a series of queries addressed to them regarding their respective districts.

In presenting the details collected it is proposed to follow the natural geographical and geological sequence of grouping rather than the route pursued in visiting different localities, which route, or line of travel, was sometimes purposely made to cross the formations, as affording, although more laborious for man and horse, better opportunies for inspection than when following the strike or continuous ridge level.

Although a county may, by this arrangement, be described as belonging chiefly to a certain formation, yet it must be borne in mind that at least two formations may exhibit themselves in a county and even be modified besides by drift. Therefore in assigning any one to a particular section it is only meant that the described county presents *chiefly* to view the formation under which it is ranked.

Again, a county may afford fossils of a certain geological formation, as the Upper Silurian, particularly in the bed of water courses, and yet the soil may have mainly resulted from the decomposition of the originally overlying Devonian shales, thereby imparting to the county agricultural and other features more nearly allied to Devonian than to Upper Silurian regions. Thus, although when the water is low on the Falls of the Ohio, we find unmistakable Upper Silurian fossils, it would be wrong to describe either Clark or Floyd county as in the Upper Silurian formations, the former deriving its character from Devonian limestone, the latter from sub-carboniferous rocks. For these reasons, although Upper Silurian rocks are found on the Wabash, at Delphi and Logansport, yet Carroll and Cass counties are embraced under the head of Devonian, because in the former the black shales constitute the great plateau and the upland of the latter Devonian limestone is abundantly indicated by its fossils.

On the other hand, Devonian fossils being found at Pendleton, Madison county, would seem to justify its being described under that head; but, as the majority of the county has Upper Silurian rocks, it was deemed best in that and similar cases to classify the county by the prevalent formation.

In a region chiefly covered with drift, there may be, as on the Little Menon, at West Bradford, sufficient rock exposure to identify that part of White county as Upper Silurian, overlaid by Devonian, and so with other drift counties; yet the disintegration of the quaternary materials furnishing the chief elements of the soil, that county is embraced among the counties of the drift formation. These examples may serve to explain many similar cases. In accordance with the skeleton index of reference given at the commencement it will be perceived that, before describing the counties in detail which belong to any geological system, it was considered best to speak—

1. In general terms of that geological formation and its prevalence in our State.

2. Of the soil usually resulting from its decomposition; its adaptation for different agricultural products and stock; also, of the materials, if any, wanting to render it highly productive.

3. The coal, if any exists, would next be enumerated; otherwise the quarries would come in order, such as those affording building rock, materials for roads, grindstone and whetstone quarries, &c.; also, deposits of marl, hydraulic limestone, gypsum, clays for pottery, firebrick, and the like.

4. The metals would next demand attention, as iron, lead, zinc, gold, or any others found in that geological system.

5. The growth of timber and leading vegetation, whether suitable for exportation in the form of veneers, hoop-poles, tanning material, medical roots, &c.

6. Mineral springs, artesian wells, and similar subjects.

7. Miscellaneous facts, such as the prevalence of milk-sickness, potato rot, hog cholera, &c.

8. Specific enumeration of the fossils found in Indiana imbedded in that geological formation, by an inspection of which it may be recognized.

9. A detailed description of each county in the formation.

Commencing, then, acording to the above plan, with the oldest geological deposits found in our State, we have first to describe, in detail, those situated in the Silurian System.

SEC. I.—COUNTIES IN THE LOWER SILURIAN FORMATION.

Sub-Section 1.—General Charcter of the Lower Silurian Formation and its Prevalence in Indiana.—Eight of our south-eastern counties are situated in this lower sub-division of the Silurian system, viz.: Wayne, Union, Fayette, Franklin, Dearborn, Ripley, Ohio and Switzerland. Several adjoining counties exhibit at deep natural or artificial cuts this Lower Silurian formation, especially Jefferson county, also the eastern parts of Decatur, Rush, and probably of Henry, besides the southern portion of Randolph, as nearly as could be determined under the heavy drift.

The New York Geologists have distinguished the Lower Silurian formation into seven different groups, besides subordinate members of some of those groups. The portions which seem most prevalent in the West are the Trenton Limestone and Hudson River Group, extending from our State into Ohio and Kentucky, under the name, usually, of the Blue Limestone, and constituting the hills around Cincinnati and Frankfort, as well as appearing at Nashville, Tennessee, all of which localities afford good fossils.

These middle and upper groups of the Lower Silurian formation are chiefly beds of limestone, with intervening spaces in which a deposition of clay predominated and gave rise, by compression, to intercalated beds, sometimes called mudstones, more frequently, argillaceous shales. The limestone is often of a deep blue color, passing into gray, crystalline, sometimes hard and compact, usually rich in fossil remains. The mudstones, although at first apparently solid, rapidly attract moisture, in consequence of their argillaceous composition, and soon disintegrate.

The fossils most abundant in these groups will be found specifically enumerated at the close of this section; but in general terms it may be remarked that the Trenton Limestone, estimated in New York to be about 400 feet thick, and the Hudson River Group occupying, according to the same authority, 700 feet, are both characterized not only in the United States, but also in Europe and elsewhere, by several genera of Trilobites, (singular crustaceans, not unlike some of our crabs,) which seem to have frequented the shores of the Silurian seas. Besides these, Brachiopods, remarkable bivalve shells of which a few genera, not however identical in species, are still to be found in the Mediterranean Sea and elsewhere, were exceendigly abundant, indicating, according to Prof. Forbes, a depth in the ocean of at least four or five hundred feet. Of the living genus Terebratula, one fossil species of which is very common in the Indiana Blue Sandstone, Prof. Forbes dredged samples from the nullipore mind of the Mediterranean, at the depth of about 250 fathoms or 1,500 feet.

The junction between the Lower and Upper Silurian is recognized in our Western States, by a coral assigned to the Hudson River Group. This, resembling a wasp's nest petrified, and called by Milne Edwards, Columnaria alveolata, (formerly Favistella stellata,) has been found in Indiana at various part of the confines between the Lower and Upper Silurian formation, as that at the Madison cut, not far from the top, also at Enochsburg, about the junction of Decatur and Franklin counties, at places in Wayne, Randolph, and even Delaware counties. The specimen from Enochsburg, now deposited in the State collection at Indiananapolis, weighs 153 pounds, and is evidently a single mass, originating from one parent stock, the apex around whose nucleus successive generations grew, in constantly increasing ranges of concentric communities.

SUB-SECTION 2.—CHARACTER OF THE SOIL RESULTING FROM THE DISINTEGRATION OF LOWER SILURIAN ROCKS, AND ITS IMPROVEMENT.—It was ascertained during the process of the Kentucky survey that the soils in this formation, judging as well by the analytical proof as by the evidence offered in the crops, were rich in the lime and phosphoric acid so necessary for the growth and filling out of small grain and grasses. So much so that it was considered such lands were more likely to remain permanently productive than some rich black soils, deficient in these inorganic ingredients. The rocks of this system usually abound in fossils, and in the third volume of the Kentucky report, the axiom is laid as an established fact in Agricultural Chemistry that "*the more replete the rock has been with fossil organic relics, and the more earthy and easy of decomposition the calcareous rock, the more productive the soil derived therefrom.*"

Of course where these these limestones and marlites exist, there is not likely to be any deficiency in calcareous matter, and usually the intercalations of agrillaceous materials furnish as much alumna or clay, say 10 to 30 per cent., as is desirable for mechanical* mixture with the average per centage of sand and insoluble silicates. This sand, even in aluminious soils, amounts to from 70 to 85 per cent. or more, while in

---

*Alumina or clay has very rarely, if ever, been found in the ashes of plants, although abundant in the soil from which that plant grew, and very necessary to prevent the water charged with nutrition, from filtering through to rapidly.

some formations, made up greatly of sandstones, the silex or sand rises to considerably over 90 per cent. In addition to the lime and phosphoric acid, the soils resulting from the blue limestone and mudstones, contain considerable quantities of sulphuric acid, potash, soda, magnesia, iron and manganese.

If we examine carefully the best works on agricultural chemistry, we find that any soil, in order to be fertile, must have, besides a certain amount of organic matter, which may vary from 5 to 40 per cent., several inorganic substances, at least in small quantities, these are lime, potash, soda, magnesia, phosphoric and sulphuric acid, with, usually, some iron and manganese, the whole diffused through a mechanical mixture of clay and sand, which should never exceed 60 per cent. of the former, nor over 92 or 93 per cent. of the latter. On the other hand the same works will show that barren soils are usually especially deficient in potash, soda, phosphoric acid, and perhaps also lime.

By an inspection of the soils derived from the Lower Silurian formation, analyzed by Dr. Peter for the Indiana survey, or the still more extensive sets of the same for Kentucky, it will be seen that these blue limestone soils have almost invariably, at least in the virgin soils, a fair proportion of all the essentials of fertility. Therefore all that is necessary to preserve them in good heart is to restore to the soil a fair amount of the ingredients taken from it, or to retain on the farm, as far as practicable, by stock grazing, the materials raised. This, however, is by no means the case with soils originally deficient in some essential inorganic element, which may require us to resort to an expensive dressing with lime, plaster or bone dust. When a soil in the Lower Silurian formation has, through ignorance or neglect, been exhausted by cropping, it would also be much more easy to return to it, than to an originally defective soil, the lost ingredients, by limited amounts of lime, plaster and the like, applied in the hills, or even by rolling the soaked seed grain, in such a manner in plaster and the like, as to cover each grain with a moderate coat. Perhaps the same result may be obtained even at a less cost by sub-soil plowing, bringing up for intermixture with the surface soil, a few inches at a time, if previous analysis has proved that the sub-soil contains the necessary ingredients, as was demonstrated to be the case in many of the Kentucky sub-soils of Blue Limestone origin.

The fact was already alluded to that soils such as these, containing considerable amounts of lime and phosphoric acid, are well adapted for the growth of cereals and grasses. Hence the cause why we find this blue limestone portion of Kentucky, comprising the middle counties of that State converted into great stock farms. For the same reason the south-eastern counties of Indiana have gone largely into grazing, wheat and hay raising; the counties, such as Wayne, Fayette, &c., apparently preferring the two former, because the stock and small grain can be more readily shipped by railroad, while Dearborn, Ohio and other counties on the Ohio river, find it profitable to bale and ship their hay to the great river cities; as well as to grow Indian corn extensively in the river bottoms, a crop that requires, and there finds, considerable amounts of potash, magnesia and phosphoric acid.

Portions, however, of this formation are well adapted by local circumstance for other besides gramineal crops; thus parts of the blue limestone of the Ohio have been most successfully cultivated in vineyards, and that culture seems gradually and favorably extending itself into some of our river counties in this as well as other geological formations. Especially when the steepness of land, or its mechanical character, renders it subject to wash on a hill-side, under plow cultivation, some such crop as the above, which may even be grown in terraces, is worthy of being considered when we are making our decision as to the adaptation of land to certain agricultural products.

More than thirty years since the Swiss settlers, at Vevay, Switzerland county, produced a red, light wine, resembling claret, which sold readily in Cincinnati, not at so high a price as the Catawba, but of a quality better adapted to general use, where only a mild stimulant is desirable.

As another recommendation of the blue limestone soil, we must not omit to mention that the clay resulting from the decomposition of the aluminous shales, already described as intervening between the beds of harder limestone, is sufficient to give the tenacity necessary for the retention of manures or other fertilizers, when the agriculturist considers it expedient to employ them.

SUB-SECTION 3.—QUARRIES OF MATERIALS SUITABLE FOR BUILDINGS, ROADS, GRINDSTONES, WHETSTONES, AND FOR BURNING QUICK LIME OR HYDRAULIC LIME; ALSO, DEPOSITS OF MARL, GYPSUM, CLAYS FOR POTTERY FIRE BRICK, &c.—There being no prospect for finding coal in a formation geologically below the carboniferous, such as this, we next proceed to examine whether any of the minerals enumerated in the heading to this third sub-section show themselves, in Indiana, within the sub-division of the Silurian system. As a general rule, the older the deposit, and the more it has been submitted to the compression produced by subsequently deposited materials, the more compact the limestone, sandstone or shale is likely to prove. Thus the limestones of the palæozoic period are usually harder and more compact, consequently more capable generally of sustaining vertical pressure in the form of foundation stone, or of resisting cross-fracture, as in door and stair steps, door and window sills, &c., than the limestones of very recent date, such as the tertiary limestone of Mexico. These I saw the natives, at Monterey, dressing with broad axes, when first quarried, but after the evaporation of the quarry water the rock appeared to consolidate into pretty fair building materials for superstructures, not involving great strain.

Such being the case, we would naturally expect to find the Blue Limestone of Indiana and elsewhere furnishing good building materials. This is actually the case if care and judgment are exercised in the selection of the rock. I observed some years since, when in Kentucky, that a bridge near Georgetown, Scott county, had been built of blue limestone rocks, in which the aluminous materials formed so prominent an ingredient that the bridge was rapidly crumbling. Yet many portions of Kentucky, Ohio and Indiana furnish from these groups (the Trenton Limestone and Hudson River Group,) some layers of solid crystalline limestone, less replete with fossils than other adjoining layers, which by this selection will be found well adapted for building, especially for foundations, where great sustaining strength is required.

As will be seen by an examination of the sub-section giving in detail the counties in Indiana of this geological formation, there is no lack of rock affording strength and durability sufficient for ordinary building purposes; the presence occasionally of organic remains prevents some varieties from receiving a regular face in dressing; yet shells are somewhat abundant in the beautiful and durable marble of Jefferson county, fully described in the former report of the late State Geologist. The recommendation of that marble as a good building material belongs here, although the county is described among the Upper Silurian, because the bed is of Lower Silurian age, being found on the Ohio river beneath the Upper Silurian formations, which mainly characterized the upland of that county.

The same blue limestones, above alluded to, can be found abundantly through the counties of this section, for the construction of turnpikes, and, by selection, the rock burns also into a good, strong quick lime, well adapted for mortar, although not so much sought after for hardfinishing or whitewashing as the lighter colored varieties furnished from the sub-carboniferous limestones. No extensive beds of hydraulic limestone have yet been found in the Lower Silurian, although a deposit near Connersville partakes of that character. Neither is this the formation in which we are most apt to find beds of marl; yet some of the decomposing marlites have to some extent the properties of the less calcareous marls. In this connection we must remember that the term marl is rather indefinite, as Prof. Johnson, in his agricultural chemistry, giving the analysis of one variety having only about 8 per cent. of carbonate of lime, while another has over 85 per cent.

Gypsum and potters' clay are not so usually found in this as in some other geological formations, and our south-eastern counties form, apparently no exception to the rule, for no such deposits were brought under our notice in the Lower Silurian formation.

Clay for fire brick, demanding a freedom from lime, magnesia and iron, would not be apt to occur in regions where the rocks furnish, by their disintegration, those ingredients detrimental to fire clay, but beneficial, as already remarked, in soils otherwise suitable for the growth of corn, small grain and grasses.

SUB SECTION 4.—THE METALS IN THE LOWER SILURIAN FORMATION OF INDIANA.—Iron, the most truly valuable of all metals, is more generally found in the Carboniferous than in the Silurian system, consequently we do not find any important deposits in these counties of Indiana.

Gold, hitherto washed in our State only from the Quaternary deposits, cannot be expected to be in any considerable quantity in the Lower Silurian formation of Indiana, because the Drift has been mainly arrested somewhat north of them, as we shall see hereafter. Silver, in the United States, has been found native at a few localities, but is more usually associated with native copper, as in Michigan, or with galena, to the amount sometimes of three per cent., in those lead ores of the Western States. Therefore, it is not impossible that some might exist in the galena of these counties. Galena, or sulphuret of lead, is found most abundantly in the Silurian system of Iowa and Wisconsin, therefore we might not unreasonably expect to find this ore in these counties of Indiana. We were shown several localities in Ohio county where at former periods galena had been taken out; but the indications did not promise a large yield, so far as we could judge from a hurried inspection in the midst of very heavy rain. Copper is generally found in the United States either native, as about Lake Superior, or in the form of oxides, sulphurets or carbonates, as in Pennsylvania, New Jersey, North Carolina, Tennessee, and other States. In some of those localities it occurs in Trappean dikes or walls of upheaval. Such igneous rocks not coming to the surface in Indiana, veins of metallic copper are not very likely to be found. Some masses of copper picked up in Indiana were evidently rounded by attrition, and had been transported by the Drift probably from the region of Lake Superior.

No other metals have as yet been detected in considerable quantity in our Lower Silurian formation.

SUB-SECTION 5.—THE GROWTH OF TIMBER AND OTHER LEADING VEGETATION.—As might be expected from the considerable amount of clay, derived from the marlites, Beech timber is very abundant in the Lower Silurian counties; perhaps it may be correctly represented as the prevalent forest growth. By an analysis of the ashes of beech, that tree evidently requires for its growth considerably more lime (42.6 per cent.) and of phosphoric acid (5.7) than the coniferous trees, such as the Pitch Pine, stated by Prof. Johnston, in his Agricultural Chemistry, to afford 27.2 of lime, and 1.8 of phosphoric acid. These inorganic matters, as already shown, are abundant in our blue-limestone. Besides Beech, however, Sugar Maple, Oaks, and Poplar or Tulip tree are not uncommon. In one part of Union county, grey and blue Ash, and some Black Walnut, were added to the above list; and on the celebrated Walnut Plains, near Jacksonburg, Wayne county, White Walnut and some Wild Cherry combined with the majestic Poplar (Liriodendron tulipifera) and stately Beech to beautify the landscape. In the details of this county will be found mention of a species of Locust tree resembling the Black Locust, in addition to the abundant Sugar Maple, White Oak and Black Walnut.

Everywhere along the lines of railroads traversing this section of country, piles of staves, hoop-poles, &c., evinced the fact that, notwithstanding our well-known lavish destruction of the primeval forest, a dense growth of fine timber still blesses this as well as many other portions of our State.

The apple and most other fruit trees prove, by analysis of their ashes, that they demand also considerable supplies of lime. As might be expected, therefore, orchards thrive well here, as indeed in most parts of Indiana, (except that the peach trees in the northern parts were killed some years since by severe frost;) several fine nurseries were also noticed, and the osage hedges testified that only correct culture, close cutting the second and third years, with some trimming afterwards, is wanting to make these supply good enclosures when timber becomes scarce.

SUB-SECTION 6.—MINERAL AND OTHER SPRINGS, ARTESIAN WELLS, &c.—
Some of the counties in the Lower Silurian formation are probably topographically higher than any other portions of Indiana, portions of Wayne near Randolph and Henry, as well as of Fayette near Rush county, also the region near the line of junction between Decatur and Franklin, being, according to my barometrical estimates, made at various times, fully 1,000 feet above high tide in the Gulf of Mexico. That this region is the highest, is further indicated by the fact that our largest streams take their rise in this section of our State. The Wabash heads in Ohio near Randolph, and the west fork of White River has its source in that county. The east branch, or Driftwood fork, of White river, heads in Henry county near Randolph, and Whitewater takes its origin chiefly in Wayne.

But, to prove this matter incontestibly, I would refer the reader to the table of altitudes in Indiana, reported to the Legislature, Jan. 20th, 1836, by Messrs. H. Stansbury and J. L. Williams, Civil Engineers, a copy of which report was politely furnished me by the latter gentleman.

These tables give two hundred and eight different altitudes, from surveys made. 1. Along Whitewater valley; 2. From Indianapolis to Lawrenceburg; 3. From Indianapolis to Madison; 4. From Indianapolis to Evansville; 5. From Terre Haute to Evansville; 6. From New Albany to Vincennes; 7. From New Albany to Crawfordsville; 8. From Indianapolis to LaFayette; 9. From Indianapolis to Wabash and Erie Canal; 10. From State Line to Terre Haute; 11. From Lake Michigan to the Wabash and Erie Canal; 12. Levels in various sections of the State.

The level at three places in the Lower Silurian, near its junction with the Upper, exceeded 1,000 feet above high tide in Hudson River; the summit between Sand and Salt Creeks, near the eastern line of Rush county, and only a short distance from three other counties, Fayette, Franklin and Decatur, being the highest point in the Table of Altitudes, 1057 feet above high tide. But one other level in the State, examined by these surveys, attained over 1,000 feet; that point is the summit between the head waters of White Lick and Eel River, in Hendricks, near Boone and Montgomery counties, part of the Knob or Sub-carboniferous Sandstone formation.

As might be anticipated, although some portions of the extreme summit levels are deficient in full supplies of good water, yet the majority of the blue limestone region is well watered. Portions of the northern counties in this formation are partially covered with Drift, and, by digging through this, or even only a portion of those quaternary deposits, sometimes only twelve to twenty feet, good water is often reached, which has filtered through these more porous materials, and been arrested by an impervious substratum, either of quaternary clay or of Lower Silurian rocks.

With regard to Artesian wells in this region, the theoretical probabilities are rather against their success, because greater heights of land are not likely to be found in the vicinity dipping their beds beneath these localities, and sending water along an impervious, inclined stratum of rock under the artesian boring, thus furnishing head enough to raise the water in the tubing to the same height from which it originally descended. In other words, success is more likely to occur in beds having a synclinal axis, or converging slopes, than in strata having an anticlinal axis, or parting slopes.

Several Artesian wells are to be found in Rush county, west of these described heights, and usually chalybeate in their character; they will be more fully described in speaking of the Upper Silurian formation, and are only mentioned here to show that they occurred where theoretically we might expect them. As the expense of verifying this matter is not very great, probably one dollar per foot for a boring of moderate depth, when the object of thus finding artesian water is very great, the most advisable plan seems for those interested to unite and test it practically. The water generally is, as we might expect, hard from the limestone; and mineral springs, some sulphurous, some chalybeate and some saline, are not uncommon; one which I examined, in Union county, is strongly chalybeate, in other words, impregnated with iron.

SUB-SECTION 7.—MISCELLANEOUS FACTS REGARDING DISEASES, &c.—In former years there was a considerable amount of milk-sickness in Franklin county; but as usual under cultivation, it is disappearing. The member from the agricultural District in which several of these blue limestone counties are situated, reports the ratio of the diseases to which the inhabitants of those regions are subject, to be about the following:

Typhus and typhoid fevers, about..20 per cent.
Bilious, remitting and intermitting, about..............................40 per cent.
Consumption, about ....20 per cent.
Rheumatism and other inflammatory diseases, about............20 per cent.

Little or no hog cholera has shown itself; potatoes, however, the same gentleman describes as frequently diseased, and insects injurious to agriculture quite abundant.

SUB-SECTION 8.—SPECIFIC ENUMERATION OF THE FOSSILS MOST COMMON IN THE LOWER SILURIAN COUNTIES OF INDIANA.—The fossils found in these early formations throughout the Globe are chiefly Corals, Mollusks and Trilobites, occasionally a star-fish or a stone-lily (crinoid;) also, some marine plants. The most common in our Indiana* Lower Silurian are the following:

A. RADIATES.—
    a'. *Amorphozoa*: Syphonia (Scyphia,) digitata, (Owen.)
    a. *Corals.* Chætetes petropolitanus, (lycoperdon,)
          C. rugosus,
          C. frondosus, [C. Pavonia,]
          C. mammulatus,
          Ch. trigiri,
          C. ramosus,
          Protarœa, (Porites, Hall,) vetusta,
          Streptelasma corniculum,
          Columnaria alveolata, (Favistella stellata, Hall,) [Constellaria antheloidea,]
          Fungia corrugata.
    b. *Acalephs.*
    c. *Echinoderms*: subdivision, *Crinoids.*
          Glyptocrinus do. decadactylus, [Hemicystites parasitica, Hall.]

B. MOLLUSKS.—
    d. *Molluscoid Bryozoa.* Escharopora (Ptilodictya, Lonsd.,) recta.
    e. *Brachiopods.* Rhynchonella, (atrypa) increbescens,
          Orthis occidentalis,
          O. testudinaria,
          O. (Spirifer) biforatus, var lynx,
          Strophomena (Leptæna) alternata,

---

* Some characteristic fossils, found in Ohio or Kentucky, very near some of those Indiana counties, are given between brackets, as being provisionally extra-limital. Doubtless, when time permits a more extended search, they can be found in Indiana also; the names in parenthesis are those formerly used instead of the name immediately preceding, which is selected as being now most generally approved.

       Leptæna sericea,
       Strophomena planumbona,
       S. tenuistriata,
       S. alternistriata,
       S. deltoidea.
 f. Conchifers. Ambonychia radiata,
       Modiolopsis modiolaris,
       Orthonota parallela,
       O. contracta, [Tellinomya (Nucula) levata,]
 g. Pteropods. [Conularia Trentonensis.]
 h. Gasteropods. Bellerophon bilobatus,
       Plenrotomaria percarinata,
       P. (Cyclonema) bilix,
       Murchisonia bicincta and bellacincta,
       Cyrtolites ornatus.
 i. Cephalopods. Orthoceras vertebrate,
       [Cyrtoceras constrictostriatum,]
       Trocholites ammonius.

C. ARTICULATES.—
 k. Worms.
 l. Crustaceans: subdivision *Trilobites.*
    Calymene senaria,
    Asaphus canalis, (Isotelus gigas) [Trinucleus concentricus, Ceraurus pleurexauthemus.]
 [Cy-therina Baltica.]
 m. Insects.

D. VERTEBRATES.—
 n. Fishes.
 o. Reptiles.
 p. Birds.
 q. Mammals.

SUB-SECTION 9.—DETAILED DESCRIPTION OF EACH COUNTY IN THE FORMATION.—

WAYNE COUNTY.

The greater part of this fine county is of Lower Silurian formation. Dr. Plummer, of Richmond, who, in Silliman's Journal, gave a minute description of the vicinity of that town, informed me he found Upper Silurian fossils, not many miles off, on Elkhorn. I obtained them also at Macksville, in Randolph county, and therefore consider the out-crop of the Upper and Lower Silurian junction to be at no great distance from the line uniting Randolph and Wayne.

The soil of the latter county is generally sufficiently rich, very durable, and well adapted to the growth of grasses. Hence this is a great grazing county, and sends remarkably fine stock to our fairs and markets.

The calcareous ingredients derived from the disintegration of the Blue Limestone in this county, are intermingled, to a considerable extent, with the Drift, which somewhat modifies and even improves the soil. In portions of the county bowlders are abundant, in others the Quaternary is represented chiefly by gravel, from a few feet thick to 40 and 50 feet, furnishing materials for their excellent turnpikes, the coarser gravel being laid on first, and the finer serving to consolidate the road into a smooth and hard surface. From beneath this gravel-drift numerous springs flow out after reaching the clay, rock, or other impervious substratum, thereby gradually excavating small valleys of denudation at intervals, sufficient to produce the gentle undulations so important to thorough drainage of the country, the whole thus conducing admirably to unite all the requisites for a fine grazing region.

As bearing on the Drift phenomena of this section of country may be mentioned here the remarkable natural channel, exposed during road excavations. Crossing White Water on the bridge at Richmond, and ascending the opposite hill, by the Dayton turnpike, (National Road,) under the polite guidance of Mr. Dennis, now Secretary of the State Board of Agriculture, and of Mr. Bennett, member for that agricultural District, we had an opportunity of ascertaining that this channel, so far as exposed, bears in a direction west of north and east of south, and must be more than a hundred feet above the present level of the river.

At the office of the *Palladium*, through the kindness of Mr D. P. Holloway, member of Congress from that District, I obtained some interesting samples of soil for analysis: one of rich black meadow muck, the other poor unproductive upland, with a view to ascertaining whether they might be advantageously mixed. Our limited means have prevented these as yet from being reached in analysis, as selection had first to be made by Dr. Peter of a few, such as were more especially characteristic of each geological formation.

As might be anticipated the scenery is pleasing, fine farms succeeding each other in rapid succession, as we pass along the line of the Indiana Central Railroad, through Dublin, Cambridge City, Germantown and Centreville to Richmond. Also, afterwards in going South, with our camp, through Milton towards Fayette county. Good barns, meadows, shocked corn, orchards, a nursery near Centreville, occasional Osage Orange hedges and small fields of Sorghum, and woods' pastures with fine cattle and sheep, all indicated a high state of agricultural development, due, doubtless, partly to natural advantages, but greatly to the intelligence and industry of the Quakers or Friends, who extensively colonized this county by emigration from Ohio.

Good building materials are by no means deficient. By judicious selection out of the fifty of sixty feet of Blue Limestone which was measured from White Water up to the town of Richmond, solid, durable rock can be obtained, although usually only from six inches to a foot thick. Nearly the same observations would apply to the quarries of Mr. Eli Henby and of Mr. Burgess, which were examined near Cambridge City, as well as probably to other localities in Wayne county; the drift, as already mentioned, furnishes fine materials for road making.

This county is well timbered, Beech prevailing to some extent, also Sugar-Maple, Black Walnut and Hickory. A few miles north of Cambridge City we observed a small swamp-muck prairie, interspersed with willows, flags and boneset, (Eupatorium perfoliatum,) and skirted by slopes made up chiefly of gravel.

There is an interesting water fall, about twenty feet in height, some five miles south of Richmond, which we regretted not having time to visit.

The Lower Silurian rocks of this county afforded, both at Richmond and Cambridge City, fossils from the middle and upper groups of that formation, being partly characteristic of the Trenton Limestone, described in the New York Geological Reports, but more especially of the Hudson River group, such as I had found near the south-eastern limit of this same Silurian upheaval, while I resided in Nicholas county, Kentucky, near the well-known Blue Lick Springs. The organic remains mostly abundant near Richmond are Chætetes mammulatus, Rhynchonella (Atrypa) increbescens, Ambonychia radiata, Cyrtolites ornatus, Bellerophon bilobatus, Pleurotomaria (Cyclonema) bilex, and similar associate fossils.

The drift phenomena near Cambridge City are highly interesting, furnishing a fine opportunity for examining and studying the polished and grooved surfaces of rocks, supposed to have received their polish and peculiar marking partly from the gyrations incident to the somewhat arrested progress of the hard, movable, superincumbent masses, partly from more direct linear transportation over them of the hard bowlders of other materials of the drift period.

Mr. Henby's quarry, above alluded to, furnished fine polished and striated slabs for the State collection. The location is on the east half of the north-east quarter of section 33, township 16 north, range 12 east, about a mile west of White Water, in a bottom partially denuded and partially covered with the Quaternary bowlders and gravel; a foot or two of clay usually intervening between the rock and the gravel. The bed rock has a slight westerly dip, the seams or joints are about five to seven feet apart, and the surface of the rocks quarried in this region is smoothed and polished as if by the long grinding under water of hard materials. These polished slabs, when closely examined, are found frequently to have grooves running along them, sometimes only a few inches apart, and commonly, according to Mr. Henby's observations, these grooves and also the seams run a little north of north-east, consequently also south of south-west.

This would perhaps indicate that the drift which, judging from the general phenomena in Indiana, usually pursued a direction somewhat west of north to south of east, had been arrested by the Silurian hills, probably constituting the highest ground, partly above the water level of that period, partly beneath it, and that the stranded materials, now arrived in a latitude warm enough to melt portions of the transporting ice, were carried over the underlying rocks so long as to rub them smooth, then to send some of the harder and not yet detached bowlders in a direction a little south of south-west, to pass around the Silurian elevation. The clay of abrasion derived chiefly from the mudstones or argillaceous materials of the Silurian rock, would be apt to settle first somewhat unconformably on the gently inclined rocky bed, and superimposed on this clay, finally rested the gravel and other quaternary materials, as the quiet waters gradually receded, leaving layers of coarser and finer sand, gravel and silt, so horizontally deposited as to furnish no angle appreciable by a delicate clinometer compass.

Dr. Johnson, of Cambridge City, who had kindly directed us to the above interesting quarry, also showed us a slab from Jacksonburg, about five miles north-east of the other locality. The same gentleman also gave us fragments of some rare bowlders from the drift of that region.

UNION COUNTY.

This county, like several others in the Blue Limestone, is topographically high, nearly as much so as any portion of Indiana, parts of it being at least a thousand feet above high tide.

By reference to the report of Dr. Peter it will be seen that the soils of this county, two samples of which have already been analyzed, sustain the character given to those of the Lower Silurian formation generally, having in the virgin soil a fair per centage of those essentials to fertility, lime, potash, soda, magnesia, phosphoric and sulphuric acids.

The main agricultural products of this county are corn and hogs; gradually, no doubt, more meadows will be introduced into their farming system.

Near Liberty, the county seat of Union, measuring from Silver Creek up, are found exposed about thirty-five feet of blue limestone, overlaid by forty to fifty of drift. Judicious selection here, and in other portions of the county and formation, generally affords substantial, if not thick, materials for many building purposes, as well as abundant "metal" hard enough for turnpiking.

The county is also well timbered, the prevailing forest growth being Beech, Oak, Sugar-Maple, Poplar* and Black Walnut.

Notes were taken for the future examination, if opportunity occurred, of a Chalybeate spring, near Liberty. The diggings from a well in the same locality afforded first gravel, then sharp sand, sometimes compacted into absolute rock.

The fossils in this portion of the Lower Silurian formation are very similar to those found at Richmond, Wayne county. The most predominant are Chætetes petropolitanus, Protaræa vetusta, Steptelasma corniculum, Rhynchonella (Atrypa) increbescens, Orthis occidentalis, O testinaria, Strophomena (Leptæna) alternata, and Leptæna sericea, Ambonychia radiata and Cyrtolites ornatus; but no trilobites that we could find during the short time we were able to devote to the search.

---

*In the West, where we use the term poplar as applied to trees or plank, we always mean to designate the Tulip tree or Liriodendron tulipifera, not any of the family of *populus* which includes the Cottonwood, Aspen, &c.

FAYETTE COUNTY.

Passing from the northern line of this county towards Connersville, the county seat, we traveled through parts of the valley of White Water. Extensive bottoms rising into gentle and undulating drift elevations, exhibited fine farms and the prospect of abundant corn crops. Near town the osage hedges betokened high cultivation, and the mill race, with extensive buildings, indicated where a part, at least, of their staple product, wheat, receives its preparation for the flour market. Pork and beef are also largely produced in this county. Although the soil in places appeared clayey, indicated by the ponds along the road sides, yet it was susceptible of fine pulverization by the harrow, and the wheat which, on the 19th of September, the day we passed through Connersville, had already been put in on several farms, was, much of it, drilled in excellent order. This system of drilling wheat appears to be rapidly gaining in the estimation of our farming community, as rendering it less liable to freeze out, besides saving seed and distributing it more rapidly than even a long experience in broadcast sowing can possibly secure.

The prevailing timber is Oak and Beech, occasionally thinned out so as to form fine woods pastures, in which the blue grass (Poa pratensis*) thrives kindly.

Building materials are abundant, rock being extensively quarried in tolerably heavy layers at several places near the county line of Franklin, and across the line at Somerset, as well as on Williams' Creek, near which locality they also manufacture hydraulic cement from limestone.

Adjoining Williams' Creek, two or three miles west of Connersville, we found, in about twenty-five feet, vertical thickness, of blue limestone, interspersed with marlite, abundant samples of the following fossils: Chætetes petropolitanus, Streptelasma corniculum, Rhnychonella (Atrypa) increbescens, Strophomena (Leptæna) alternata, S. planumbona, Leptæna sericea, Orthis testudinaria, portions of Calymene senaria, and of Asaphus canalis, (Isotelus gigas.)

In traveling towards the extreme western limit of Fayette, about four and a half miles from the Rush county line, we found, at a deep natural cut, a fine exposure of the upper members in the Lower Silu-

---

*The less common blue grass of Botanists is Poa compressa.

rian formation, surrounded by a reddish silico-calcareous rock, apparently of Upper Silurian age, although we failed to find any fossils in it. The natural section* furnished in this hundred and ten feet exposure or valley of denundation, gave the following succession of strata:

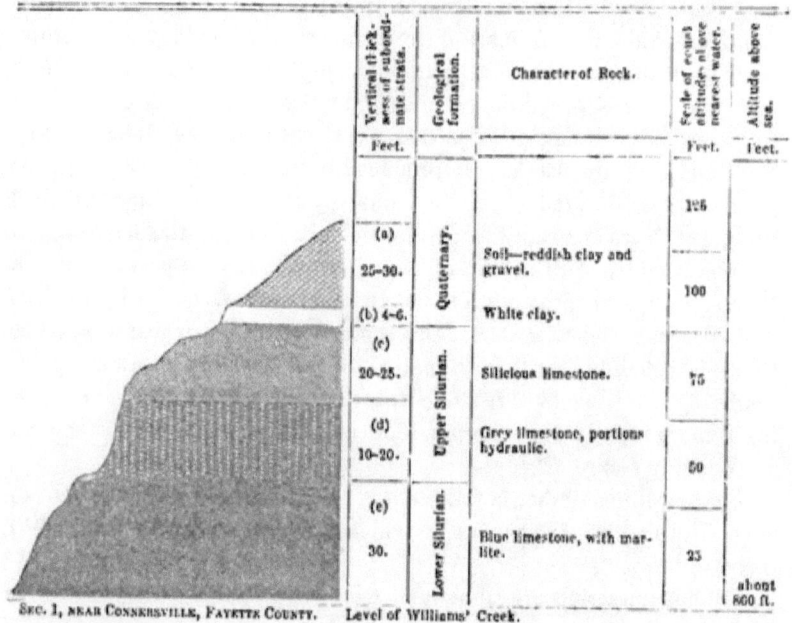

Sec. 1, near Connersville, Fayette County. Level of Williams' Creek.

Nearer Connersville the member (a) in the above series had a greater vertical thickness, comprising in the descending order three feet of soil and subsoil, twenty-five feet of gravel, ten feet of sand, six feet of blue clay and twenty feet of bowlders intermingled with gravel.

Soon after passing the locality which exhibited section 1 we ascended still higher, over coarse gravel and bowlders, to about the highest land in the State; the Bar. at 2 P. M. falling to 28.97 inches, although it stood a few hours before at 29.28 at Connersville. Allowing that it had fallen, as it often does in the afternoon, about 2-100th, still we had ascended two hundred and seventy feet after leaving Connersville. We continued sometime on this elevated plateau with but little variation in the Bar., passing some very fine farms and a dense growth of large

---

* In order that the general reader may more readily understand the sections given in this Report, it is thought useful to subjoin the following description and illustration of the markings herein adopted, as the Normal or

OF INDIANA. 47

Beech, Sugar-Maple and Oak timber, with Papaw undergrowth, even beyond Vienna, the western limit of the county, that town being built in Rush up to the Fayette line.

We readily perceive from observation that a great portion of the surface soil in this county is Drift, amounting sometimes to fifty or sixty feet in vertical thickness, which has thus greatly modified the soil from that of pure Upper Silurian detritus.

TYPICAL FORM OF THE SECTION.

Vertical thickness of subordinate strata. Feet.	Geological formation.	Character of Rock.	Scale of equal altitudes, counting upwards, usually from or nearest water level. Feet.	Altitude above high tide in the ocean. Feet.
		Soil, sand or fine gravel, chiefly recent Quaternary; also, tufas, shell marls and bog ores.	450	
		Coarse gravel and bowlders of the Drift Period, also conglomerates of any age.	400	
		Shales—Silicious, aluminous or bituminous.	350	
		Coal. Iron ore.		
		Sandstones.	300	
		Clays or marlite.	250	
		Domolite Rock. (Magnesian limestone.)	200	
		Silicious or aluminous limestones, or silicious Domolite.		
			150	
			100	
		Crystalline limestones, of any age.		
			50	
		High water in nearest stream or lake.		

DESCRIPTION OR EXPLANATION OF THE NORMAL OR TYPICAL SECTION.—In the column on the extreme right will be placed, when known, the level above the sea, obtained partly from

On portions of this plateau there is a deficiency of running water for stock, although a supply is obtained on many portions of the elevation by digging ten or fifteen feet through bluish clay, when they reach gravel and usually find water in that or the sand overlying an impevious substratum. It is commonly hard because during filtration through the superincumbent Drift, the water encounters fragments of limestone. Notwithstanding some inconvenienceon, this score of a scarce supply in dry seasons, there are farm houses on the plateau in Fayette and the adjoining county of Rush, as fine as any we saw in the State. Some of them could not have cost less than four or five thousand dollars. The style of architecture is elaborate and sometimes highly ornamental.

Ellet's levels, referable to high tide in the Gulf of Mexico, partly from the levels of Messrs. Stansbury and Williams above high tide in the Atlantic, where it ebbs and flows at the mouth of the Hudson river, New York.

The next column, on the left, gives the heights above the nearest water course, say usually *high* water, sometimes however *low* water in the Ohio river, if near it, or of a lake or rivulet. If none is near, the unit or starting point, or bench from which we level up, is the lowest adjacent part of the valley or deundation, which exposes the section. This scale saves the trouble of adding together the separate strata beneath, to know how high any one bed is in the section.

In the middle column, the lithological and palæontological character of the rocks is given, and in the next column on the left the relative Geological age or stratigraphical formation.

The extreme left column denotes the thickness of each separate or subordinate bed, and has a letter affixed for convenience of reference, should a more extended description in the text be necessary.

The markings in the wood-cut attached to these descriptive sections, although arbitrary are designed to be uniform, and to some extent appropriate, for the similarity of lithological character, thus: soil, sand, fine gravel, shell marls and calcareous tufa, or bog iron-ore deposits, usually of recent Quaternary age, are indicated by short horizontal lines or dots; coarse gravel and bowlders of the Drift period, as well as conglomerates of any age, by light rounded fragments in a shaded ground; shales by close, waved, continuous lines; iron ore by short dark marks; coal by a solid black band; marls and clay by heavy, straight lines; sandstone by light, continuous straight lines; magnesian limestones (Dolomites) by fine lines running diagonally from right to left downwards, and crossed by short, light marks; silicious and aluminous limestone, or silicious magnesian limestones, are indicated by similar lignographing, except that the diagonal lines bear from left to right downwards, and the faint crop marks are omitted; crystalline limestones, finally, are indicated by the imitation or semblance of blocks or bedstones.

The indication of an iron-ore stratum might appropriately have been placed just above the coal and the clay bed immediately beneath; but, as the representation of a normal section is only designed to convey an idea of the wood-cutting employed to facilitate the more ready understanding of lithological character in a rock, irrespective of age or superposition, the *relative* placing of the typical strata is unimportant. Sometimes the columns, instead of occupying the exact order indicated above, may, for convenience, be reversed; but the heading of any column will always define the use it is intended to subserve.

FRANKLIN COUNTY.

We were unavoidably prevented from seeing the greater part of this county, although it was included in our projected route for the autumn of 1860. The western portion, which we visited, appeared a new and thriving region of country, settled to a considerable extent by Germans, who are erecting, at Enochsburgh, a substantial Catholic church of stone. The material is obtained from quarries in that vicinity, some of which are Upper Silurian; but on Salt creek, immediately adjoining town, the formation is the extreme upper part of the Lower Silurian, characterized by beds composed of the coral designated by Prof. Hall, in his New York Palæontology, as Favistella stellata, from Favus, honey-comb and stella, a star; but considered by Edwards and Haime, in their fine Monograph on corals, as generically and specifically identical with the Columnaria alveolata of Hall himself, as well as of Goldfuss, Brown and Edwards.

The specimens which we collected at this place, at Madison, Jefferson county, and which are abundantly found on University Hill, Nashville, Tennessee, all bear the characters exhibited by Prof. Hall in plate 75, fig. 1, of the New York Palæontology, vol. 1, while a sample I possess from Bourbon county, Kentucky, by the striated cell walls and less extension towards the centre of the vertical lamellæ, is more nearly represented by fig. 1, in plate 12 of the same volume.

A splendid mass of this coral, found at the above locality on Salt Creek, in Franklin county, already alluded to as weighing 153 pounds, was conveyed to the railroad and thence sent by Adams Express to the State collection at Indianapolis, where it can now be seen.

Associated with this coral were Rhynchonella (atrypa) iucrebescens, Ambonychia radiata, Orthis occidentalis, and similar fossils, in a bed of Blue Limestone, eight or ten feet in thickness; then about twenty feet of non-fossiliferous rock over the Favistella bed, with several intercalations of chert.

This rock, rejecting the chert, is burned into a fair quality of lime; it seems, judging, in the absence of palæontological evidence, from its lithological character and its position, to be of Upper Silurian age. Passing westward a short distance, into Decatur county, to the town of Rossburg, we found, in ascending a hill, fifty to sixty feet of this crystalline limestone, surmounted by about a hundred feet of encrinital limestone, and fifteen to twenty feet of Drift. This brought us to the general level of the country at New Point, on the Indianapolis, Lawrenceburg and Cincinnati railroad, which station my barometer made more than one thousand feet above the level of the ocean. The railroad survey makes the summit between Sand Creek and Salt Creek, which can not be far from the station, one thousand and fifty-seven feet above high tide in the Atlantic, estimated near the mouth of Hudson river, New York.

In speaking of the quarries near the line of Fayette and Franklin, allusion was already made to those of Somerset, in this county.

There is said to be considerable diversity in the character of the soil in Franklin county, which may arise from the fact that portions are derivable from the Lower, others from the Upper Silurian, and others again from the Drift.

The timber must be fine, judging from the large quantities of lumber shipped to Cincinnati, and the piles of shingles and staves observed along the line of railroad.

We were informed that, in the early settlement of the country, milk-sickness was not uncommon in portions of the county; which may have been partly due to the saline contents of the shales, before exposure by the plow.

Brookville, the county seat, is remarkable as having been the residence of Hon. O. H. Smith, Gov. Wallace, Gov. Noble, and other distinguished men. The county seems also to have been a favorite resort of the Aborigines: at least Indian mounds are found in several parts of it.

DEARBORN COUNTY.

This county is generally of at least average fertility, and even towards the river, where the surface level is broken, we observed, near where the railroad, in approaching Lawrenceburg, the county seat, had cut first through the hills, covered with a few feet of Drift, and had afterwards passed with a rapid down-grade, through rocky cuts, that the industry of the inhabitants had piled the loose rocks in many places into heaps, like the cairns of our British ancestors, and had plowed, probably with the hillside, movable mould-board, and planted in corn, the steep mountain sides, while other bluffs, a hundred to a hundred and fifty feet high, afforded nourishment for vineyards, which even the rapid motion of the cars permitted us to observe bore abundant and heavy clusters of grapes, as we neared the picturesque Ohio, showing her capabilities of emulating her twin sister in Europe, the Rhine, with her vine-clad hills.

As this and other limestone regions in Indiana seem well adapted for this department of agricultural industry, a few words calling attention to the soil best adapted to this growth, as well as other points bearing on the culture of the vine in our State may not be out of place, when commenting on the soils of Dearborn, which result chiefly from the Blue Limestone, similar to that around Cincinnati.

On this subject I ask permission to extract briefly from the 250th page et seq. of a work on Fruit and Fruit Trees, by the late great American Horticulturist, A. J. Downing, where he makes the following forcible and appropriate remarks: "*Vineyard Culture.*—While many persons who have either made or witnessed the failures in raising the foreign grapes in vineyards in this country, believe it is folly for us to attempt to compete with France and Germany in wine-making, some of our western citizens, aided by skillful Swiss and German vine-dressers, emigrants to this country, have placed the fact of profitable vineyard culture beyond a doubt, in the valley of the Ohio. The vineyards on the Ohio, now covering many acres, produce regular and very large crops, and their wine of the different characters of Medeira, Hock and Champagne, brings very readily from seventy-five cents to one dollar a gallon in Cincinnati. The Swiss, at Vevay, first commenced wine making in the West, but to the zeal and fostering care of N. Longworth, Esq., of Cincinnati, one of the most energetic of western horticulturists, that district of country owes the firm basis on which the wine culture is now placed. The native grapes, chiefly the Catawba, are entirely used there, and as many parts of the middle States are quite as favorable as the banks of the Ohio for these varieties, the much greater yield of these grapes leads us to believe that we may even here pursue wine-making profitably.

"The vineyard culture of the native grape is very simple. Strong loamy or gravelly soils are preferable—limestone soils being usually the best—and a *warm, open, sunny exposure* being indispensable. The vines are planted in rows, about six feet apart, and trained to upright stakes or posts as in Europe. The ordinary culture is as simple as that of a field of Indian corn, one man and horse, with the plow and the horse cultivator, being able to keep a pretty large surface in good order. The annual prunning is performed in winter, top-dressing the vines when it is necessary in the spring; and the summer work, stopping side shoots, thinning, tying and gathering being chiefly done by women and children. In the fermentation of the newly made wine lies the chief secret of the *vigneron*, and, much as has been said of this in books, we have satisfied ourselves that careful experiments, or, which is better, a resort to the experience of others, is the only way in which to secure success in the quality of the vine itself."

We may have occasion to revert to this subject again in connection with other counties equally well adapted for the culture of the vine.

Main street in Lawrenceburgh is 473 feet above the level of the sea, and high water in the Ohio, at the same place, is 482 feet. The hills back of Lawrenceburgh average from 300 to 325 feet above high water in the Ohio at that place, consequently are about 800 feet above the ocean. The upper forty or fifty feet of rock, immediately below the soil, comprise layers of Blue Limestone, six to ten inches thick, with alternations of an occasional foot or more of aluminious shales and indurated clays, occasionally rather silicious, with very few fossils. From the upper layers of limestones just mentioned, which the workings of quarries near there afforded a fine opportunity to inspect, we obtained Chætetes ramosus and C. pavonia, a small variety of Orthis, (Spirifer) biforatus, also, Orthis testudinaria, Ambonychia radiata, &c., while the lower beds, some seventy-five or eighty feet below the upper quarry, afforded some six to ten-inch limestones, alternating with aluminous shales, Orthis occidentalis, Strophomena (Leptæna) alternata, and a large variety of Orthis (Spirifer) biforatus, viz.: O. (Spirifer) lynx.

RIPLEY COUNTY.

Not having had an opportunity, as we anticipated, to visit this county, we would recommend a thorough examination to be made hereafter, particularly of the quarries five or six miles north-east of Versailles, the county seat, from which building rock is obtained, said to be the best that can be found in the Blue Limestone region and much resembling that of Xenia, Ohio.

The soil is represented as being, in portions, not so fertile as that of adjoining counties; perhaps this may be the case where the silico-calcareous rock of Upper Silurian clate has furnished by its disintegration proportionally more silex and less alumina and lime than the detritus of the blue limestone, with intervening mudstones and marlites, usually affords.

I was informed that near Versailles petrified wasps nests were found. This is probably the Favistella stellata of Hall, and would indicate that the junction of the two sub-divisions constituting the Silurian System reached the surface somewhere not far distant from that vicinity.

OHIO COUNTY.

Although somewhat hilly, this county is productive. The chief agricultural products are hay and Indian corn, besides considerable crops of Irish potatoes.

Mr. Rabb, member of the State Board of Agriculture for that District, informs me that the Mormon hay press, costing about $200.00, is the one generally employed for baling. It does not require any one to tread down the hay, as is the case with most other presses, and enables the farmer, in consequence of the power with which it compresses, to put from one hundred to one hundred and twenty-five tons in a flatboat, when with a common press he could only put about fifty tons. His yield is about one and a half to two tons per acre, and the price about $14.00 per ton.

This gentleman, according to the report given in the second number of the Indiana Farmer for the year 1860, raised, in the year 1857, an average of eighty-three and a half bushels of corn to the acre on his lands, without manure; in the year 1859, an average of one hundred and twenty bushels; and in the year 1859 an average of one hundred bushels. His wheat crop, for the past three years, averaged successively thirty-one and a quarter, twenty-two and twenty-one and three-sevenths per acre; it sold at $1.20 per bushel. The net profits, after allowing interest on investments, were $21.52 cents per acre, on the land actually under cultivation, or nearly $18.00 per acre on the whole farm, including the woodland, lanes and fence corners. This profit would entirely pay for the farms in seven years, besides allowing an annual income or expenditure for the purchaser of $482.00 during that period; or, supposing the land paid for, would be yearly over seventeen per cent. investment on the capital. Yet merchants (with all the risk of prices and wear and tear of their constitutions from anxiety and indoor confinement,) often consider they are doing a thriving business, when their net profits average ten or twelve per cent. per annum. The above exhibit is encouraging to the farmer, and to the young man selecting a profession for life. The whole article alluded to, detailing the expenses, mode of culture, &c., is well worthy the attention of the community.

Back of Rising Sun, the capital of this county, the hills rise more than four hundred feet above low water in the Ohio at that place. The large Orthis (Spirifer) lynx is common near the top of the hills, also

Leptæna alternata and Orthoceratites, while lower down are found vast quantities of Lower Silurian corals, particularly of the genus Chætetes.

In this region considerable quantities of lead ore (sulphuret) have been taken out, associated with sulphate of Baryta, (Heavy spar.) In the eastern States this matrix is sometimes ground up for white paint.

The abundant lead deposits of Iowa and Wisconsin are, according to Prof. Hall, in rocks termed Galena limestone, an upper member of the Trenton limestone group, the same as part of our Blue Limestone; but there the lead ore is usually in a matrix of magnesian limestone, more easily worked than the Baryta, while in Gallatin county, Illinois, the "lead-blossom" is Fluor Spar, (Fluate of lime.) The lead ore of Missouri is sometimes in Baryta, at other time associated with calcareous spar (carbonate of lime) and occasionally loose in the soil, a stiff reddish clay.

It is not improbable that further search may develop more extensive deposits of lead ore in this and other portions of Indiana, the geological formation, as just remarked, being the same as that of the productive lead regions of Iowa and Wisconsin.

On these hills, along with Beech, Oak, Sugar Maple and Walnut, a tree flourishes abundantly, (although somewhat infested by the borer,) which the inhabitants consider a variety of the Black Locust (Robinia pseudo-acacia) and call *Yellow Locust*. The rain was falling in torrents when we made our examinations here for the lead or we would have endeavored to ascertain whether or not this is the *Yellow Wood* of Tennessee, described by Michaux under the Latin name Virgilia lutea, although he thinks it properly belongs to the genus Sophora. It grows as far north as latitude 37°, the flowers forming white pendulous bunches, a little larger than those of the Black Locust, but less odoriferous; the leaflets are also, as in that tree, borne on short petioles, supported by a common footstalk, and the seeds contained in a pod, resembling in size and shape that of the above mentioned valuable Robinia. The wood is yellow, affording a dye, which, according to the same author, only requires some mordant ensuring permanence to be of importance to the arts.

The flying weevil (Anacampsis cerealella) has not troubled the county for the last twenty years; the Hessian fly (Cecidomyia destructor) but little, nor have they any potato rot. Apples, peaches and grapes all thrive well; the two former fruits are cultivated somewhat extensively.

Rising Sun is said to be remarkable for its health, which doubtless is partly owing to its being well drained. It is situated partly on a late

Quaternary gravel, which, in places near here, and extending across the river into Kentucky, has been consolidated, and afterwards fissured by large vertical seams, so as to receive the name of *split-rock*. The fragments of this conglomerate being, to a great extent, limestone, would appear to assign its origin rather to the breaking up of the rocks forming the beds of the Silurian seas, near that region, than to the debris and detritus derivable from the more northern crystalline rocks, such as scattered bowlders so extensively over our prairies, and even left some south of thirty-eight degrees north latitude.

No milk-sickness exists nearer than the Miami bottoms of Ohio; typhoid and intermittent fevers are rare; consumption somewhat more prevalent; but the three physicians of the place are said to find most of their practice on the opposite side of the river.

SWITZERLAND COUNTY.

This county is, in many respects, very similar to the county just described, therefore most of the observations would apply to it also. Nearly the same agricultural products are raised, which were enumerated for Ohio county, except that probably Switzerland does not export quite so many potatoes.

In former years, the vine was largely cultivated by the hardy and laborious Swiss who settled around Vevay, the county seat. More than 20 years since, I tasted some of their wine, which retailed in Cincinnati at twenty-five cents per bottle, and very much resembled claret in taste and appearance. As a general beverage it seemed more suitable than the stronger and more expensive products of the Catawba. The variety cultivated at Vevay was a native sort, described by Downing under the head of Alexanders' Grape, or the Schuylkill Muscadell, "not unfrequently found as a seedling from the wild Fox grape, on the border of our woods. It is quite sweet when ripe, and makes a very fair wine; but is quite too pulpy and coarse for table use." To this Mr. Longworth adds, in a letter to Mr. Downing, in 1845: "The cultivation of the grape at Vevay is on the wane, as they cultivate only one variety—the Cape grape—a native sort, otherwise known as the Alexanders' or Schuylkill Muscadell. From it they make a rough, red, acid wine."

In addition to what was already said, under the article Dearborn county, regarding the soil best adapted for the growth of the vine and development of the grape, it may be useful here to add a few more extracts on this subject from the same letter of Mr. Longworth: "The grape requires a good soil, and is benefited by well-rotted manure. For aspect I prefer the sides of our hills, but our native grapes would not succeed well in a *dry* sandy soil, particularly the Catawba, which is a cousin german to the old Fox grape, that prefers a spot near a stream of water. The north sides of *our* hills are the richest, and I believe they will, as our summers are warm, in the majority of seasons produce the best crops." * * *

"I believe our best wine will be made in latitudes similar to ours. A location further north may answer well if the ground be covered with snow all the winter to protect the vine. It is to this cause that they are indebted for their success in the cultivation of the grape in the Jura mountains, in France. There is little doubt that the grape will bear better with us, and (judging from samples I have had from the first grower at the South) will make a better wine here than in Carolina." * * *

"One favorable year, I selected from the best part of one of my vineyards, the fourteenth part of an acre, the product of which was 105 gallons—at the rate of 1470 gallons per acre. The best *crop* I have ever seen was here, at the vineyard of Mr. Hackinger, a German, about 900 gallons to the acre, from the Catawba grape."

When now we consider that Switzerland, as well as most of the counties described in this section, are in precisely the same geological formation on which Mr. Longworth and others have for many years so successfully and profitably cultivated the grape and manufactured wine, besides being nearly in the same latitude, it seems well worthy of consideration whether the culture might not be advantageously extended in these, and other counties to be mentioned hereafter, particularly on hill sides too abrupt for easy grain-tillage. To the consideration of profit may be added the forcible argument that, with the increase of light wines, there will be, in all probability, a proportionate decrease in the production and consumption of those fiery alcoholic beverages, over-stimulating even in their purity, but truly poisonous, health and moral-destroying beverages, in their too frequently drugged adulteration.

Further observations on this county must be reserved for future examinations, as the corps was unable to visit it personally.

SEC. II.—COUNTIES IN THE UPPER SILURIAN FORMATION.

SUB-SECTION 1.—GENERAL CHARACTER OF THE UPPER SILURIAN FORMATION AND ITS PREVALENCE IN INDIANA.—Eighteen counties may be assigned to this second section, as deriving the character of their soil chiefly from the disintegration of the Upper Silurian rocks, viz.: Adams, Wells, Huntington, Wabash, Miami, Jay, Blackford, Grant, Howard, Delaware, Madison, Randolph, Henry, Hancock, Rush, Decatur, Jennings and Jefferson.

The same formation extends in a north-west direction under the Drift to Lake Michigan, probably through Lake, Newton, White, Porter, Stark, Pulaski, Cass, LaPorte, Marshall, Kosciusko, Whitley and perhaps others; inasmuch as Upper Silurian fossils have been found at low exposure on a northern tributary of the Tippecanoe river, and on the Kankakee, as well as in adjoining parts of Illinois, and at Milwaukie, Wisconsin; besides the overlying black shales of Devonian age in the Grand Prairie. Still, as the Drift has to a great extent formed the soils of these counties, most of them are described under that head; while others, receiving their character from the black slate of Devonian age, are assigned to that period.

The strata of Upper Silurian formation which seem most developed in Indiana, belong rather to the upper and middle than to the lower groups of the same, especially to one which, from its prevailing between Lakes Ontario and Erie, around the Niagara Falls, has been designated in the New York Geological Surveys as the Niagara group. Some of the fossils found seem assignable to yet higher groups, the Onondaga Salt Group of the New York geologists, and their Lower Helderberg Limestones.

As in the Lower Silurian formation, so in this, Corals and Brachiopod Mollusks, with Trilobites, constitute the prevailing fauna in beds which in the Western Hemisphere reach a thickness of about 4,000 feet, consisting, in British Wales, of sandstones, more or less calcareous and argillaceous, resting upon earthy limestones, with intercalations of shales, locally known as "Mud stones."

The New York geologists assign to this formation, in the eastern portion of the United States, a thickness of two thousand, four hundred feet, the upper members of which are characterized by argillosilicious limestones, the middle by shales and limestones, the lower by shales and sandstones. In Indiana, the thickness of this formation may be placed provisionally at about 1500 feet, comprising, as already stated, chiefly the middle and upper groups. This is confirmed by the borings from Dupont's Artesian Well, through part of the Devonian and all the Upper Silurian rocks at the Ohio Falls. Prof J. Lawrence Smith reports the fragments as being doubtless from Lower Silurian rocks, when the borings had reached a depth of sixteen hundred feet from the surface, while the Devonian rocks first passed through amounted to about one hundred feet or less.

SUB-SECTION 2.—CHARACTER OF THE SOIL RESULTING FROM THE DISINTEGRATION OF UPPER SILURIAN ROCKS, AND ITS IMPROVEMENT.—Prof. J. F. W. Johnston, the distinguished Agricultural Chemist and Geologist, characterizes the soils derived in Wales from the shales and earthy layers of the sandstones and limestones, as less productive than the red marls and clays of the Old Red Sandstone in Hereford, while the muddy, impervious soils of the lower Ludlow and Wenlock rocks, he says, "subjected to the drainage of the upper beds, form cold and comparatively unmanageable tracts. It is only where the intermediate limestones come to the surface, and mingle their debris with those of the upper and lower rocks, that the stiff clays become capable of bearing excellent crops of wheat. This fact, however, indicates the method by which the whole of these cold, wet clays might be greatly improved.

By perfect artificial drainage and perfect limeing, the unproductive soils of the Lower Ludlow and of the Wenlock shales might be converted into wheat lands more or less rich and fertile."

Prof. Hall expresses a very favorable opinion regarding the fertility of some groups in this formation on our continent. Speaking of the Onondaga Salt group, he remarks: "The soils derived from the decomposition of the rocks of this group, and of those of the Niagara group, are among the most fertile in the United States."

The above opinion is fully sustained by the fertility of some Indiana counties situated in the Upper Silurian, while, in portions of others, where the cold shales or the argillo-silicious limestones preponderate over the purely calcareous rocks, some remedy may in time be necessary, similar to those pointed out above by Prof. Johnston, as the best means of rendering the land highly productive. A sufficient number of the soils from this formation, in Indiana, has not yet been subjected to analysis to furnish a safe criterion for deductions of practical value. Those four results given in Dr. Peters' report, as well as the table of some Upper Silurian soils analyzed by him for Kentucky, exhibit many of the inorganic ingredients necessary for fertility, in fair average proportions. Where such is the case, the same crops would be suitable as were indicated for the Lower Silurian formation, the cereals and grasses.

On this subject, Mr. Fisher, member of the State Board of Agriculture from Wabash county, who embraces in his District several other counties of Upper Silurian soil, remarks: "Our staple products are corn and wheat; our lands produce timothy, clover, blue grass, potatoes, &c. The reason why corn and wheat are our staple articles, is because we have not a market always reliable for other farm products; much of our corn is fed to hogs, and finds a market in the shape of pork and some in beef. It is not because our soil will not produce other things equally well; but we always have a cash market for wheat and corn. We have nearly every variety of soil except the high rolling prairie. * * The soil in the northern portion of Wabash, in what is called "barrens," quite lightly timbered and covered with hazel bushes, is of a sandy character, and produces good wheat and good corn, if the season is not too dry. In much the largest portion of the District, the land was covered with a heavy growth of timber, having a strong clay soil, which, when underdrained, will produce abundantly."

SUB-SECTION 3.—QUARRIES OF MATERIALS SUITABLE FOR BUILDINGS, ROADS, GRINDSTONES, WHETSTONES, AND FOR BURNING QUICK LIME OR HYDRAULIC LIME, ALSO DEPOSITS OF MARL, GYPSUM, CLAYS FOR POTTERY, FIREBRICK, &c.—Good quarries adapted to the purposes of building are very abundant in this formation. Beginning with the more northern, we have extensive quarries described more in detail under their separate counties, in the neighborhood of Huntington, Huntington county, several in Wabash, especially that of Mr. Fisher, member of the State Board above alluded to, convenient to the railroad and canal; also, one at LaGro; again near Peru, along the Mississinewa for some distance to Miami county, rock is abundant; although at places it does not quarry in large slabs, having a tendency to cross fracture, and near Logansport, Carroll county, many good quarries extend along the Wabash.

Somewhat further south we have stone quarries at Bluffton, Buena Vista and other places on the banks of the Wabash, in Adams and Wells counties; also not far from Marion, Grant county; at Macksville, in Randolph; at Yorktown, Delaware county, besides extensive beds at the junction of the Upper Silurian and Devonian strata of Pendleton, Madison county, also some quarries at Strawtown and other points in Hamilton. Still further south, the Upper Silurian affords good building materials on Flat Rock, in Shelby county; at the St. Paul and Jordan quarries, in Decatur; also the well-known and extensively shipped Vernon rock of Jennings county, and the beautiful materials from the Dean Marble quarry of Jefferson, which, although in upper layers of Lower Silurian strata, fall under this section because the soil of the county, as remarked previously, is derivable chiefly from Upper Silurian detritus. Ripley county, we believe, has abundance of rock, but we were unable to visit the localities at which we understood it was obtained, and therefore cannot speak definitely regarding it. Limestones in thin layers, suitable for turnpiking, is also found in Henry county.

From the above it will be evident that in this formation there is no deficiency of limestone suitable for the manufacture of lime, although some layers have occasionally been rejected as too silicious to be perfectly burned. Hydraulic limestones of excellent quality are abundant in this formation, particularly in Wabash county, and near the eastern line of Rush adjoining Fayette county.

No marls, nor grits for grindstones or whetstones, were especially brought to our notice in the Upper Silurian strata, but we were informed that there was no scarcity of clays suitable for pottery, &c.; and near Madison, in Jefferson county, a fine clay is extensively dug for the use of foundries.

SUB-SECTION 4.—THE METALS IN THE UPPER SILURIAN FORMATION.—Considerable quantities of Bog Iron ore (designated by Prof. Dana as hydrous peroxide of iron) are found in the more northern counties of this formation, as Wabash, Delaware, Miami, Henry, and probably others. Gold has been washed on Blue River, Henry county; but no workable quantities of other metals have yet been examined by the corps, though a considerable amount of zinc blende found may lead to the discovery of large quantities.

Many of the general observations made regarding metals, &c., in the Lower Silurian formation would apply here also, as both belong to the same System, and have, consequently, some characters in common.

SUB-SECTION 5.—THE GROWTH OF TIMBER, AND OTHER LEADING VEGETATION.—On this head, Mr. Fisher, whose District embraces several of the counties in this formation, and whose remarks on the soils we quoted above, makes the following observations: "The timber is as various as the soil, and is, as in all other places, a complete index to its quality. I think much the largest portion of my District is what would be called, in common parlance, 'Beech and Sugar-tree land,' that is, these two kinds of timber abound more than any others, yet you will find few quarter sections entirely destitute of Black Walnut, Bur Oak,* Ash, Hickory, &c." In Henry county we noticed also White Walnut, Poplar and Wild Cherry.

Although perhaps not peculiar to these geological strata, it may be mentioned that the Boneset (*Eupatorium perfoliatum*) and the Stickseed (*Echinospermum lappula*) were abundant in several of the counties of this formation.

SUB-SECTION 6.—MINERAL AND OTHER SPRINGS, ARTESIAN WELLS, &c. The well and other water in this formation is, as might be anticipated, chiefly hard limestone water; the supply in most places is abundant.

A few sulphurous and saline springs exist, and chalybeates are numerous, especially in the northern and middle counties of the formation, as Delaware and Rush.

Centrally through the latter county, in a line north of east and south of west, embracing Rushville, the county seat, at most places where wells have been dug or borings made, the water rises to the surface and by inserting pipes may be carried even higher; it is almost invariably chalybeate in character.

SUB-SECTION 7.—MISCELLANEOUS FACTS REGARDING DISEASES, &c.—On this subject I quote again from Mr. Fisher: "The prevailing diseases are intermitting fever, or fever and ague, bilious fever, &c., in winter there is occasionally a case of pneumonia, and various other diseases throughout the year, but not of sufficiently frequent occurrence to mark them as diseases generally prevailing in the country. In former years our country had some reputation as a good ague country, but it is fast losing it; for the last three or four years there have been but few cases, and those of the mildest form. Bilious diseases are supposed, I believe, to be superinduced by the malaria arising from stagnant pools of water and decayed or decaying vegetable matter, and as the county improves, these causes are removed and the disease ceases to exists."   *   *

"There has been in years past some little 'potato rot,' but not to an alarming extent, nor has it been general. Some complaint has been been made of injury to wheat in the fall by what is called the 'fly,' but this is not very extensive. The injury is done by destroying the young plants; some persons say they are eaten by the 'fly,' others by 'black crickets.' I incline to the latter conclusion."

---

*This is the overcup White Oak of Michaux, (*Quercus macrocarpa*,) the chene a gros gland of the early French settlers in Illinois. Prof. Gray considers the Mossy cup White Oak (*Quercus olivæformis*,) only a variety of the above.

"Plums (except wild plums) are almost invariably destroyed by the curculio; other fruits are not much injured by insects."

"I have never heard of a case of milk-sickness or hog cholera in my District."

SUB-SECTION 8.—SPECIFIC ENUMERATION OF THE FOSSILS MOST COMMON IN THE UPPER SILURIAN COUNTIES OF INDIANA.—The fossils of the Upper Silurian are still, as in the Lower, chiefly corals, mollusks and trilobites; but frequently specifically distinct from those found in the lower part of the Silurian System. In our State they, to a great extent, belong to the Niagara Group of New York Geologists, and part of the Wenlock and Dudley of Europeans; the most common are these:

A. RADIATES.—
- *a' Amorphozoa:* Stromatopora constellata, (Hall,)
  S. concentrica, (Hall,)
- *a. Corals:* Heliolites pyriformis,
  H. macrostylus,
  Plasmopora follis? (Edwards and Haime,)
  Favosites Niagarensis,
  F. favosa,
  F. Hisingeri, (Edwards and Haime,)
  Astrocerium parasiticum;
  Halysites catenularia, (Catenipora escharoides or chain coral of Hall,)
  Halysites sexto-catenatus,
  Syringopora (?) multicaulis, (Hall,)
  Columnaria inæqualis,
  Cænites (Limaria) laminata, (Hall,)
  Zaphrentis turbinatum,
  Streptalasma calicula and strombodes pentagonus.
- *b. Acalephs.*
- *c. Echinoderms:* sub-division Crinoids,
  Cariocrinus ornatus,
  Stems of Eucalyptocrinus decorus,
  Eugeniacrinus costatus.

B. MOLLUSKS.—
- *d. Molluscoid Bryozoa.* Fenestella tenuis,
  Clathropora frondosa,

  e. Brachiopods. Atrypa reticularis,
        A. ———, n. sp., [Terebratula Wilsoni,]
        Peteramerus occidentalis,
        P. oblongus,
        P. Huntingdonensis,
        Orthis elegantula, Leptæna depressa.
  f. Conchifers.
  g. Pteropods.
  h. Gasteropods. Platyostoma Niagarensis, Hall,)
        Bucania euomphaloides, (R. Owen,)
  i. Cephalopods. Orthoceras undulatum, (Hall,)
        O. imbricatum, (Hall,)
        Gyroceras rhombolinearis, (R. Owen,)
        Nautilus Wabashensis, (R. Owen,)

C. ARTITULATES.—
  k. Worms.
  l. Crustaceans: sub-division *Trilobites.*
        Calymene blumenbachii, var. senaria,
        C. blumenbachii, var. Niagarensis, (Hall,)
        Bumastus Barriensis.
  m. Insects.

D. VERTEBRATES.—
  n. Fishes.
  o. Reptiles.
  p. Birds.
  q. Mammals.

SUB-SECTION 9.—DETAILED DESCRIPTION OF EACH COUNTY IN THE UPPER SILURIAN FORMATION.—

ADAMS AND WELLS COUNTIES.

As there is considerable similarity in the character of these two counties, it has been thought best to give the description of both under one head, especially as some of the most important palæontogical investigations were made on the Wabash, very near the junction of the two counties.

The Drift which spreads so extensively over counties north of these two, has reached them also, and has considerably modified the soil, not so much by the disintegration of bowlders, which are comparatively rare here, as by the beds of gravel and quaternary clay overlying the silico-calcareous Upper Silurian rocks that pervade these counties. In most cases the resulting soil, although fertile, inclines occasionally to be tenacious, and the surface of the country being rather level than hilly, the character of the land may be designated as frequently too retentive of moisture, except in very dry seasons; requiring thorough drainage to develop its best qualities.

We saw, especially in the northern portion, where gravel rather predominates, some fine farms, with good barns and fair crops, the counties being traversed in part by plank roads, and to some extent settled by a thriving German population, with railroad transportation at no great distance. Clover seems to thrive very kindly in this region, and Sorghum is cultivated to some extent. Corn and wheat are staple farming products, and at Decatur we observed some excellent grist and saw mills. Cattle, horses and hogs are raised extensively; and meadows, both for hay and pasture, evidently furnish heavy crops.

In portions of these two counties rock for building materials is abundant; it is also well adapted for the construction of turnpikes; and, although in places silicious, the majority of it burns into a good lime.

At Buena Vista, (near Newville, in Wells county,) where they are building a bridge over the Wabash, (and otherwise exhibit enterprise and energy in the erection of mills, &c.,) they have quarried rock somewhat extensively for building purposes, and, although some layers have a disposition to cross fracture, and calc spar fills occasionally large cavities, yet by selection, particularly from the lower layers, a tolerably fair material can be obtained here, as well as, at intervals, some distance up the Wabash to the Ohio line; also down the river from this point, commencing two miles above Bluffton and continuing for twenty miles below, then disappearing, until three miles from Huntington it again shows itself. The rock is highly bituminious in places, emitting a strong odor, when freshly broken. The dip here was found to be, as usually noticed also by us, south-west.

The forest trees chiefly observed in passing through Adams county were Beech, Maple, Hickory, Ash, Elm, Black Walnut, White Oak and Dogwood, while in Wells county the timber, of large size, comprised Hickory, Poplar, (or Tulip tree,) Maple, Sycamore, and some Beech. A considerable amount of smart weed, rag weed, stick weed, (Echinospermum lappula,) and boneset, (Eupatorium perfoliatum,) indicate their growth by a cool, moist soil, very fine ferns, including some species not before seen by us, also extend in Wells county over considerable tracts of land, besides other rather unusual vegetation.

Milk-sickness is said to prevail to some extent in portions of Adams county; but my informant, Mr. Robert Zimmerson, says it was never so troublesome here as in Miami county, Ohio, where he formerly resided. In that county he knew of a lick, in a sugar camp, to which the cattle resorted for the saline matter and near which they died. On fencing up the lick, the cattle were not affected with milk-sickness for several miles around. Having worked in lead mines on Fever River, he bore me out in a view, formerly expressed by me, that the disease in many of its symptoms resembles poisoning from lead.

Still we must admit that the disease often occurs where we see no surface indications of that metal; indeed, so far as I could learn no metal of consequence has been found in these counties; the nearest being the zinc-blende of Huntington county. The general nature of the soil in parts of Adams and Wells, partakes of the characters often mentioned by my late brother as favorable, by a tenacious substratum, sometimes of aluminous shales, for the retention of water charged with salts and their subsequent efflorenscence in the form of licks; near such regions he generally found that milk-sickness had prevailed at some period.

Typhoid fever is of somewhat frequent occurrence in this part of our State; indeed it is becoming in most parts of the United States more prevalent than formerly, while higher grades of fever are diminishing.

Springs are not very abundant in these counties; but, on the north side of the Wabash from Buena Vista, water is readily reached in digging wells without encountering rock; on the south side they obtain good water at from sixteen to twenty feet below the surface, among sand and gravel overlying the Upper Silurian rocks: a mile and a half further south-west, and also north-west, on Three-Mile Creek, a bed of rock is encountered at from two to four feet, consequently no water can be obtained without blasting and penetrating these beds.

Fossils are not abundant in the rocks of these counties at the localities examined; but the chain coral, Halysites catenularia, as well as the Heliolites pyriformis (Hall) assign these strata to the Niagara Group of Upper Silurian age.

HUNTINGTON COUNTY.

The soil of this county is generally deep and fertile, argillaceous in places, in others more arenaceous, the result partly of decomposing Upper Silurian rocks, partly of the Drift, which covers a considerable area to some depth; the Drift hills around the town of Huntington consist, in places, of forty-five feet of sand and gravel deposited on the Upper Silurian rocks which, with some alluvium, form the Wabash and Little river bottoms.

Wheat and corn grow well in this county, and are cultivated extensively; the latter is partly fed to cattle and hogs, which also form staple articles of exportation. Rock is quarried abundantly near the town of Huntington. Near one of the quarries the following section exhibited itself:

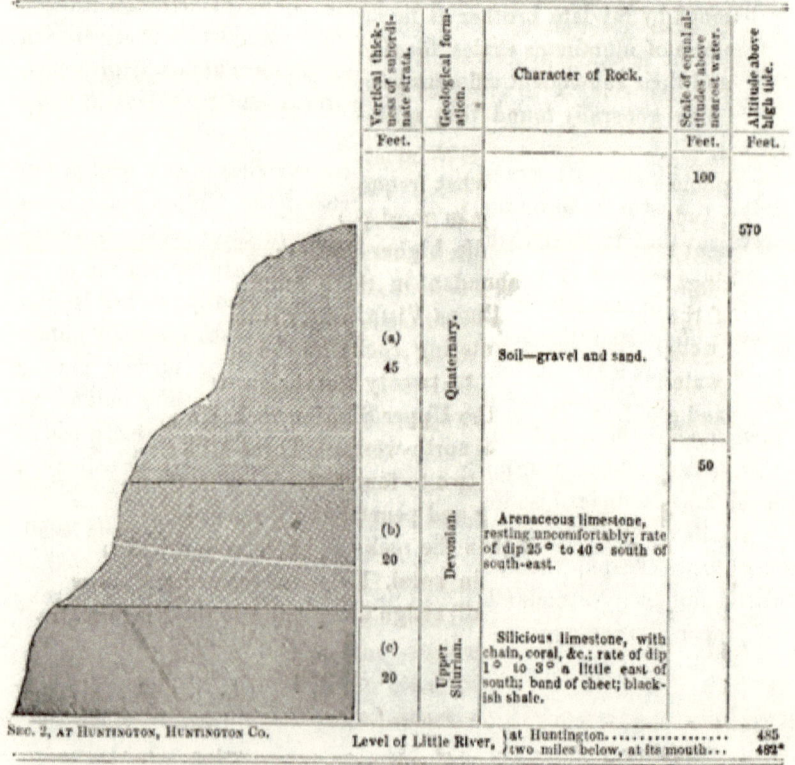

SEC. 2, AT HUNTINGTON, HUNTINGTON CO.

*On authority of Messrs. Stansbury and Williams.

These quarries afford abundant and fair materials for the construction of buildings and roads, also for the manufacture of lime. Considerable quantities of zinc-blede (the black-jack of miners) are obtained about two miles from the town of Huntington on the north side of the Wabash, near the feeder dam; the locality merits close examination, which in our rapid visit in 1859 we were unable to give it, expecting to return to that county. No other metals were heard of as having been found in this county.

The country is well timbered, Beech and Sugar-tree being prevalent in parts of the county passed through by us.

Under the very efficient direction of Mr. Luzon Warner, politely recommended to me by the Rev. Mr. Skinner, of Wabash, as having a fine collection of fossils, we were conducted to the quarries near the Huntington fair grounds, and enabled to obtain the following fossils, chiefly belonging to the Niagara Group: Halysites catenularia, Heliolites pyriformis, (Hall) and H. macrostylus, (Hall) Favosites Niagarensis, Favosites Hisingeri, (Edwards and Haime) Stromatopora concentrica, Stromatopora constellata, (Hall) Syringopora multicaulis, (Hall) Strombodes pentagonus, Streptelasma calicula, (Hall) Zaphrentis turbinatum; Cænites (Limaria) laminata (Hall); Stems of Eucalyptocrinus decorus; Atrypa reticularis, Pentamerus occidentalis; Platyostoma Niagarensis; Bucania ——— ? (Hall,) Nautiloceras (of D'Orb. or Gyroceras of DeKoninck) species undetermined; Calymene Blumenbachii, var senaria; Caly. Blum. var. Niagarensis, (Hall) Bumastis Barriensis.

By drilling through 20 feet of rock in the lower part of town they obtain water; in the hills, by sinking wells from 40 to 60 feet deep, good water is reached.

WABASH COUNTY.

Under the polite guidance of Mr. Fisher, of your State Board, we had a fine opportunity of examining parts of this county. Not far from this gentleman's beautiful farm and quarry there is a railroad cut, rendered somewhat noted as being close to the scene of a whloesale murder committed a few years since by Mr. and Mrs. Hubbard. They killed with a shoe hammer, as proved on trial, and buried under the floor of the cabin, the family with whom they boarded, consisting of a man, his wife, and, I think, five children, circulating the story that the family had sold out to them and moved away. The subsequent murder of an Irish canal laborer led to their detection and conviction.

The evidence of natural convulsions exhibited in the adjoining railroad cut by disturbed and dislocated strata, were almost as inexplicable as these outbursts of a perverted human nature. Entering at the west end we find beds inclined to the west of an angle of about 45°; approaching the centre an anticlinal axis partakes rather of the character of curved or folded strata, with huge masses of the purest crystalline calcite, partially covered by a crust of tufa. This is doubtless derived by infiltration from the calcareous matter of the superincumbent Drift, as somewhat farther east we encounter gravel, sometimes consolidated by this cement into a hard conglomerate, resting now on beds that occupy the railroad level, although at the centre of the cut these strata were nearly thirty feet over our heads. Beneath this bed we discern chert, sometimes pure and detached, sometimes apparently the result of silicious filtration into the cavities of the limestone. Emerging from this remarkable section at the eastern end, we find shales with an easterly dip at the rate of about 25°.

Drs. Ford and Winton of the town of Wabash were kind enough to drive us out to Linn's Mill, on Treaty Creek. Here we again found evidence of the convulsions and uncomformable stratification noticed at the Fair Ground quarries of Huntington, and at the railroad cut described above. On the west side of the creek, opposite to the mill and close to the dam, a hill is formed by an anticlinal axis, the beds dipping northward and southward about 43°. But the extreme summit of the hill has evidently been subsequently denuded and abraded by water, until a hollow affords a channel for a rippling rivulet, while in the bed of the main stream, beneath the axis, the undisturbed strata are visible. Of this interesting locality I made an outline sketch, from which the engraver executed the subjoined lignograph.

No. 2.—DEVONIAN, RESTING ON UPPER SILURIAN, AT THE MILLS ON TREATY CREEK, WABASH COUNTY.

As the quarries at LaGro are stated to be very fine, we regretted not having an opportunity to visit them. From Mr. Fisher's quarry large quantities of excellent building rock are shipped by canal and on the Fort Wayne Railroad. The layers usually furnish very large and thick slabs, one bed has hydraulic properties and formerly was burnt in that neighborhood for cement.

Near the town of Wabash the bluffs ascending from the Wabash River, which is here 638 feet above high tide, furnished the subjoined section:

	FEET.
Loose and thin limestones	15 to 20
Chert and flag stones	8 to 10
Aluminous shales	15
Silico-calcareous rock	15 to 20
Good building rock	20 to 25
Hydraulic limestone	5 to 8
Good building rock, thickness (beneath general surface of ground)	unknown

SEC. 3, NEAR WABASH, WABASH COUNTY.

From some of the various localities in Wabash county above enumerated we obtained Halysites catenularia, Astrocerium parasiticum, Orthoceras imbricatum and fragments of other large Orthoceratites, an undetermined species of Nautilus and Calymene Niagarensis.

Bog iron ore occurs in the northern part of the county; no other metals have as yet been seen or reported as found; nor gritstones suitable for grindstones or whetstones.

The soil is various, producing usually good crops of corn, wheat, timothy, clover, bluegrass and potatoes.

The northern portion of the county is more sandy and the timber lighter, than in the south, where clay predominates,* with large Beech trees and Sugar-Maple abounding, besides some Black Walnut, Bur Oak, Ash and Hickory.

---

*Nos. 8 and 9 of Dr. Peter's reports furnish the analysis of a soil from Mr. Wm. T. Ross' farm near the centre of this county. The remarkably red soil, of Mr. D. Ross' farm, highly impregnated with iron, has not yet been analyzed. The locality is one-fourth of a mile southwest of Somerset, near the corner of three counties—Wabash, Miami, and Grant.

MIAMI COUNTY.

The weatherings of silico-calcareous rock (or Magnesian limestone as it may also be termed,) have mingled with the Drift which has reached this latitude to form the soils of the county. They are also often charged with iron which has filtered, while held in solution by water, into many of the rock cavities, and been deposited there until again mingled with the soil, after decomposition of the stony matrix or mineral nidus, if the term be admissible. This union has given rise to a soil of varied character, but usually of sufficient fertility to produce good crops of the staple agricultural grain, maize or Indian corn. Although most of the farms came into market only sixteen or eighteen years since, (the land forming this county having been reserved for the Miami tribe of Indians) yet they already exhibit proof of good culture and enterprise. Forty dollars per acre is by no means an uncommon price for the land, and when we camped near the farm of Mr. William Godfrey, (16th and 17th of June, 1860,) we found corn selling at 35 cts. and wheat at $1.25 per bushel.

Close to this farm, on the banks of the picturesque Mississinewa, about three miles from Peru, are the celebrated "pillars," resulting from the unequal disintegration produced by the waters of this river on the harder and softer portions of the silico-calcareous rock, chiefly forming its banks.

In describing this locality among other scenery of Indiana, as some of the readers of the "Indiana Farmer" may remember, I used language, the coloring of which, I think, does not over paint the scene : "Again on the Mississinewa, a tributary of the Wabash, we find, close to the residence of Godfrey, a son of the Miami chief, whose tribe left these fine lands only eighteen years since, bluffs in which the rocks have been weather and water washed into fantastic pillars and natural cornices, which might serve to inspire the genius of a Michael Angelo with some new architectural design, to rival his St. Peter's at Rome."

These bluffs, or pillars, are here about 25 feet high, while nearer the ford they rise to 40 and 50 feet above low water.

The bed of this interesting stream was, during our visit at this locality, full of confervæ, (simple jointed water weeds) and had more crawfish dashing, with their peculiar, quick backward movement, from under the rocks into the sunshine, than I ever before saw in one stream. Various species of unio, cyclas, paludina (chiefly dead) and melania, were also common; the latter leaving a track in the sand resembling that of a worm. Besides these, numerous specimens of the larva of the Phryganea, or water moth, were seen dragging their wooden habitation of cemented sticks along the bottom of the shallow fresh water coves formed by the river*.

In this camp we noted, besides the usual timber, abundance of the Ohio Buckeye or American Horse Chestnut, (Pavia Ohioensis,) the buds of which, eaten in early spring by the cattle, frequently produce in them symptoms resembling an attack of " trembles or tires," in man called milk-sickness.

Mr. Godfrey showed me where he had partially opened a quarry of fine grained sandstone, very similar to that in the bluff below Logansport; and, (as was the case there,) here also overlying the yellowish silico-calcareous rock of the Upper Silurian age. Further excavations are required to prove its extent; but there seems every prospect that from this bed and the underlying limestone abundance of material can be obtained for the various purposes of construction.

A short distance above Peoria, where there is a dam across the Mississenewa, affording good water power, they are likewise quarrying rock of fair quality and thickness. It seems usually not to extend more than eight or ten feet above low water level in the river, the superincumbent deposit, consisting of quaternary bowlders, gravel, and red clay, and amounting sometimes to seventy-five or one hundred feet in thickness. Near Somerset,† where this county adjoins Wabash and Grant, on descending some twenty-five or thirty feet, we found them quarrying the lower limestone beds, which were even more solid than the upper, and from which materials a fine woolen factory has been erected; a kiln close by exhibited a good quality of lime, burnt from the same strata. About Peru the limestone seems to have been abraded and denuded to some extent by the Wabash, and replaced by later quaternary deposits. In digging their wells they generally pass through about twelve feet of gravel and then through a few feet of tough blue clay before finding water.

---

*It was here also that we captured a bull-frog for camp provision and found, on dissection of its intestinal canal, that it contained a pebble weighing at least an ounce.

†This town in Wabash county must be distinguished from Somerset in Franklin county mentioned above; indeed we have frequently, in order to avoid error, to be careful, inasmuch as there are unfortunately in Indiana two "Salt Creeks," two "Eel" rivers, two "St. Joseph" rivers, two "Bloomfields," more than one town of "Liberty," and numerous "Buena Vistas," &c., &c.

Half a mile east of town, in digging a ditch, considerable quantities of bog iron ore were thrown out, of which the following analysis exhibits the chief constituents:

One-tenth of a gramme operated upon, became very red when heated to 300° F., and lost by thus drying	0.006
It contained of insoluble inlicates	0.030
Sesqui, or peroxide of iron ($Fe_2 O_3$)=37.1 per cent. of iron	0.053
Alumina	a trace
Carbonate of lime	0.004
Magnesia, alkalies (undetermined) and loss	0.007
	0.100

At Peru Mr. Wilson gave me a fine sample of very pure sulphuret of lead, said to be found in that neighborhood; he also promised to investigate and report further on the facts, as the finder was unwilling then to indicate the locality, although claiming that the metal was very abundant, even to the amount of many tons.

At a quarry about a mile west of town, where the rock again appears, but too silicious in character to make good lime, we found a slight dip to the south-east, and obtained the following fossils: Columnaria inæqualis, Fenestella elegans, Clathropora frondosa, Orthis elegantula? (Hall,) and an indistinct fragment of a trilobite.

Beech and Sugar tree appeared the prevailing timber in the southern part of the county, with some Elm and Oak. Good barns and school houses indicated a prosperous condition of the inhabitants.

JAY AND BLACKFORD COUNTIES.

Blackford, unfortunately, we were unable to visit, but it is stated to be a well-timbered, rather level and fertile county, with no rock showing itself near the surface, so far as we could learn; Lick river and Salamanie produce, by their valleys, gentle undulations for drainage; the former meandering past Hartford, the central county seat, to empty into the former, which, after flowing with the geological strike of the country, parallel to the Mississinewa and the head waters of the Wabash, discharges into that river at LaGro, soon after its curving, with a sudden bend, to flow south-westerly *with* the dip, until it reaches the coal field of our State.

In Jay county, we observed also a considerable quantity of level clay land, which could undoubtedly be readily improved by drainage. In places there is a decided prevalence of gravel, with some bowlders. Occasionally the corn, on these ridges, seemed not so heavy, but the potato crop looked remarkably well. Sorghum, also, seemed abundantly cultivated with success in parts of the county, and the flocks of good sheep near Bloomfield indicated an improving system of husbandry. The steam mills in and around the county seat, Portland, evinced enterprise, while two newspapers, in a population short of 400, might well arouse the astonishment of regions settled long before the axe of the white man had felled a single "monarch" in this forest.

Our route through West Liberty, Bloomfield, Portland and Bluff Point, formerly called Iowa, exhibited Beech, Sugar Tree, Coffee Tree, (Gymnocladus Canadensis) and Black Walnut timber, all indicating rich land.

Further north, about the junction of this county with Wells, we occasionally passed, on corduroy causeways, small swamp-muck prairies, luxuriating in large asters, daisies, golden rods, the American aspen, (Populus tremuloides) willows, the indigo plant, (Indigofera caroliniensis,) smart-weed and boneset, (Eupatorium perfoliatum;) these lower lands are flanked, on the surrounding gravel ridge, by scrubby oaks, a few hickories, hazel bushes and abundant ferns, among which the delicate and graceful Maidenhair (Adiantum pedatum) appeared pre-eminent.

Some of the springs in this county are stated to be strongly impregnated with Magnesia.

GRANT AND HOWARD COUNTIES.

The new and fertile county of Howard, formed from part of the Miami National Reservation, we were unable to visit; but, judging from its position and surroundings, I should suppose that, in this county, the junction of the Upper Silurian and Devonian would be found, if the rocks could be reached beneath the Drift, which undoubtedly covers the greater portion to a considerable depth.

The soil is represented as being well adapted for corn, wheat and the grasses; timber is abundant, the farms being yet comparatively new.

The soil of Grant county, modified sometimes by Upper Silurian limestones and shales disintegrating, but composed chiefly of heavy quaternary deposits, exhibits alternations of a light-colored soil on some of the higher undulating grounds, resting on a more productive reddish clay, with bowlders, presented to view wherever the lighter soil has been washed away. Occasional prairies, of rich, black soil, are interspersed with the woodland, and orchards were noticed to be abundant and thrifty. Considerable flocks of sheep were seen, and the barns were observed to be very substantial; many of them built on the Pennsylvania plan of selecting a hillside for the foundation.

The rock quarried in this county is chiefly on or near the Mississinewa, seldom showing itself, however, higher up the river than Marion, the county seat. Near this town, it is taken out in considerable quantities for buildings and similar purposes. We noticed some excellent doorsteps quarried and dressed, for the dwelling of Mr. Wallace, brother to the late Gov. Wallace. Some of the layers afford slabs, which ring very clearly when struck, and from which we obtained a few indistinct fossils, chiefly small orthoceratites.

The Woolen factory of Marion is situated on this limestone bluff; but in many parts of the town gravel is thrown out in grading, or excavated in digging cellars, even at a level somewhat lower than the limestone exposure; the latter doubtless having been denuded in places, which then received the bowlders and gravel of the Drift, as well as later quaternary deposits in hollows and "pockets."

On the road from Marion through Jonesboro, leading toward Madison county, especially near Fairmount, the clay mingled with the gravel thrown on the roads, is so highly impregnated with peroxyd of iron as to give to the whole a strong yellow-ochre color.

The timber on the same route consists chiefly of Beech, Sugar Tree and Oak.

A mill, near Jonesboro, is on Deer Creek, which empties into the Mississinewa. The clover and wheat in the red soil appeared heavy; oats and corn were, however, occasionally somewhat thin on the lighter upland.

DELAWARE COUNTY.

Toward the western limit of this county, we observed some hills of gravel and sand, with oak timber; the farms and cultivation, however, were good, as may be inferred from the fact that a large amount of the wheat crop had been put in with the drill machine, and that hay stacks as well as flocks of sheep were numerous. Nearer Muncietown, the county seat, there is a greater admixture of clay, giving rise to a growth of Beech and Sugar tree. The quantity of hoop-poles at the depot would also indicate an abundant growth of hickory. Mr. Kirby, near Muncie, keeps a dairy of about one hundred cows, and manufactures cheese. Land is worth about one hundred dollars per acre in this vicinity, although some milk-sickness still prevails in portions of the county, where formerly it was common.

Some very fine and large specimens of bog iron ore were shown us at Muncie, obtained in that vicinity, where it is reported to be found in considerable quantities.

Close to the town of Muncie are four large and three smaller hills of Drift, bearing north-west and south-east from each other; one of the larger, opened for gravel, exhibits coarse bowlders on the north side, gravel on the south side, with fine sand beneath it. One hill, south of this, is entirely composed of fine sand.

Rock is obtained from at least two quarries near Muncie; although not in very thick slabs, it is suitable for many purposes. The upper layers are somewhat cherty, the lower more solid and pure. The dip here seemed a little east of south, while, at the other quarry, on White river, about a mile from town, it appeared rather west of south. In both quarries chain coral (Halysites caterularia) and Favosites favosa were found, although characteristic fossils were somewhat rare.

At Squire Gilbert's quarry, near Yorktown, where Buck Creek discharges into White River, about ten feet of rock were exposed, continuing downwards to an undetermined depth. The upper layers are here also rather thin, with numerous imbedded orthoceratites, (chiefly orthoceras imbricatum, of Hall;) the lower, bluish beds, although somewhat silicious and hard to work, as well as difficult to burn into lime, can be dressed into slabs of from a foot to fifteen inches in thickness. Here, too, the dip was found to be southerly.

In digging wells at Yorktown, they encounter rock at fifteen feet below the surface; the water is hard, and some springs, near there, are chalybeate.

MADISON COUNTY.

The northern part of this county, forming the summit level between the tributaries of the Wabash and those of White river, is indicated by our barometrical observations to be nearly 1,000 feet above high tide. The highest portion passed over by the party running the level for a projected canal from Indianapolis to the Wabash and Erie canal. is laid down as being 942 feet above the sea.

In this region we still found the ashen soil, spoken of in describing Grant county, forming the more elevated undulations, with the red, clayey soil in the bottom, resting on gravel and bowlders, these being frequently rudely stratified into curved beds, thus:

Sec. 4—Curved Quaternary Beds in Madison County.

In the lower part of the county, where the upper layers of rock are of Devonian age, the soil is somewhat modified. For the analysis of samples taken from the field of Mr. Irish, near Pendleton, where, in parts of his field, the plow often encounters solid rock, with scarcely enough of earth to cover it, see No. 13, 14 and 15 of Dr. Peters' Report.

Chemical research evinces no lack in this soil of the elements necessary to the production of fair average crops, and experience corroborates the correctness of the theoretical assumption, as wheat grows well and corn only exhibits short crops after injudicious cropping has exhausted some of the inorganic requisites.

The fine growth of Beech, Sugar Tree and Black Walnut, with some Ash, in the adjoining forest, amply corroborates the original fertility of the virgin earth. Wheat, clover and timothy; also, hogs and horses, are staple articles of farm profit; we also observed a fine breed of cattle near Alexandria.

Beneath the upper layers of rock here, some aluminous shales disintegrate, and exhibit to those digging cellars or wells a metal which attracted some attention. It proved, on examination, to be sulphuret of iron.

At a quarry belonging to the same Mr. Irish, not far from his farm, where rock is excavated in considerable quantities, we obtained the following section:

	FEET.	INCHES.
1. Slabs of limestone, susceptible of a fine polish, others with a silicious grit, containing Devonian fossils, as a Conocardium, Favosites Polymorpha, &c., 1 to......................	6	
2. Good sandstones..	3	
3. Another similar bed, 4 inches to...................................		6
4. Another similar bed, 4 inches to...................................		6
5. Sandstone hardening by exposure.................................	1	6
6. Yellowish, rather hard and somewhat aluminous, fine-grained, silico-calcareous rock, to bed of stream, 2 to......	4	
Total..	15	6

SEC. 5, NEAR PENDLETON, MADISON COUNTY.

Nos. 2, 3, 4 and 5, distinguished by the quarrymen, do not present sufficiently distinctive characters to be separated. At the falls of Fall creek, the same section exists with the minimum figures given above.

Four or five miles south-east of this, occasional cases of milk-sickness occur; but chill and fever, with some typhoid, are the prevailing types of disease.

On the dividing ridge above alluded to, water is obtained, about 20 feet below the surface, by passing through bowlders and gravel to a quicksand.

Near Alexandria, on Big Pipe creek, I found, at Mr. Calloway's mill, rather thin layers of silico-calcareous rock, in the upper beds, excavated for building chimneys and walling cellars; beneath these, near the surface of the water, solid slabs, six feet long by four wide, or larger, if necessary, and ten inches thick, can be quarried, by drilling and blasting. The lower layers are a bluish crystalline limestone, and appeared, from the fragments of fossils obtained, to be of Upper Silurian age.

The bowlders near here were chiefly granite, greenstone and hard sandstone.

Approaching Andersontown we saw some small prairies; and a cut made for the turnpike exhibited the following:

Decomposing granite and limestone bowlders, red clay, &c.,......$2\frac{1}{2}$ feet.
Pure, clear gravel, hen-egg size..2 feet.
Small, fine gravel and sand......... ....4 to 6 feet.

Calling on Mr. Henry, Editor of the "*Gazette,*" at Andersontown, we were directed to the quarries, about a mile and a half south, which we found owned by Mr. Davis and Mr. Moss. At both places considerable excavations had been made, usually to the depth of about 6 feet, affording slabs, which dress to eight inches in the upper layers, and to one foot thickness in the lower. Large orthoceratites prevail throughout, and, from between the superincumbent Drift and the rock, springs flow, descending into the adjoining bottom of White river.

RANDOLPH COUNTY.

The soil around Winchester, the county seat, is chiefly clayey, producing good clover and fine wheat crops, that average 20 bushels, and sometimes attain a maximum of 37 bushels, to the acre. On Mr. Irwin's place, about 4 miles south-east, only a few miles from the source of White river, an efflorescence was observed to form on the earth thrown out to the depth of several feet, in digging a ditch, to drain a few acres of craw-fish, swamp-willow land. By some persons this salt is supposed, from the taste, to be alum; but, as the effect in the mouth was said to be rather cooling than astringent, it may be nitre; when we were there, no appreciable amount could be obtained, and a sample promised to be sent has not yet reached the laboratory.

In the northern part of the county, near Dearfield, although there is much quaternary, consisting chiefly of bowlders and gravel, yet the aluminous material desseminated through it, on land already level, causes it to be somewhat too retentive of moisture, giving growth to flags and ferns, and requiring corduroy or log causeways for winter use.

From Messrs. Monks, Neff and Garrett useful information was obtained regarding the neighborhood of Winchester. Among hand specimens exhibited by them and represented as being found within a few miles around, the coral, Columnaria alveolata, (Favistella stellata of Hall,) often familiarly termed "petrified wasps nests," proved the junction of the Upper and Lower Silurian to be at no great distance.

Mr. David Heaston has bog iron ore on his farm near town, three-quarters of a mile up Sugar Creek, and his well, which we examined, proved a strong chalybeate; Mr. Monks says bog iron ore is also found on Cabin Creek, eight miles south-west from Winchester, as well as at other localities in that direction.

This gentleman showed me a rock somewhat rounded by attrition or weathering, which he had hauled to his garden, estimated by the teamsters to weigh one ton; it is made up of a mass of very hard, conglomerated materials; some of the oval fragments being as large as the egg of an ostrich, many more the size of a goose egg. Another curiosity submitted to our inspection by Mr. Monks, consisting of a very finely and regularly marked water-worn material, nearly two inches square by from one-eighth to a quarter of an inch thick, was politely lent for closer inspection. The markings being all on one side, exhibiting no tubes, floors, septa or side pores, and the specimen effervescing freely with acid and being soft enough to receive a decided impression from the knife, and the edge showing successive lines of deposition, it seems most probable that it is a fragment of a hard shell, ornamented with markings by the Indians; and the only wonder is that, without the aid of machinery, so much regularity should be obtained. But this seems a case similar to that of the carved footprints described by my late brother in Silliman's Journal, and which had been pronounced by some English writers, who saw them in the possession of Frederick Rapp, the founder of New Harmony, "almost too perfect to emanate from the chisel of a Chantry." When, however, as really happened, various specimens are found near the same locality, of similar character, but of various grades from very rude to fine sculpture, we are led to agree with the argument of my late brother, that in all probability they are the workmanship of the Aborigenes, in whom doubtless sometimes reside the elements to form a Phidias, should circumstances arise to call forth the dormant power.

Some quaternary ridges near Winchester are composed of coarse gravel on the westerly sides, of finer gravel with beautiful, sharp sand on the eastern slopes. At Macksville, about four miles west of Winchester, a limestone characterized by abundance of Pentamerus oblongus, affords, from beds having a westerly dip, material for building purposes and for kiln-burning into lime.

In the low swamp-muck grounds of middle Randolph, which, however, are being rapidly drained, we saw, luxuriating in the fertile humus, the Chestnut White Oak (Quercus prinus palustris of Mchx.) with several of the willows, (I think Salix candida and S. eriocephala,) with flags and swamp dock? (Rumex verticellatus.) On somewhat higher portions are Beech, Sycamore, Elm and Hackberry, with silk or milkweed, (Acelapias cornuti,) smart-weed, iron weeds, (Vernonia fasciculata and sometimes V. Noveboracensis,) and a thistle-like weed, not the Canada thistle, replacing the hitherto abundant Boneset.

At our camp near Winchester, on the west fork of White River, we observed four Sycamores growing from one root; and, apparently emanating from the same bases with these, although doubtless separate under ground, a Red Elm and a Hackberry completed the group. Buckeyes also, at this locality, grew around us.

Chalybeate springs are very abundant in this county; and among the swamp-muck, near Salt Creek, from which chalybeate water oozed abundantly, we observed in places numerous dead shells of the genera paludina, planorbis and physa. Large bones have also been dug from these swamps, but we had no opportunity to see any. We heard also of Indian bones being common in aboriginal mounds and forts around here, saw likewise some beaver dams, and learned that muskrats are very abundant.

For the analysis of the very red and productive soil obtained from Mr. James Clayton's farm, three miles west of Winchester, see No. 7 of Dr. Peter's report. There is also, near here, some of the soil termed "mulatto."

After passing over clay lands, eight miles south of Winchester, the soil, on approaching Huntsville, becomes gravelly and drier, the natural undulations, which amount to twenty-five or thirty feet, serving well for drainage. Large bowlders are still numerous, and, judging from saw-mills, timber must yet be abundant in this region.

HENRY AND HANCOCK COUNTIES.

We regretted that our visit to the former was necessarily very brief and the examinations impeded by heavy and continuous rain.

After passing, on the straight-line "Cincinnati and Chicago Railroad," the Walnut Plains described when speaking of Wayne county, we crossed Flat Rock about four miles south of New Castle, and Blue River about half a mile from town.

The soil generally is rather clayey, usually good for timothy and clover. Their staple products are wheat, corn and grass; the general market for which is Cincinnati, although a considerable trade is also kept up with Richmond. On the 12th of November, 1859, when we were at Newcastle, the average price of wheat was $1.00, and of corn 30 cents per bushel; hay eight to ten dollars per ton.

Cattle and hogs are raised abundantly in Henry, likewise some sheep; the horses bred seem of good quality and the mule raising is, gradually, also becoming more common.

The Blue River country, being finely watered, is especially well adapted for grazing farms.

Gold is washed abundantly from the quaternary gravel drift near the mouth of a small stream and a mill-race, emptying about eight miles form Newcastle into Blue River; and these localities, on closer examination, may prove worthy of being more extensively worked.

Bog Iron ore was thrown out while ditching Blue River about twelve or thirteen miles north of Newcastle, on Mr. Raymond's farm, section 1, township 19 north, range 10 east, and appearances indicate that the deposit is somewhat extensive.

Their building materials of rock they usually obtain from Williams' Creek in Fayette county; but shelly limestone, suitable for road making and burning into lime, can be found in nearly all the branches. From these materials several good turnpikes have been completed in this county. Besides the above Cincinnati and Chicago railroad, another railroad, denominated the Southern Cincinnati and Chicago road, has been graded, passing through Newcastle to Cambridge City, Liberty, in Union county, &c.

Newcastle, the county seat, numbers about 1,300 inhabitants and sustains in and around it many excellent schools, some of high grade, kept up the entire year and giving courses in the languages and mathematics.

Knightstown, the largest town in the county, is surrounded by much agricultural wealth; and Raysville is quite a thriving place.

About seven or eight miles west of Newcastle, a number of Indian skeletons were disinterred in constructing a turnpike, and about the same distance south of town some remarkable human bones and skeletons of giant size were dug out, with other relics, during the making of a road.

Sugar-Maple, Oak and Walnut were observed to be abundant after leaving the Walnut Plains and entering Henry; the county is considered a good fruit region, although here as elsewhere in middle and northern Indiana, the peach trees were killed some years since.

They have had some hog cholera; and we were informed that the prevailing diseases, among the inhabitants of the county, are bilious-remittent in type, with some typhoid and occasional recent cases of milk-sickness in portions of the county. Sometimes the disease was apparently traceable to certain springs, inasmuch as the cases ceased to appear in that region after these springs were fenced in from access of cattle, through which the poison is most generally communicated.

Dr. Reid informed us that post-mortem examinations usually disclose a thick tarry matter in the stomach and intestines, the evacuation of which generally would effect a cure. The constant nausea, abdominal tenderness, and want of action or engorgement of the liver, render it, however, often difficult to produce sufficient quietus for the retention of the necessary cathartic. Blistering on the stomach sometimes alleviates the irritability, and the softened condition of the bowels indicates the propriety of diffusible stimulants, which are consequently often successfully employed. In some cases there is an apparent tendency to mortification.

In this county there are several sulphur and chalybeate springs; one of the latter made its appearance suddenly in the dead of winter, about two years before our visit.

Well-water here is hard and reached usually at from twenty to forty feet, without encountering any rock, after passing through 1. blue clay; 2. gravel; 3. sand.

Of Hancock county also we did not see as much, while passing through, as was essential to the formation of correct inferences, on many important points.

The eastern part of the county seems modified by the proximity of both Devonian and Upper Silurian rocks, and the western brings the Devonian shales into the region of the knob-sandstones that underlie parts of Marion and Johnson.

The continuous predominance of Beech woods as we passed successively through Charlottsville, Greenfield, Philadelphia, Cumberland, &c., indicated an aluminous soil, although the staves, oak poles and barrels at depots, pointed to more arenaceous ridges; the corn, standing in large and close shocks, the meadows, the numerous small grain and hay stacks, some orchards and a nursery near the eastern confines, all seemed to justify favorable conclusions regarding the general fertility of the soil and the improved state of agriculture in the county.

The railroad cuts, usually low, evinced undulations favorable for drainage of the country, while the streams crossed showed no want of water for the stock; and an occasional flouring mill demonstrated the correct use of water power. The cuts being usually through gravel, exposing occasional bowlders, proved the quaternary drift to have spread over the subjacent rocks, and thus probably to have aided as is often seen, in adding fertility to their detritus.

Further particulars must be reserved for minute explorations, where instead of general reasoning from a few striking facts, ascertained details can be given in full.

RUSH COUNTY.

The remarks already made regarding the western portion of Fayette apply in a great measure to the eastern part of Rush. The great Lower Silurian plateau spoken of, with a growth of large Beech, Elm, Sugar Tree, and Oak, and undergrowth of Papaw, continues somewhat into Rush; the town of Vienna being in that county, close to the dividing line. Thence we gradually descend on to the Upper Silurian formation. A superficial deposit of quaternary gravels and clays still modifies the soil, and the same character of fine farms, good barns and dwelling houses continues—some of the latter in Rush county are evidently both costly and ornamental; the osage hedges, clover fields and improved farm implements of this county indicate likewise high agricultural prosperity, notwithstanding the apparent scarcity, when the geological corps was there, (29th September, 1860,) of corn and water for stock. The county generally is represented as possessing soil as fertile throughout as any other county in our State.

In Rushville they obtain their materials for works of stone-masonry partly from Vernon, Jennings county, and partly from the Flat Rock quarries near the Decatur line and in that county.

From Drs. Helm and Sexton we received valuable information bearing on our researches; and, through the politeness of Mr. Shaddinger, had an opportunity of seeing the collection of Mr. George C. Clark, at the Court House, as well as of examining the very interesting artesian wells in and adjoining the town.

Some of these are simply dug in the usual manner and the water, rising rapidly to the surface, flows over, others are dug seven or eight feet, then bored and tubed, enabling the water to be drawn off from pipes some feet above the surface. In all probability it might be raised even to the second story of the buildings or higher, as the rise seems due to the fact of these springs deriving their head from water which has flowed along an impervious substratum, off the high plateau described above, and which by my barometer is in places nearly a hundred feet higher than the level of Flat Rock creek at the bridge, near Rushville.

These wells are almost all strongly chalybeate; and, when dug to the usually depth of twenty to twenty-five feet, pass through soil, gravel, ordinary clay, and lastly through tough blue clay. Some of those bored here are now running about one pint per second; others somewhat less.

Near the head waters of Mark Creek, where a saw mill is established, a spring broke out, twenty years after the first settlement of the place. The artesian chalybeate at Slabtown, in Walker township, at the mill near the railroad crossing, flows freely, winter and summer, from a depth of twenty-six feet, dug through gravel and clay.

A remarkable feature connected with these artesian wells is that all which have thus far been dug successfully are situated nearly on a direct line, passing through the county, by Rushville, bearing from ten to fifteen degrees north of east to the same amount south of west.

The celebrated springs near Knightstown, Henry county, I was informed, are situated in Rush county, not far from the dividing line.

A considerable amount of bog iron ore was taken out about five miles east of Rushville, on the "Tiner" farm, also on that of the "Alexanders." About a mile above there, near the south bank of Little Flat Rock, a space, fifteen or twenty rods long by some eight rods wide, was so impregnated with oxide of iron that nothing would grow; and sulphate of iron was also found. Not far from this neighborhood salt was formerly made, by boiling and evaporating the spring water, according to the primitive usage of frontier settlements and times; whereas now it is found more profitable to reach by expensive shafts a stronger, subterranean brine.

According to data kindly furnished by the County Surveyor a point in Fayette county, about two miles east of Vienna, is eleven feet higher than the highest point in Rush; and this point is, if I understand him correctly, according to the levels run for the Junction railroad, 386 feet above Indianapolis. By the table of altitudes given by Messrs. Stansbury and Williams, the bottom of the canal at Indianapolis is 337 feet above high water mark at Evansville, and the latter place, according to Mr. Ellett, is 320 feet above high tide, although according to Messrs. Stansbury and Williams it is 361. Assuming the former, then the highest point in Rush would be 1,043 above high tide, by the latter, 1,084 feet.

DECATUR COUNTY.

In the north-west part of this county, approaching the county seat, Greensburgh, from the Shelby county line, we passed alternations of level and undulating, rather clayey land, affording growth for fine

Beech woods and extensive meadows, besides the usual staples of corn and wheat. Cattle, hogs, mules and sheep were abundant; the latter browsing on the short sweet herbage of gently rolling, quaternary hills.

Further south, Poplar and Oak timber are more abundant, and on "Narrow-bone ridge" we observed splendid trees of Black Walnut, Poplar, Sugar Tree, Beech and Ash.

The county furnishes excellent building rock at the quarries of Flat Rock Creek and Jordan; with these materials they have constructed a magnificent Court House. Some samples of the St. Paul or Flat Rock stone, dressed for Cincinnati, were twenty-two inches thick and weighed one hundred and eighty-eight pounds to the cubic foot; from these quarries fifty hands daily ship from twelve to fifteen car loads, besides burning large quantities of lime. The stone usually averages from six to twenty inches, and when placed in the foundation of heavy buildings, as the Court House, evidently proves itself well adapted to sustain great pressure. From an examination of the tombstones in an old grave-yard, both the Flat Rock and Jordan quarry stones stand the weather well. Some of these had almost the appearance of marble, and had preserved their edges sharply. With the pleasure and advantage of being piloted by Mr. Bonner, member for this agricultural district, we visited the Jordan quarry, passing two miles south of town by the strong chalybeate spring of Dr. Wheeldon. At six miles we crossed a fork of Sand Creek, the whole bed of which is rock, and six and a half miles a little east of south from Greensburgh we came to the Jordan river and quarries, the lower beds of which are there a few feet above the level of the stream, capped by some layers almost marble. Following the meanders of Jordan down stream, we found these upper harder layers sometimes standing out in bold relief as table rocks, twenty-five feet above the stream, the Jordan having a rapid fall, while the dip is very slight. The softer magnesian limestone with its orthoceratites and lituites has often disintegrated under the united action of air and water, sometimes forming cavernous sinuosities large enough to tempt boyish adventure in exploration, with the hope of developing a new cave to rival Wyandot or the Mammoth of Kentucky. Sometimes even the upper strong material, like the table rock of Niagara, can no longer sustain its own weight and is precipitated into the bed of the stream. One of these large masses was pointed out to us, which from its accidental resemblance, has been denominated "the coffin."

JENNINGS COUNTY.

This county is justly celebrated, judging from the samples shown us, for the excellent quality of the building materials obtained at numerous quarries. Mill-stones are also taken out, which are well adapted for grinding corn. Near North Vernon they claim to have marble; but we were unable to obtain a sample for examination. As the corps was accidentally prevented from including this county in the fall survey of 1860, although it was embraced in the original route laid down, it is exceedingly desirable that localities so full of geological and mineral interest should be fully examined and reported upon.

Mr. Legg, of Vincennes, proprietor of coal mines in Daviess county, informed me that in digging a well two miles north of Vernon, the county seat of Jennings, they found, about twenty-eight or thirty feet below the surface, a considerable amount of two metallic ores, one supposed to be antimony and the other not known. Had these specimens been immediately forwarded to the laboratory, they would have been analyzed and the results embodied in this report.

Part of Jennings is said to be rather too broken for arable purposes; but would make excellent grazing land or vineyards, and afford valuable water power. Probably the western part of the county will be found chiefly characterized by Devonian limestones, as, judging theoretically, the junction of these with the Upper Silurian rocks must crop out about that region.

JEFFERSON COUNTY.

Ascending the hill back of Madison through the deep railroad cut, we found in the lower and middle portions abundance of Lower Silurian fossils, whereas after passing North Madison and ascending to the general level of the country the Upper Silurian gives character to the soil. In the lower part of the cut, Strophomena (Leptæna) alternata, a few Trilobites, (chifly Calymene senaria,) Ambonychia carinata, and A. radiata, Modiolopsis modiolaris, Pleurotomaria percarinata, and Orthis (Spirifer) lynx are readily obtained; somewhat higher up are Streptalasma corniculum or Orthoceratites; with Avicula demissa, Atrypa increbescens, Leptæna sericea, Murchisonia gracilis, and Asaphus canalis, (Isotelus gigas,) Orthis testudinaria and O. occidentalis extend a long distance, succeeded, at about 350 feet above low water, by a dark-

FALLS OF EEL RIVER, OWEN COUNTY.—(Coal Measures.)

D. D. Owen, Del.

*This sketch of the Falls of Eel River, in Owen county, is inserted instead of the promised view exhibiting the cut at Madison, which the Engraver failed to complete.

colored coraliferous limestone, with a continuous band or layer, over two feet thick, chiefly made up of Columnaria alveolata (Favistella stellata of Hall,) considered as often marking the upper limit of the Lower Silurian. With this were some large nodules of calcite and pearl spar.

Above this a silico-calcareous rock, with some cherty layers, makes its appearance, in which we failed to detect fossils, but from which Mr. Thurston, of Madison, presented some fine Trilobítes, (Calymene Blumenbachii.) Undoubtedly these various cuts for the railroad, forming the inclined plane of ascent from Madison to the interior, have developed for the palæontological collection a rich field, which should be thoroughly explored and described.

Above the Columnaria bed the arenaceous limestone layers offer to view an anticlinal axis having a slight northerly and southerly dip; these are surrounded by eight or ten feet of aluminous shales, again capped by six or eight feet of cherty, porous limestone. The mingled Upper Silurian, with some quaternary detritus of North Madison, succeed these stratigraphical layers and develop on the adjacent farm of Mr. Elias Stapp a fine clay extensively used in the foundries near there. We obtained a sample for analysis.

The view from this section of country through the cut to Madison is truly magnificent, and we have endeavored to give a faint idea of it by a sketch; it is equaled, we were told, by the river prospect in coming from South Hanover (the location of a thriving college, which we regretted not having time to visit,) up the river to Madison.

In this county, two or three miles north of Bethlehem, on the Ohio river, close to the Clark county line, is situated the celebrated "Dean Marble Quarry" so fully described in the first report of the late State Geologist, alike suitable for good building materials and for ornamental purposes, such as table, bureau, and wash-stand tops.

In cutting and polishing these, as well as in the rough quarry rock, numerous fossils exhibit themselves, chiefly gasteropods of the genus Murchisonia, (species bellacincta and bicincta.) From low water in the Ohio here to the base of the marble quarry, is about two hundred feet by my barometer, above which a silico-calcareous rock, or perhaps more properly an aluminous and silicious limestone extends, for twenty or thirty feet, with cherty limestone surmounting the whole.

Jefferson county also affords the beautiful, Lower Devonian coral, Pleurodictyum problematicum; indeed the uplands, near the line of junction with Floyd county are chiefly Devonian.

SEC. III.—COUNTIES IN THE DEVONIAN SYSTEM.

The greater portion of the following counties were found so characterized as to rank them in the Devonian system, viz.: Cass, Carroll, Tipton, Hamilton, Shelby, Bartholomew, Jackson, Scott and Clark.

SUB-SUCTION 1.—GENERAL DESCRIPTION.—The geological assemblage of rocks often termed the Devonian system consists with us, in the lower Pleurodictyum problematium beds of Jefferson and Scott counties; and in the middle sub-divisions, of beautiful limestones, alternating with others more cherty and aluminous, replete with splendid fossils, such as those found so abundantly about Charlestown and Utica, Clark county, and the Ohio Falls, chiefly large masses of coral, belonging especially to the families Favositidæ and Cyathophyllidæ, ranging from the delicate tabulated Syringopora to the gigantic Zaphrentis, also Brachiopodous and Conchiferous mollusks, from the curious slipper-shaped bivalve of the family Orthidæ (Calceola sandalina) to the snouted Conocardium and delicate Lucinia proavia. This is probably the equivalent to the great Eifel limestone or calceola schists of Europe. The upper sub-divisions are more aluminous in character, being especially represented in Indiana at various points from New Albany, on the Ohio, to Delphi, on the Wabash, and in the same north-west line towards the lakes, by a bituminous, argillaceous black shale, (often containing beautiful lingulas,) which has been frequently mistaken for a coal shale, but which is yet geologically far beneath the true carboniferous deposits. To these upper beds, the equivalent apparently of the Rhenish Goniatite schists, as well as of the Gennessee slate of the New York Geologists, we assign Jackson county with its Goniates and Orthoceratites.

In Great Britain, where this system frequently receives the name of the "Old Red Sandstone," Prof. Johnston describes it as being in its upper part made up of red sandstone and indurated sandy gravel, its middle of clayey marls and impure silicious limestones, and the lowest of mottled sandstones, sometimes wholly silicious, at other times partially calcareous in character.

In Europe, too, fishes of remarkable type, sported in the Devonian seas, such as those described by H. Miller in his fascinating "Red Sandstone," the Cephalaspis, with its external buckler of bone to protect its cartilaginous interior, the Holoptychius with tubercled scales, the winged and horned Pterichthys, and its equally remarkable congener the Coccosteus, all of them fishes more resembling our sharks, rays, gars and sturgeons, than our perches or salmon.

Sub-Section 2.—Soils, Agricultural Products, &c.—The soils resulting from the disintegration of the Indiana Devonian rocks, we found in many instances of excellent quality; although occasionally varying somewhat in character from the proximity of other formations, as may be more fully traced in the description of the separate counties of this formation.

Some English writers have characterized the "Old Red" as giving rise to some of the finest agricultural regions in Great Britain; but others assert that this is the case only where there are intercalations of marl.

The agricultural products in this geological area of Indiana are rather small-grain and grasses, than Indian corn; some of the counties, however, cultivate the latter also successfully, and raise considerable droves of hogs.

Sub-Section 3.—Rock Quarries, &c.—Some of the limestones of Devonian age furnish beautiful building materials; and as Cass county is assigned to this formation, although some of the lower beds are Upper Silurian, we may here cite the numerous quarries in and around Logansport. The building materials in the quarries of Hamilton and Shelby counties are also from near the junction of the Upper Silurian and Devonian; but the pure lime shipped so abundantly from Clark county is entirely burnt from Devonian limestones, while the western part of Bartholomew furnishes rock derived from the sub-carboniferous sandstones.

Sub-Section 4.—Metallic Ores.—With the exception of the gold washings to be spoken of in describing Carroll and Clinton counties, and some carbonate of iron found locally in the ash-colored shales over the black slate in some knobby regions, few metals of importance were brought to the notice of the corps in the counties of this formation; yet, although metallic ores of good quality are not usually so abundant in this stratigraphical section, as in some others, it is quite possible that more extended research may develop some minerals worthy of the practical operative's attention.

Sub-Section 5.—Timber and Predominant Vegetation.—Beech timber seems the prevalent growth, particularly on the clay soils, resulting from the disintegration of the aluminous shales; but other valuable trees are also abundant in various parts of the formation, such as Sugar Tree, Black and White Walnut, Ash, with some Buckeye and Wild Cherry.

Sub-Section 6.—Springs, &c.—Mineral springs are probably not so common in this as in some other formations; but we have, back of Jeffersonville, a chalybeate which was formerly much frequented as a watering place and summer resort. The analysis of this water will be found in the description of Clark county. Several medicinal springs are spoken of when describing Carroll county.

SUB-SECTION 7.—MISCELLANEOUS FACTS REGARDING PREVALENT DISEASES, &c.—Although bilious and remittent fevers are not so common perhaps as in the alluvial bottoms, nor milk-sickness so prevalent as in parts of the coal measures, yet on the cold clays of the black slate, the latter has been reported in places, and some chill and fever observed; as well as the ever prevalent typhoid and not uncommon winter fever or pneumonia.

SUB-SECTION 8.—CHARACTERISTIC FOSSILS:—

A. RADIATES.—

    a. Corals: Favosites Goldfussi,
                  F. basaltica, (Gothlandica,)
                  F. polymorpha,
                  F. dubia,
                  F. reticulata,
                  F. mammillaris,
                  F. maxima,
                  F. fibrosa,
                  Emmonsia hemispherica,
                  E.? cylindrica,
                  Syringopora tabulata,
                  S. tubiporoides,
                  Zaphrentis corniculum,
                  Z. Rafinesque,
                  Z. gigantea,
                  Amplexus Yandelli,
                  Aulacophyllum sulcatum,
                  Hadrophyllum Orbignyi,
                  Cyathophyllum rugosum,
                  Heliophyllum Halli,
                  Acervularia Davidsoni,
                  Eridophyllum strictum.
    b. Acalephs.
    c. Echinoderms: sub-division Crinoids,
                              Olivanites Verneuili.

B. MOLLUSKS.—
    d. Molluscoid Bryozoa. Fenestella antiqua, (Lonsdale,)
                           F. tenuiceps, (Hall,) [This species, found on the Falls, is probably Upper Silurian.]
    e. Brachiopods. [Spirigerina reticularis, Caleeola sandalina,]
                    Atrypa aspera,
                    Athyris concentrica,
                    [Spirifer mucronatus,]
                    S. cultrijugatus,
                    S. euritines,
                    Terebratula ——— ?
                    [Pentamerus galeatus,]
                    Lingula prima?
    f. Conchifers. Conocardium (Pleurorynchus) trigonalis,
                    Lucina proavia,
                    Megalodon cucullatus.
    g. Gasteropods. [Euomphalus serpens?]
    h. Cephalopods. Goniatites rotatorius,
                    G. sinuosus,
                    Orthoceras laterale?

C. ARTICULATES.—
    k. Worms.
    l. Crustaceans: sub-division *Trilobites.*
                [Phacops macrophthalma or Phacops bufo,]
                Calymene crassimarginata.
    m. Insects.

D. VERTEBRATES.—
    n. Fishes.
    o. Reptiles.
    p. Birds.
    q. Mammals.

SUB-SECTION 9.—DETAILED DESCRIPTION OF THE COUNTIES IN THIS GEOLOGICAL FORMATION.—

CASS COUNTY.

This county and even Carroll might with propriety have been described among the Upper Silurian counties, inasmuch as Pentamerus occidentalis and other fossils characteristic of the above formation are found abundantly; but as these show themselves only at low natural cuts, chiefly on the Wabash, and as the uplands of both counties derive a great portion of their character from Devonian rocks, modified by quaternary erratic deposits, it seems more appropriate to class these counties with the Devonian.

In Cass they claim to have several characters of soil:
1. South of the Wabash a rich sandy loam, with clay subsoil.
2. North of the Wabash a more sandy soil.
3. Between the Wabash and Eel rivers there is excellent wheat land, the result of a due admixture of the arenaceous soil with an aluminous gravel. All these are based on limestone. Considerable quantities of good sized bowlders are found throughout the county, particularly along the Wabash valley.

A short distance below Logansport, we found limestones in the bed of the river, with a slight dip westerly, but nearer town, not far from the mouth of Eel river, the dip was observed to be locally rather northerly. Going out along the canal towards a cut made for the Cincinnati railroad in passing through the low land, we found limestone with Devonian fossils, Favosites basaltica, (Gothlandica,) and at the foot of the hill observed large quantities of calcareous tufa, impregnated with iron, from a chalybeate spring just above. The lower portion of the cut exhibits a "hard pan" of bluish clay, with small gravel; the upper part consists of 20 to 25 feet of ordinary gravel.

Mr. Green, of Logansport, who has had extensive experience in quarrying, pointed out an encrinital limestone in thin layers on the south side of the Wabash at this place, which has been considerably used as a firestone, and exposed to high heat, without exhibiting any signs of cracking or crumbling. The same shows itself about two miles south, on the Wabash Valley road, in beds, disappearing with a westerly dip.

Four miles below Logansport, the limestone quarried contains considerable quantities of calcareous spar and sulphuret of iron, but, two miles further down the Wabash, a saccharoid limestone, apparently of good quality, hard and compact, is quarried, and was used for the construction of piers to sustain the bridge across the Wabash, for the Toledo, Logansport and Burlington railroad. For coping stones they usually import from the Fisher quarry, in Wabash county, or from Dayton, Ohio.

The hills on both sides of the Wabash, near this true limestone, and overlying it, are made up of a rock which effervesces slightly with acids, but which, from its peculiar texture and working, is denominated by the stone masons a freestone. Portions of it may be considered a true silicious limestone, standing well in walls; while, further back in the hill, some layers approach to a marble, rather disposed to be friable until the quarry water has evaporated.

Under the guidance of Mr. Wright, State Senator from this county, we visited a locality three miles above Logansport, where a layer of supposed hydraulic limestone, (of which a sample was taken for analysis,) n derlies the Favosites limestone.

From Logansport they ship large quantities of veneers to Cincinnati and New York, chiefly of Black Walnut. We were informed that these are sometimes worth nine cents per foot, and that a large, old, gnarled tree, preferred on account of the varied markings, will make from ten to fifteen thousand feet or more of veneering: one saw cutting about five thousand feet per day. Judging from these figures, this department of business seems well worthy of attention in portions of Indiana, where the Black Walnut growth is extensive.

CARROLL COUNTY.

This county, entered from the South, exhibits, near the town of Prince William, in the valley of the middle fork of Wild-Cat, as well as later, at the north fork, a considerable amount of Quaternary, comprising thirty to thirty-five feet of bluish hard-pan clay, observed on ascending from the river, then yellowish, loose gravel, with interspersed bowlders, the various layers separated by distinct horizontal lines of stratification, and springs occasionally welling from above the lower aluminous stratum. On reaching the general level of the country, sixty feet above the bed of Wild-Cat at the ford, and traveling towards Delphi, the county seat, the crops of good clover, and the rank growth of boneset, (Eupatorium perfoliatum,) in connection with extensive surface cracks in the soil, denoted a predominance of clay, requiring pretty thorough drainage.

In this region, part of which appears newly settled, crops of flax are rather abundant, raised chiefly for the seed: the farmers expressed a desire, in connection with this culture, to learn some method of working up the stalk fibre, which at present is generally thrown out as refuse, after the extracting of the seed by threshing.

Near Delphi we obtained the following section:

	FEET.
Soil and loose gravel, &c	10–15
Quaternary hard pan and conglomerate	15–20
Devonian black slate	50–60
Devonian limestone	20
Upper Silurian Pentamerus beds	20

At the lime-kiln just below town we found a local dip to the southeast, amounting to 40.°

The black slate has been washed out in the valley between Deer creek and the Wabash, and the detritus scattered over the Devonian limestones, which contain Emmonsia hemispherica, and other fossils, overlying the Stromatopora concentrica and Pentamerus occidentalis limestone.

Between two and three miles from Delphi, considerable samples of gold have been washed, from the Drift in the bed and bank of the creek, a locality well meriting further examinations.

Cases having occurred in this county which indicated milk-sickness, or which, at least, by some of the physicians, were assigned to that class of disease, several springs were qualitatively examined at their sources; but, after the employment of delicate reagents, no poisonous ingredient, at least of a metallic nature, could be detected.

On Deer creek, a few miles from Delphi, vast quantities of calcareous tufa have formed, by filtration of water through the overlying quaternary deposits, and subsequent evaporation and consolidation, while trickling slowly over the black slate bluffs of the stream. The stalactitic and columnar forms, often ornamented by distinct impressions of leaves on the soft tufa, with cavernous niches decked out in the rich profusion of cryptogamic vegetation, chiefly of the liverwort family, added to the rippling streamlets forming cascades, as they are precipitated from cedar-clad Drift hills, finally, over 30 to 40 feet of black slate, into the meanders of Deer creek, all conspire to form a highly picturesque scene, well worthy of a more extended and detailed sketch than the one which time permitted on this exploration, and which is here subjoined.

On these examinations we had the advantage of being piloted by Dr. Beck, Messrs. Holt, Milroy and Baum. Two paper mills in town, besides large warehouses for storing produce, added to other indications of industry and activity, denote this to be a place of thrift and enterprise.

QUATERNARY TUFA, NEAR DELPHI, CARROLL COUNTY.—(See report, page 38.)

R. Owen, Del.

On crossing the river below the dam at Pittsburg, about two miles from Delphi, and ascending the north bank of the Wabash, the heavy quaternary deposits are found made up of alternate clay and gravel beds. The town is from twenty to twenty-five feet above low water; and a spring flowing near the level of the streets, but deriving its head from above a clay bed forty feet higher up, has been ingeniously made to ascend inside a hollow willow tree, and to pour out in a small artesian stream, by inserting a pipe into the trunk of the tree, at a convenient height for domestic accommodation.

After attaining the general level of the country on the road to Monticello, about a hundred feet above low water in the Wabash, we traveled on Drift bowlders and gravel, sometimes forming ridges giving growth to Black Jacks, small Hickories and White Oaks, Hazel bushes, some sumachs and mulleins, while, ten to fifteen feet lower, occasionally intervened moderate areas of rick swamp-muck prairie, studded with yellow asters, numerous pink, bell-shaped corollas, contrasting with these and others of a bright blue, besides sour dock, flags, boneset, &c.

Thus were brought into juxtaposition soils, on which the industry of the farmer was carrying out quantities of manure, to supply the necessary organic and inorganic fertilizers, with other lands too rich for small grain until cropped in Indian corn. Sorghum and good corn, orchards, and flocks of sheep, with some fine cattle and substantial barns, spoke of good husbandry.

South of the Wabash, we observed in this county the swamp White Oak, (Quercus primus discolor,) which resembles the White Oak (Quercus alba) and affords posts remarkable for their durability. Indeed, Michaux considers it superior to White Oak, and recommends its encouragement, "to the exclusion of the Red Flowering Maple, Bitternut, Hickory and Hornbeam, all of which frequent swampy land, where little else would grow."

TIPTON AND HAMILTON COUNTIES.

Of the former we cannot speak from personal inspection, although we passed once near the eastern limit of the county, and at another time near the west. Judging from these and reports furnished us, it seems generally level, from heavy quaternary deposits; it is also well timbered, and favorably situated for exporting produce, as the Indianapolis and Peru railroad passes centrally through it. The soil of the east probably partakes more of the character given to Grant and Madison by the Upper Silurian limestones, while in the west it may readily have been modified by the aluminous shales found in Clinton.

Hamilton derives its chief characteristics from quaternary deposits, although the silico-magnesian limestones frequently come close to the surface between Strawtown and Noblesville.

The upper or ashen soil, forming the eminences, seems thin, and requires subsoiling and admixture with the red ferruginous clay exhibited beneath, whenever a natural section of a few feet occurs by denudation.

On a creek seven miles south of Noblesville, where drift to the amount of forty feet is exposed, we obtained the following section:

	FEET.
Ashen soil	$2\frac{1}{4}$
Reddish clay	$1\frac{1}{2}$
Hard-pan and cherty gravel	10
Very ferruginous and stiff clay	10
Gravel, hen egg size, bowlders and sand	15

Among the bowlders and gravel in the lowest stratum, granite, gneiss, hornblende-rock, greenstone, silicious limestone, and some sandstone, were observed. The sand was fine, sharp and white.

The "hard-pan" was made up of angular chert, flint, limestone and sandstone fragments cemented into a concrete gravel by a whitish clay.

The upper reddish clay kneads into a paste, and breaks when dry into small square pieces.

The white or ash-colored soil, although rather sandy, yet bakes somewhat in the sun; it exhibits a few angular cherty pebbles, diffused through it. A chalybeate water oozes from this hill side, and we noticed among the timber Beech, Sugar Tree, Walnut, Oak, Buckeye, and, in the undergrowth, abundance of iron weeds.

On the 21st of June, when passing here, corn was observed more than two feet high in the red clay hollows, when it scarcely more than showed itself on some of the light-colored eminences. In some places excellent potato crops were noticed; wheat and timothy appeared also tolerably fair.

Other sections near here presented a stratum of bowlders immediately beneath the reddish clay and above the hard-pan gravel.

Five miles north-east of Noblesville, the county seat, near the mouth of Deer creek, where there is a saw-mill, an extensive quarry has been opened, and lime is burned. A rather rapid dip, amounting to at least 30°, is well exposed here, assuming a direction rather west of south.

Near Strawtown, Mr. Guenther is successfully manufacturing pottery from the quaternary ferruginous clay described above.

SHELBY COUNTY.

The north-western portion of this county receives its character chiefly from the quaternary deposits, with proximity in places of the black slate; in the south-east, the silicious limestones, near the junction of the Upper Silurian with the Devonian, are reached at intervals in quarries, all along the valley of Flat-Rock; these, in conjunction with the erratic group, form a soil of fair quality.

Entering the county at the thriving town of Manilla, we observed, in traveling (20th Sept.) south-west to Shelbyville, good, although not large farm-houses, orchards, wheat chiefly drilled, fair average crops of Indian corn and sorghum; the country also well drained, and watered by numerous little streams. According to the "Table of Altitudes," Shelbyville is 757 feet above high tide. South of Shelbyville, we observed large fields of corn, on somewhat flat, clayey soil, occasionally varied by gravel and large bowlders, with a predominance of Beech timber. Yet further south, the soil in places indicated a black, sandy loam, with high iron weeds, Beech forest not so exclusively preponderating.

The silicious limestone in the bed of Flat Rock did not afford many fossils, nor did it offer as fine slabs as in the quarries already spoken of near the junction of Shelby with Decatur and Rush, where stone is worked so extensively. The best are said to be on Conn's creek. A southerly dip was quite perceptible in the bed of Flat Rock; at an old dam, the rocks were much rippled, marked and grooved by the wearing effects of the water running here with the strike. Immense quantities of chert and confervæ covered some portions.

At the office of the "*Republican Banner*" we were informed that, about 2½ miles south-east of Shelbyville, 'Squire Allen Sexson has an artesian chalybeate well: this is almost on a direct prolongation of the W. S. W. line of artesian chalybeates described as being found in Rush county. A natural chalybeate also flows out below Freeport, in Hanover township, Shelby county, on Big Blue river, a branch of White, to be distinguished from Great Blue river, mentioned hereafter as forming the boundary between Harrison and Crawford counties.

BARTHOLOMEW COUNTY.

The rich, sandy loam of southern Shelby county continues into the northern part of Bartholomew, particularly in the valley of Flat Rock. The uplands, averaging fifty to sixty feet higher than the bottoms, are generally composed of from twenty to thirty feet of quaternary gravel, with bowlders, overlaid by soil. On the route to Columbus, the county seat, we observed some good barns, full average corn crops, and luxuriant clover—Beech and Poplar timber prevailing. On approaching Taylorsville, level land, heavy timber, small swamp-muck prairies, extensive fern tracts, in which Adiantum, the maidenhair genus, appeared most abundant, and corduroy roads tried our springs, all pointed out the propriety of thorough drainage. These appearances are readily explained when near by, at Tannehill's mills, on the east fork of White river, we find the Devonian black slate, and learn that the knob sandstone of sub-carboniferous date is encountered five miles west.

Other fine rich bottoms show themselves also along the valley of Sugar creek, in this county, a stream not to be confounded with the Sugar creek of Fountain county.

Traveling westward towards Brown county, and ascending about 360 feet above the water level at the Tannehill White River Mills, we were shown the materials passed through and thrown out by Mr. Thomas Brown, when digging a well at this elevation. After excavating the yellowish clay, with mingled gravel, a bluish clay was reached; then, on digging with difficulty through some feet of "hard pan," water was obtained in the gravel, 36 feet below the surface. In "pockets" above the clay, he found sand, iron particles and mica, probably the detritus of decomposed bowlders. On the highest Bartholomew ridge, near Brown county line, 390 feet above White River, and consequently little short of a thousand feet above the sea, the water level near the mouth of Cliffty, at White river, being 596 above tide, the Fall grapes were observed growing in clustered profusion, on the gracefully pendant vines of native, unpruned luxuriance; promising well for success in artificial cultivation, under judicious selection and management.

JACKSON AND SCOTT COUNTIES.

The portion of this county of which we saw the most was around Rockford, (visited by us partly on account of the fine fossils obtained there at low water,) and the central portion presented to view in passing by railroad, chiefly along the White River valley from Seymour, through the county seat, Brownstown, and past Vallonia, towards the knob sandstone near the junction with Lawrence county.

Unfortunately, the river was rather high for our purposes at the Rockford dam; however, we found, in thin layers of an aluminous limestone, portions of fine goniatites, belonging to at least two species, and some orthoceratites were presented by the owners of the mill. One species of the former so strongly resembles DeKoninck's Belgian species, rotalorius, as probably to be identical; and the other seems closely allied to the G. sinuosus of Hall. The best orthoceras is provisionally assigned to the laterale.

The true hard, bituminous black slate, often with lingula spatulata (Hall) is here scattered profusely along the banks of the Driftwood fork of White river, derived chiefly from a locality about four miles above, where the black slate constitutes the east bank. The rock used at the Rockford mill is from Silver creek, six miles north of Jeffersonville.

At Seymour they employ the Vernon stone, having no rock near the surface: the water, which they obtain by digging wells, being almost soft, probably filters through the fine alluvial sand, interspersed with some gravel, here forming the adjacent hills, of ten to twenty feet, and, after such filtration, rests on the knob sandstone beneath; hence the absence of calcareous and saline impurities.

Dr. Monroe, of this place, informed us that the county exports chiefly corn and hogs, Lawrenceburg being their main market. Somewhat further west, towards the county seat, we observed (21st Nov.) numerous straw stacks, and a considerable amount of new wheat, that promised well, of cattle and pastures, also of sheep grazing on the quaternary hills, generally from 25 to 50 feet high, and at places even 75 or 100 feet above the railroad level. Beech timber predominated, but piles of staves and hoop-poles along the road were not uncommon. The rich Beech, Papaw and Iron weed land, north of Langdon, seemed to require drainage.

In Scott county we find the same black slate, and some deposits of marl. From the heavily timbered lands of Austin, so low as occasionally to permit chill and fever miasmata, and from other points in this county, vast quantities of staves are shipped.

In the region of Vienna the Beech land is higher, as we pass from the Devonian clay shales to the sub-carboniferous or knob sandstone.

Of this county, however, we did not see as much as would justify a detailed description. It ought, therefore, to be more thoroughly examined hereafter.

CLARKE COUNTY.

As we near the northern limit of this county, in the region between Summit, (the ridge dividing the Ohio and White River tributaries,) and Henryville, Oak and Hickory become very prevalent; but farther east, about Utica and Charlestown, Beech, luxuriating in the furruginous clays, again makes its appearance.

A cut between Memphis and Sellersburgh, and various places on Silver Creek, exhibit the same bituminous, aluminous black slate seen in Jackson and other counties, usually assigned to the Upper Devonian, but by some writers placed in the lowest beds of the carboniferous system.

About Utica large quantities of lime are burned from Devonian rock, in beds about twenty feet thick, with fossils only in the lower layers; surmounted by ten to fifteen feet of chert and reddish clay. As nearly as we could ascertain they ship annually from this place 100,000 barrels of excellent white lime, chiefly burnt in fire kilns, some of which hold 350 barrels and are charged fifty or sixty times a year.

Near Utica, on "Clarke's Grant," No. 26, owned by Mr. Jacob Ruddell, a Devonian soil, for the analysis of which see No. 10, 11 and 12 of Dr. Peter's report, grows, even on upland after thirty years cultivation, fifty to fifty-five bushels of corn and twenty-five of wheat, also good clover and potatoes. On this farm we found abundant samples of Lucina proavia and other Devonian fossils. The timber is Sugar Tree, Black and White Walnut, Elm, Ash, Buckeye and Cherry. But it is chiefly in the neighborhood of Charleston that the splendid and magnificent coral characteristic of this age, and already gracing many a European cabinet, are found in profusion. On the farm of Mr. Frank Weller, Clark's Grant, No. 26, where they were quarrying rock for the turnpike, we obtained samples of Zaphrentis gigantea two feet long, Cyathophyllum rugosum, Favosites Basaltica, (Gothlandica,) F. mammillaris, besides many others; and, on the road from here to Jeffersonville, Spirifer cultrijugatus, &c. Along the entire rapids or falls of the Ohio, from Jeffersonville to New Albany, are found numerous beautiful Devonian fossils, such as Syringopora tabulata, Favosites mammillaris, Emmonsia hemispherica, Amplexus yandelli, Zaphrentis cornicula, Z. gigantea, Olivanites Verneuilli; beautiful Bryozoa, chiefly of the genus fenestella; trilobites, as Calymene crassimarginata, &c., &c.

Associated with these beds is the stratum, about fourteen feet thick, from which the celebrated water lime* is quarried to be burnt, barreled and sold as "hydraulic cement." Above these limestones are the black slates, which at places near New Albany, have been ascertained by Dr. Clapp to occupy 104 feet in thickness. Prof. J. Lawrence Smith states that in boring on Main street, Louisville, for the artesian well of Messrs. DuPont, they passed through thirty-eight feet of shell limestone and forty feet of coralline limestone, both of Devonian age, before reaching the Upper Silurian, which was supposed to occupy the succeeding twelve hundred feet.

Back of Jeffersonville the saline chalybeate spring, already spoken of as a summer resort, afforded on qualitative analysis:

Bi carbonate of the protoxide of iron;
Bi carbonate of lime;
Bi carbonate of magnesia;
Chloride of sodium, a small quantity;
Chloride of potassium, a trace;

Consequently it is tonic, slightly aperient, and perhaps alterative.

At New Washington the landlord of the hotel informed us he struck the black slate in digging his well.

Seven miles south of that place, on descending a natural cut, we observed about thirty feet of saccharoid limestone, over thirty to thirty-five feet of chert, with reddish soil and Devonian fossils, then eighty to eighty-five feet of silicious limestone beneath. The same was verified on ascending the next hill, where near the summit we found beautiful Devonian corals.

On approaching the Dean Marble Quarry, at the eastern edge of Clark county, we observed near Bethlehem more than two hundred feet of Lower Silurian formation, overlaid by thirty to thirty-five feet of silico-calcareous rock, twenty to forty of Devonian limestone, and fifteen to twenty-five feet of chert, disintegrating into ferriginous clay. Here at least then, the Upper Silurian seems to have thinned out considerably.

---

*Dr. Peter's analysis for the Kentucky report of this rock will be found in the appendix.

SEC. IV.—COUNTIES IN THE SUB-CARBONIFEROUS SANDSTONE FORMATION.

SUB-SECTION 1.—GENERAL OBSERVATIONS.—Some English writers subdivide the carboniferous system in ascending into the Carboniferous or Mountain limestone formation, the Millstone Grit, and the true Coal Measures. In this country, in consequence of there being a considerable thickness of strata beneath the productive coal seams, or carbon-bearing beds, some Geologists have named these lower strata sub-carboniferous; and as great diversity of scenegraphic and agricultural character exists between the upper and lower strata of this formation, it is frequently again sub-divided into the sub-carboniferous or knob sandstone, and the overlying sub-carboniferous limestone, or cavernous limestone.

As this latter arrangement seems convenient for Indiana, we begin with those counties in which the prevailing rock is a series of sandstones assuming various tints of yellow, green and gray, with intervening shales. Prof. Hall is of opinion that in denominating rocks of this lithological character, in the State of Ohio, the "Waverly Sandstone," the strata were not always carefully distinguished from Devonian beds of the Chemung and Portage groups.

The following counties in Indiana are considered as chiefly characized by the sub-carboniferous sandstone and shales: Tippecanoe, Clinton, Boone, Hendricks, Johnson, Morgan, Brown, Washington and Floyd counties.

SUB-SECTION 2.—SOIL, AGRICULTURAL PRODUCTS, &c.—The soil resulting from the disintegration of sandstone, and somewhat aluminous shales, might naturally be expected to be rather cold where the shales predominated, as well as too thin and light where the detritus of the sandstone was the chief ingredient of the soil. This is undoubtedly to some extent the case, but generally speaking the two are blended, and sometimes the modifying proximity of the not-far distant limestone, or the natural top-dressing of quaternary deposits bringing clay, gravel, decomposed bowlders, some of them rich in magnesia, lime, the alkalies and oxide of iron, forms a varied soil well adapted for most agricultural purposes.

Where there is a defect in the soil of this formation an artificial top-dressing of lime or plaster would well merit a trial, which at first may be on a small and inexpensive scale. Tobacco has been advantageously raised in portions of this formation; and, on the high knobs, if some laws could be devised for the protection of sheep, laws* equally just in their bearing on the sporting "Nimrods" who love pointer and setter, and on the pastoral "Abels" who see in these latter only the sheep-killer, then we may hope to have those hills covered with Southdowns and Cotswolds, furnishing excellent food and clothing to the consumer, as well as profit to the farmer. As a general rule small grain and perennial grasses would return a better remuneration in this formation than corn or clover; and among these cereals, where there is a market, barley will be found most productive in a sandy soil. Potatoes too, if light showers favor, would here produce abundantly.

SUB-SECTION 3.—ROCK QUARRIES, &c.—It is rather in the upper limestones than in these lower sub-carboniferous sandstones that we should look for the best building materials; still we saw localities where, by judicious selection, they had quarried out good rock for various purposes. The details will be given in describing the separate counties.

Near Bedford they claim to have hydraulic limestone, and in some places probably also grindstones might be manufactured; although those brought to our notice were higher up in the series, among the Millstone Grits and true Coal Measures. We saw no marls or clays which were considered commercially important.

SUB-SECTION 4.—METALLIC ORES, &c.—This is not the formation in which we usually find extensive metallic deposits in situ; but as will be seen in the detailed description of counties, iron ore was found in Marion, lead in Tippecanoe, and gold has been extensively washed at several places from the drift deposits that have lodged in favorable portions of this geological rock stratum, where hollows had formed, whether by natural ridging from the dip of the beds or more usually by subsequent denudation of intercalated strata, less hard and weather proof than the purely silicious peaks.

SUB-SECTION 5.—TIMBER AND PREDOMINANT VEGETATION.—On the higher sandstone knobs, Oak, Elm and Poplar are most common; while occupying less elevated or more clayey portions, we still see Beech, Sugar Tree, Black Walnut and Ash. In yet different regions of this formation, where rich quaternary marls and other deposits have con-

---

*It is said that in some States a tax has been levied on dogs, (and who that is fond of sporting would thus begrudge a dollar or two each year,) and the resulting proceeds from a fund for the repayment of losses to farmers whose sheep have been killed or worried by dogs. The plan is reported to work well.

tributed to fertilize, the growth of timber will again vary from the above; indeed the flora of the sub-carboniferous soils is exceedingly various, comprising in parts the plants that love to revel in cold, wet, aluminous plains of our latitude, in part those whose habitat inclines to high, dry and arenaceous ridges. Some of the latter, as will be seen in the detailed description, rise abruptly three hundred and fifty feet, in places even four hundred feet, above the small streams which form the valleys of drainage for the country.

Sub-Section 6.—Springs, &c.—Except on the above high and dry ridges, springs and well water are sufficiently abundant throughout this formation for all practical purposes, and are usually less apt to be hard than in most other formations, where calcareous beds generally form at least one layer in the stratigraphical series constituting the system or formation.

It is only on the westerly boundary of this geological zone that we find the overlying limestones again charging the waters of filtration with the soluble alkaline earths.

Sub-Section 7.—Miscellaneous Facts.—From the facts above adduced of the varying physical geography and development of vegetable life, we are naturally led to expect a variety in the climatal and pathological results. In this we are not deceived, for the diseases incident to the undrained flats are usually more bilious and remittent in type, while on the dry ridge, dysenteric epidemics are more common. Pneumonia and typhoid are perhaps scarcely bounded by geographical areas or subject to restrictive lines of altitude.

Sub-Section 8.—Fossils Characteristic of this Formation.—Palæontologists have long observed that during the progress of the earth's history and modification, whenever the chief deposits, washing from the higher portions of land to raise the lower grounds, whether dry or under the bed of the ocean, partake of the character usually assignable to arenaceous deposits, that of loose drifting sands, organic remains in these are comparatively rare. It seems to require something more coherent in the form of aluminous, silicious or calcareous particles, oozings of peroxide of iron or some other soluble metallic oxides, or sometimes a salt, to imbed or surround the delicate organism with such a nidus as shall, by a chemico-geological metastasis, replace organic matter by its less perishable organic representative. This applies to the Knob-sandstone, for the fossils found in it are rare; and, as they are chiefly, generically and specifically the same as those in the superincumbent limestone, also of sub-carboniferous age, will be described under that head.

SUB-SECTION 9.—COUNTIES IN THE KNOB-SANDSTONE:—

TIPPECANOE COUNTY.

In the north-eastern portion of this county we encounter the Devonian, black, aluminous, bituminous shales, and a few miles further east find them, and thin Devonian limestones, reposing on the Pentamerus limestones of Upper Silurian. In the south-west, a short distance from the county line, we meet with the sub-carboniferous limestones, which overlie the sub-carboniferous sandstones; and near the western junction with Warren county the Millstone Grit and true Coal Measures show themselves.

From these facts, although the greater portion of the county is covered with quaternary deposits, forming the beautiful Wea prairie, and stretching to the celebrated " battle ground," we feel justified in assigning the sub-strata of Tippecanoe mainly to the Knob-sandstone subdivision.

As, however, we found these drift deposits in many places over a hundred feet thick, and as, in boring the Lafayette Artesian Well, they passed through 170 feet of superficial deposits, including the bowlders, before reaching the black shales, we may readily imagine that the soil of Tippecanoe county chiefly derives its character from this geological deposit. The great valley of the Wabash has denuded the lighter portions of these older quaternary deposits, exposing long lines of the heavier bowlders, especially between Lafayette and Americus, and has replaced the older by newer alluvial quaternary, thus still further modifying the soils of the Wabash bottom, and rendering them more sandy than the general level of the older quaternary prairie soil, from 120 to 150 feet above low water in the Wabash at the Lafayette bridge. These bottoms, here as elsewhere, sustain their high character for large crops of maize, while the uplands are well adapted for cereals and grasses. The upper portions of these drift deposits are frequently loose sand and gravel, intermingled with decomposing bowlders, crumbling oxides, and other detritus, that furnish some calcareous and ferruginous matter. This, washed by the filtering rain water down to the impervious quaternary clays and gravels, is then arrested, and serves to cement them into a hard, pebbly conglomerate, not unlike the artificial concrete of civil engineers, although with component fragments more rounded by previous attrition.

About three-quarters of a mile east of the Lafayette Court House, on Mr. L. B. Stockton's farm, also at another point, southerly from town, in Taylor's hollow, Rochester's addition, we saw fine sections of this denuded Drift conglomerate. In company with Mr. Wagner, President of the State Board, we traced, under the great advantage of his thorough acquaintance with the country, the same line of exposure at various localities, going out to Warren county, particularly six miles west of town, about high water mark, the other at the mouth of Indian creek, near the western limit of the county.

In the above hills near Lafayette, the hard conglomerate has again partially disintegrated in places, leaving rude columns supporting a roof of concrete, with ample cavities below, giving to the whole at some distance a picturesque effect, not unlike the representations of the ancient Druidical temples of our British ancestors, during their early savage life.*

In one of the hollows bearing north, near Indian creek, we were told that extensive deposits of lead ore had been found; but, as usual with these discoveries, the finders threw a mystery around the description of the exact locality, which precluded the probability of finding it readily, and prevented any extended search at that time.

The celebrated Artesian Well at LaFayette, (bored by Mr. McKay, the water analyzed and full details of the whole work ably given by Dr. Chas. M. Wetherill,) shows that the head waters are derived probably from the Upper Silurian rocks, extending from Huntington to Delphi, and then dipping under the black slates. The tubing passed through 170 feet of Quaternary, then nearly 30 feet of Devonian shales, and lastly more than 30 feet of Devonian coralliferous limestones.

The water of this artesian well resembles that of the Blue Licks, Kentucky, and may be used either for beverage or bath, in various chronic diseases. It is reported as beneficial in many cases of chronic rheumatism, some forms of dyspepsia, scrofula, &c.

It is scarcely necessary to add, that, in the large and prosperous city of LaFayette, every accommodation can be furnished to invalids visiting this Artesian well for health or pleasure.

---

*A sketch of this locality was made, in case it should be called for as illustrative of quaternary conglomerate beds, but is omitted, as well as other illustrations, in consequence of the engraver and writer both having entered the army before fully completing the work.

CLINTON AND BOONE COUNTIES.

As a calcareo-aluminous rock (becoming sometimes silicious, and even passing occasionally into Lydian stone, a dark, banded quartz) destitute of fossils, shows itself on Sugar creek, about 15 miles south-west from Frankfort, the county seat of Clinton, this sub-carboniferous substratum probably underlies the western portion both of this county and of Boone, at no great distance from the general level. Aluminous shales, however, show themselves on Prairie creek, near Frankfort, and, the black bituminous shales of Devonian age being abundant at Americus and Delphi, the eastern parts of Clinton and Boone are probably underlaid by similar strata. Still, the chief character of the soil is derived from the Drift, which averages about 60 feet on the western limit of Clinton county, near Colfax or Midway. Although in portions of Clinton the soil is white, with a substratum of gravel and bowlders, and a growth of hazel, yet in others there are miles of stiff quaternary clay, requiring drainage, and affording abundance of elder bushes and boneset, with rank ferns, particularly near the edge of Twelve-mile prairie. The crops of corn and clover appeared, however, excellent, and orchards were not unfrequent.

At Frankfort, Mr. J. N. Simms, in having a well dug, on the general level, after passing through $8\frac{1}{2}$ feet of soil and subsoil, penetrated 28 feet of blue clay and reached a quicksand, through which the water rapidly rose until it stood 20 feet deep. Following the creek up a short distance from town, we found, near its bed, several feet of stratified bluish aluminous shales, covered by drift, from which iron ore is obtained, and in which they claim to have Cobalt. This merits thorough chemical examination.

In the northern portion of the county, on the middle fork of "Wild Cat," iron ore is yet more abundant; and at the Maxwell mill, now owned by Mr. Kemp, there is a strong chalybeate spring, and bog iron is found. Also at Grey's mill, on the south branch of "Wild Cat," the water is strongly impregnated with iron, and heavy ferruginous deposits show themselves.

Accompanied by the editors and several other gentlemen from Frankfort, we visited the gold locality on the Kilmore branch, which heads on Indian prairie and runs into the south fork of "Wild Cat." We found it, as expected, in a pocket of Drift; and partly from the yellow and blue clay in the bank, a foot or two above the stream, and partly from the later quaternary arenaceous deposits in its beds, one of our party, who had washed gold in California, and another, who had worked at the Pike's Peak "Diggings," panned out several fragments of pure gold, chiefly in flat scales. When the matrix is not near, and the deposit solely in drifted materials, the question of its profitable yield could best be answered by frequent washings, in portions somewhat distant, combined with the tracing up of the Drift for some miles, and an inspection of the prevalent character in the accompanying bowlders. This subject will be again discussed in connection with Carroll, Bartholomew and Brown counties; gold was also found, as before remarked, in the Drift of Henry county.

Some cases of milk-sickness were reported as occasionally occurring on the Seminary land.

In the south-west portion of Boone county we found the land new and rich, in places rather wet, the junction of sub-carboniferous sandstone and limestone being near by in Montgomery county; the timber near Zionsville is good, chiefly Oak, Beech, Walnut, Sycamore, with Sugar Tree, Elm and Hickory; other parts afforded a light-colored sandy loam, with good meadows of red-top.

Among several flocks of sheep we observed some fine Merinoes; the horses and cattle also seemed of an improved breed.

Among the bowlders of the Drift, in this county, were noticed felspathic granite, syenitic granite, and gneissoid granite; greenstone, quartz-rock, metamorphosed sandstone, and one of limestone. One granite bowlder here must have weighed at least half a ton.

On some of the waters of the Big Racoon, small unios are found with numerous shells of the genus clyclas and melania; on another creek the genera physa and planorbis were more abundant. Corduroy roads, dog fennel, (Matricaria foetida,) smart weed, (Polygonum hydropiper,) and elder, showed the necessity in some places of attention to drainage.

MARION COUNTY.

This fine and flourishing county in which our State Capital is situated constitutes a geographic midway, a central heart, whence radiate the great iron roads serving to keep up the product-circulation and harmonious intercourse, which should exist between the capital of a great State and its more distant members. The geological formation of Marion county corresponds to her functions and her mission. Based on sub-carboniferous rocks, with Devonian at no great distance, these solid foundations are yet so covered up by a varied Drift, disintegrating into a rich soil, that Marion county is like the central heart to which we have likened her, dependent in a great measure upon the counties around her for the nutritive supplies of raw material, brought on her anastomosing circulatory system of railroads to be elaborated and distributed on the iron arteries of the same diffusive civilizers, to the utmost limits of the State. Bounded on the north and north-east by a great chain of fresh-water lakes, communicating with the Atlantic seaboard, on the south and west by extensive navigable streams, which debouche into others that terminate in the Gulf of Mexico, Indiana thus forms no inconsiderable part in the great nucleus of a rich and powerful Republic.

The principal staples of Marion county, as Mr. Vinton, formerly member of the State Board for that district, reports, and our observations confirmed, "are wheat, corn, rye, oats, barley and potatoes; of stock, horses, cattle, hogs, mules and sheep." Iron, building rock and lime have to be brought from adjoining counties; but potters' clay, he says, is abundant in Marion. The timber is chiefly Oak, Walnut, Ash, Sugar Tree and Beech. They have no milk-sickness and but little potatoe-rot; hog cholera prevailed in 1857–8; latterly, however, very little. Mr. Vinton adds: "Chinch bugs," (Lygæus leucopteris of Say,) "potato bugs," (Cantheris vittata,) "weevil," (perhaps the bean-weevil or Ithycerus Noveboracensis of Schœnherr, also probably the plum-weevil or Conotrachelus nenuphar of Herbst,) "and orchard-worm" (presumed to be the Saperda candida of Fab., the S. bivittata of Say,) "are standing pests." * * * "The prevailing diseases are of a malarious character."

While experiencing the hospitality of the present member for this District, Mr. Loomis, we arranged many specimens in the State, others remaining to be deposited after serving for description. Many bowlders found in Marion are placed in the Quaternary Department, to show the characters of the Drift; a special belt or ridge, hereafter described, passing close to Indianapolis.

We also took occasion to examine the admirably arranged Rolling-Mill of this city, fitted up at a great expense with the latest improvements and appliances, which promises ultimately to be successful and eminently useful. The energy and enterprise of the citizens in Indianapolis have likewise established various other important works, such as foundries, machine shops, woolen factories, dye-houses, with many manufactories, wholesale houses for the sale of goods, &c., &c., worthy of the rank which the city now holds, that of the most populous in the State.

The fine public buildings, (such as the Asylums for the Insane, Blind, and Deaf-Mutes,) in and near the city are well worthy of a visit; and the State Fair Grounds, as well as the flourishing nurseries in the vicinity, show that the agricultural interests are duly regarded.

HENDRICKS AND JOHNSON COUNTIES.

Of Hendricks county we saw chiefly the north-eastern part, in passing from Indianapolis to Crawfordsville. Although probably the rocky substratum is in this county chiefly sub-carboniferous sandstone, judging from rock in the railroad cuts seen when passing on the cars between Fillmore and Crittenden, as well as from the section obtained on Big Raccoon, near Ladoga, in the adjoining part of Montgomery, yet here also the Drift, being heavy, gives the chief character to the soil.

Several quaternary ridges, sometimes of sand, at others of gravel and bowlders, having rather a north and south direction, were noticed. The bowlders seen were of granite, gneiss, hornblende rock and quartz rock. Near New Elizabethtown the general level of the country is from a hundred and fifty to a hundred and seventy-five feet higher than at Indianapolis, and the Danville Court House is two hundred and forty-five feet higher than the beds of the canal at Indianapolis, consequently nine hundred and forty-three feet above the sea. Although the land is newly opened in parts of the county, the houses and barns are already abundant, and of substantial character; on the 8th and 9th of June, when we passed through the county, corn, wheat and clover crops were all very promising. Near the edge of Boone county, on the head water of Eel river, not far from Jamestown, the land is new and rich, but being level requires drainage. The timber noticed was Oak, Beech, Walnut and Sycamore.

Through the Assistant, Mr. Patterson, a sample of a rather rich iron ore was sent for analysis; it is from the west fork of White Lick, near Townsend's Mill, two miles from Cartersburgh, Hendricks county. In the south-east part of Johnson county, on Sugar Creek, bog iron ore has been found to some extent. The south-western part of Johnson county exhibits the Knob-sandstone, but the eastern is probably underlaid by Devonian black slates, as these show themselves in Bartholomew county, at Tannehill's Mills, on Sugar creek, about three miles south of the south-eastern limits of Johnson county; consequently this county might, with almost equal propriety, have been described among the Devonian counties. Its chief agricultural characteristics of soil are still, as in Hendricks, due to from forty to sixty feet in thickness of the drifted quaternary materials.

Passing from Marion into Johnson county, on the dividing lands between the two branches of White river, the northern portion of the county, about Greenwood, exhibited a yellow clay soil, strongly impregnated with yellow ochre. One drift bank in that region furnished the following section, descending through thirty feet: Soil fine gravel, yellow and red clays, coarse gravel and bowlders, and hard-pan. We then passed a soil somewhat wet, as indicated by the corduroy roads; but the crops were fair and the buildings good.

Around Franklin, the county town, the clay is usually reddish, and there is fine level Beech and Sugar Tree land extending towards Edinburgh, with handsome corn, grass and wheat fields.

As we go south towards Williamsburgh we descend seventy or eighty feet below the general level of Marion county, over reddish clay and gravel, to the level of large granitic bowlders, and finally reach the Knob-sandstone, about the edge of Brown, with a prevalence of White Poplar, (Liriodendron tulipifera,) and of a close grained variety here called Hickory Poplar.

MORGAN AND BROWN COUNTIES.

On Bean Blossom, nine or ten miles south-west of Martinsville, the junction of the sub-carboniferous limestone with the Knob-sandstone is seen for the last time, when traveling north-east against the dip. At that point the sub-carboniferous sandstone has a thickness of 230 to 250 feet, capped by a few feet of geodiferous limestone.

On the bluffs of the west fork of White River, near the junction of Morgan and Johnson counties, at a locality which, in 1819, (when the site of our State Capital was under consideration,) competed favorably in the minds of some of the commissioners for selection, the sub-carboniferous sandstone forms an escarpment on the river bank of 25 to 30 feet, being close-grained and aluminous. To this succeed shales with a foot or two of hard-pan, and 60 to 70 feet of quaternary conglomerate, which in weathering has become detached and precipitated in huge masses to the foot of the bluff. Surmounting this conglomerate are quaternary hills, with a soil of black alluvium, giving root to a vigorous and varied growth of fine timber, as Buckeye, Blue Ash, Black and White Walnut, Hickory, Hackberry, &c. The above beautiful woodlands, with some adjoining farms, are now owned by Mr. Calvin Fletcher, Jun., of Indianapolis, and it was in company with him and Indiana's historian, Mr. Dillon, that I had the pleasure of examining these details. Some seven or eight miles north of this, near the edge of Marion, a sand ridge furnishes Tulip Tree, Red Oak, Beech and Sugar Tree. Part of the north-east corner of Morgan county *may* be underlaid by the Devenian black slate, but as yet we have failed to discover any, nearer than on Sugar creek, Bartholomew county.

Brown county is almost exclusively made up of a succession of hills, sometimes 300 to 400 feet above the water courses, composed of the sub-carboniferous sandstone, with heavy quaternary deposits, filling portions of the valleys but seldom found near the tops of the higher hills. On the dividing ridge between Bean Blossom and Salt Creek fruit thrives well, much better than on either of the slopes. There were also some good tobacco crops in fields recently cleared, and we observed luxuriant clusters of wild grapes among the ridge timber of Oak, Chestnut-Oak, Sassafras, Chestnut and Sumach, on the high and dry ferruginous sandstone knobs. On these there is often difficulty in obtaining water, as the materials are too porous to retain the filtering rain water. Lower down occasional layers of clay and disintegrated shales, somewhat aluminous in character, afford a sufficiently retentive substratum. Thus, near the edge of Brown and Bartholomew, at Mr. Thos. Brown's, about 350 feet above the water level in the east branch of White River, at Tannehill's Mills, water was obtained in gravel 36 feet below the surface, doubtless resting on clay, as they passed through several beds of yellowish gravelly clay and bluish hard-pan, with pockets of micaceous iron-sand. The inhabitants of Nashville, the county town, report considerable indications of iron ore; and native copper was found with gold near Spearsville.

The crops are better than might be anticipated, particularly in wet seasons. Wheat is said to produce twelve bushels to the acre. Several promising vineyards and some tanyards were observed near Nashville. These ridges seem well adapted for grazing sheep and goats, as well as for the growth of the vine and some species of fruit trees.

In the valleys Buckeye flourishes and a vast extent of ground is sometimes covered by the polypody, the Flowering Fern, (Osmunda,) and other rankly luxuriant ferns, which seem to delight in the cold and moist soils resulting from the disintegration of the silico-aluminous shales prevalent in these lower members of the sub-carboniferous sandstone.

But the chief interest of Brown county attaches to its Gold region, which, under the polite guidance of Mr. Benajah Johnson, Postmaster at Taylorsville, Bartholomew county, we were enabled to examine advantageously.

Although some "prospecting" has been done on Bear-Wallow Hill, on head waters communicating through Lick creek to Salt creek, as also in what they term the gravel of Greasy creek, a deposit of disintegrated shales, the main localities in which success has attended the washings in this county are on Hamlin's fork of Salt creek, three-quarters of a mile in a direct line from the west limit of Bartholomew, near Mt. Moriah P. O.

Here we found extensive preparations in the way of sluices and hose, rockers and "Long Toms," picks and shovels, &c. Notwithstanding the rain, we panned out enough to convince ourselves that the black sand in many of the pockets contains a considerable amount of gold particles.

Occasionally they pan out flat scales worth from a dollar to a dollar and a quarter. The chief question therefore to be determined, in order to judge of the ultimate profit, is how extensive the deposits are likely to prove. Judging from what I saw here and elsewhere in Indiana of the gold localities, I should venture the opinion that the gold is invariably associated with drifted quaternary materials, derived from a matrix, which finds its mountain home, at least from four to six hundred miles distant, and more probably double that distance, in a northerly direction.

The reason of its arrest chiefly here is obvious, when we observe that the streams, from which the inhabitants wash gold, head about two miles north, in a great ridge running somewhat east and west, the summit of which is only a mile further north. All south of this, embracing such counties as Lawrence, Orange, Crawford and Perry, is broken by hills and hollows of considerable abruptness and difference in level, whereas from the summit of the above described ridge north, a little west, through Johnson, Marion, Boone, Carroll, White, Jasper and Porter to Lake Michigan, there is little (except a few valleys of denudation, where White river, the Wabash or Tippecanoe have channeled their beds,) to break the evenness of the great northern Drift which an earlier Quaternary age strewed all over the northern counties of our

State, and extended into the lower grounds of our more southern counties sometimes as far as the Ohio river. This Drift appears not to have reached the highest elevations of Brown county; at least by our barometrical observations, no quaternary deposits rested in these gold-diggings more than two hundred feet above Sugar creek, whereas some points in Brown attain an elevation of four hundred feet above the water-courses, or about 1,000 feet above the sea, as Flat Rock, which empties into White river a few miles below Tannehill's Mills, is given by Messrs. Stansbury and Williams as 602 feet above high tide.

Another interesting fact, noticed here, and no where else, was that *two-thirds* of the numerous bowlders, seen around the gold-washings, were quartz-rock; the others were chiefly granitic.

There seemed considerable local dip, in the sub-carboniferous shales here, from 7° to 12° in a south-west direction; one section gave an anticlinal slope of 45° north-east and south-west. About fifteen feet of aluminous shales, here over the more solid greenish sandstone, have decomposed into a blue clay, very similar to that in the bed of Lake Michigan. Septaria, with indications of iron, were numerous in the aluminous shales. From some of the hills in Brown county the prospect is really magnificent, compensating to some extent for the inconveniences experienced by the horses losing shoes in the steep and rocky ascents and descents, as well as the scarcity of water for stock on the ridges. This rugged county is also fortunate in being exempt from milk-sickness, according to the data furnished us.

WASHINGTON AND FLOYD COUNTIES.

Of Washington county little was seen by us personally. Although it was included in the line of travel for the Fall Survey of 1860, we were compelled, from the illness of an employe, to omit its examination on that occasion. The western portion is doubtless underlaid by the sub-carboniferous limestones of Orange, while in the eastern half is probably found at deep cuts formed by the Muscatatuk, the Knob-sandstone, judging from the fact that but a short distance beyond the eastern limit of this county, Devonian black shales appear on the slate ford of Graham's Ford, in Scott county.

Floyd county, in consequence of the Ohio, at its rapids, cutting through the Devonian rocks, exhibits at New Albany, in very low water, some of the Upper Silurian fossils, found also on the Kentucky side near Bear-Grass Creek. Above these we find the interesting coral reef, that has been cut through by the water, exhibiting in ascending order the Coralline beds, the shell beds, the hydraulic limestone and the black slate, extending in the creek, near New Albany, thirty-eight feet above high water. Crossing on the bridge and ascending the adjoining knobs, we find the Knob-sandstone rising some 460 feet above the black slate, overlaid by limestone and again by shaly sandstone. A few miles further west a better section is obtained on the Corydon road, ascending to Edwardsville.

At the risk of being charged with a topographical solecism, or incongruity of juxtaposition, I have ventured to bring these strictly consecutive sections together, representing the one which is found a few miles from New Albany on a vertical scale half the size of the other and bringing the rapid portion of the Falls, to economize space, somewhat further from Jeffersonville than truth warrants. The facts and figures conveyed in the section otherwise are correct, and may facilitate comparative examinations.

SEC. 7.—DEVONIAN, SUPERPOSED ON UPPER SILURIAN, AT THE FOOT OF THE RAPIDS OF THE OHIO, WITH A SECTION OF THE OVERLYING CARBONIFEROUS KNOBS OF FLOYD COUNTY, IN THE BACK-GROUND TO THE LEFT, ON A SCALE HALF THE SIZE OF THAT EMPLOYED IN THE DEVONIAN SECTION.

The thickness of the limestone as we travel west continues to increase at the expense of sub-carboniferous sandstone, until near Corydon, in Harrison county, it is seen for the last time disappearing, or dipping under the sub-carboniferous limestones.

The hills of Floyd afford fine timber, Oak and Pine. Some portions on the Ohio river, as the admirably underdrained and well-managed farm of Mr. Collins, member of the State Board of Agriculture, furnish as bountiful returns as can be obtained anywhere in our Mississippi Valley, and other parts of the county are well adapted for small grain and grasses.

SEC. V.—COUNTIES IN THE SUB-CARBONIFEROUS, OR CARBONIFEROUS, OR MOUNTAIN, OR CAVERNOUS LIMESTONE.

To prevent mistake, as these synonyms are used indiscriminately by different writers for the limestone, (with some intervening shales or sandstones,) occurring *above* the Knob-sandstone, or Waverly or Sub-carboniferous sandstone, (part of which seems considered by Prof. Hall, in his Iowa report, the equivalent of the New York Chemung Group,)* and *below* the Millstone Grit or Carboniferous Conglomerate, they are all given here, although usually the term sub-carboniferous limestone will be employed to distinguish it from the limestones usually only a few feet in thickness, which occur in the Coal Measures and are hence more properly carboniferous limestones.

SUB-SUCTION 1.—GENERAL DESCRIPTION OF THE FORMATION.—Usually these limestones, formed prior to the true coal period, have obtained a considerable thickness in our State, spreading over the greater portions of Montgomery, Putnam, Monroe, Lawrence, Orange, Harrison and Crawford counties. Including the overlying shales, grits and sandstones up to the Carboniferous Conglomerate or Millstone Grit, the latter not inclusive, I have been unable to satisfy myself of a greater vertical thickness, in Indiana, than from 400 to 500 feet, which agrees approximately with the thickness of similar beds found by Prof. Hall in Iowa, viz: above the Knob-sandstone, about 100 feet of Burlington limestone, 140 of Keokuk limestone, 65 of Warsaw limestone, total 305 feet, besides the overlying sandstones included by him as below the conglomerate. On the other hand a section given by Prof. Hall, from observations by Mr. Worthen, State Geologist of Illinois, made near Huntsville, Alabama, exhibits a thickness for similar strata of about 900 feet, while in the Missouri Survey, by Prof. Swallow, over 1,100 feet have been assigned to these beds, under the names Encrinital, Archimedes and St. Louis limestones, with the ferruginous sandstones above.

With us the lower beds, immediately over the Knob-sandstone, can

---

*The Chemung Group is in the New York classification one of the higher sub-divisions of the Devonian. Prof. Hall admits that "the passage from the Chemung (Devonian) to the Burlington (Carboniferous) limestone is so gradual, both in the physical aspect and in the generic and specific characters of the fossils, that it forms no greater change than is observed between any of the subordinate groups."

be distinctly observed in Lawrence county, at Heltonsville and on Salt creek, in Monroe county on Bean Blossom, two miles north-east of Bloomington, and in Montgomery county, near Ladoga and near Crawfordsville. The first layers are often remarkably white, encrinital, and sometimes, near the junction, so hard as to strike fire with steel, yet effervescing with acids, and exhibiting numerous entrochites. Associated with these are darker beds, with Terebratula lamellosa and other Brachiopods. These with the superincumbent blue and gray limestones (which have sometimes diffused fragments of calc-spar, giving rise to the name of Bird's-eye limestone, or whose texture is so uniform as to form a good serviceable lithographic stone,) constitute what may be called the Lower or Monroe-Harrison sub-carboniferous limestone, attaining at times a thickness of from 100 to 150 feet; but more usually less, diminishing as we go west, while the ferruginous sandstones and shales increase at their expense. The fossil fish, (Palæoniscus) found in the eastern part of Monroe county, Indiana, is probably in this bed. Above these strata of limestone, occasionally silicious shales, but more frequently red clays and chert, occupy a space of from 25 to over 100 feet, often full of geodes, that have fallen from the superincumbent calcareous beds, to which the term *Middle* or Lawrence-Crawford sub-carboniferous limestone may be applied. This geodiferous limestone is from 20 to 100 feet thick, and in its upper layers is sometimes magnesian, sometimes flinty in bands like Lydian stone. The geodes, where the limestones disintegrates, fall out, exhibiting an oval, wrinkled exterior, and if broken open are seen to contain clear white crystals, pointing to a central cavity. The crystals are sometimes calc-spar or dog-tooth spar, (cabonate of lime,) sometimes quartz, (silica); occasionally we find distinct crystals of both occupying different portions of the same spherical mass; and, at other times, botryoidal chalcedony replaces the regular crystals. These seem to result from the infiltration into the limestone cavities of water holding carbonate of lime in solution, through the aid of an excess of carbonic acid, and holding silex dissolved either aided by heat or by alkalies, somewhat as flint is formed in chalk or as the splendid quartz crystals form abundantly in rock cavities of middle Arkansas, where to this day analysis shows the water to be alkaline.

The chert beds in Indiana are also usually characterized by a high red color derived from infiltration and deposition of the hydrous peroxide of iron, which shows itself abundantly in a fine crumbling condition, where some of the geodes and most of the chert blocks are broken open.

It is likewise in these strata that we find, especially in caves, Epsom salts, (sulphate of Magnesia,) resulting from the decomposition of the Magnesia, being often in the form of an efflorescence on the surface of that rock; also gypsum, stalactites, stalagmites and the beautiful fibrous satin-spar, such as that from the Pillar of the Constitution in Wyandot cave.

The Middle Limestone, occasionally replete in some strata with the coral Lithostration Canadense, also often oolitic in its upper strata, sometimes with shells of the genus Euomphalus, passes gradually into arenaceous and sometimes magnesian limestones, then to silicious shales, often with alternations as in the beds below, of Bryozoic chert and red clay, occupying in all from 50 to 100 feet. Next, in ascending order, occurs a limestone which as it sometimes constitutes one heavy bed, or member, but more frequently by the intercalations of from eight to thirty feet of silicious shales, areneceous limestones, dolomite or chert beds, is separated into two sub-members, may hence be termed the twin or upper sub-carboniferous limestone. The two sub-members constitute the first and second Archimedes limestone of some authors, and may conveniently be distinguished into "A," the sub-member above; "B," the sub-member below.

The occasional separation into two sub-members is true also of the strata forming the Middle limestone and even the Lower; but much less frequently in Indiana than is the case with the upper member.

This Upper sub-carboniferous limestone, which might be also termed the Orange-Martin limestone, from its prevalence in those counties, is often characterized, in its sub-member "B," by abundant Bryozoa, of the genus Retepora, Fenestella, Ceriopora, &c, and it is occasionally oolitic.

The sub-member "A" is more frequently compact, with few fossils and a very clear ring when struck by the hammer, sometimes more coarsely crystalline and containing Archimeidpora Archimedes, or spines and fragments of Echinites; other layers, sometimes the entire twin member, may be found so replete with several species of productus as fairly to claim the title of productal limestones. None has yet been found by us of the brecciated variety mentioned in Prof. Hall's Iowa report. The intervening silicious shales furnish the grindstone grits.

The upper limestone often forms the roof of caves, the underlying shales or argillo-calcareous sandstones and silicious dolomites washing out more or less to form the narrow passages or vaulted domes of the caves, as will be shown in detail when describing Crawford county.

Below each one of these sub-members a thin coal seam, or a fine clay, has been found in several counties of Indiana, seldom exceeding eighteen inches in thickness. Above this upper sub-carboniferous limestone we have a succession of ferruginous sandstones, with a variety of grits, embracing the grindstone and whetstone quarries of Indiana, with their Lepidodendra, Stigmariæ and other remarkable carboniferous vegetation, or an occasional thin coal seam, sometimes both, as in Orange county.

These ferruginous sandstones, occupying a space of from fifty to a hundred feet, are assigned by some writers to the Millstone Grit series, which in its true conglomerate form is often-found superposed on the finer grits to the extent of forty or fifty feet, while others consider the ferruginous sandstone as belonging to the series which we denominate the sub-carboniferous limestone series.

To render the foregoing more intelligible a consecutive section is subjoined, combining the whole, from actual sections, verified so repeatedly at different points, as to establish with considerable accuracy the correct order of superposition, average thickness, and palæontology, in descending order, thus:

System.	Formation.	Members.	Feet.	Localities.
Carboniferous.	Sub-carniferous limestone.	Ferruginous sandstone.	50–100	Whetstone quarries near French Lick and elsewhere, in Orange county; also near Greencastle, Putnam county, and near Dover Hill, Martin county.
		Upper limestone. "A."		Oil Creek, Perry county; Mt. Prospect, Crawford county; Silverville, Lawrence county; Owensburg, Green county, and Independence, Warren county; also part of Orange or Little Bluer River, Crawford county.
		Grindstone grit; intervening shales.	3–20	
		"B."	0–30	
		Aluminous and calcareous shales.	5–50	
			50–100	Upper part of Wyandot cave and similar caverns.
		Chert and red clay.		
		Silicious and Magnesian limestones.		Lowest points in Wyandot cave. Bed of Lost River, Orange county.
		Lithostrotion beds.		
		Middle limestone, with geodes.	25–50	Near Bedford, Lawrence county; Gosport and west part of Monroe.
		Chert and red clay.	40–110	
		Argillo-calcareous sandstone.		
		Locally lithographic.		
		Lower limestone.	30–100	Harrison county.
	Sub-carb. sandstone.	Green and grey shales. Yellowish sandstones.	450	Floyd county; eastern Monroe; Brown county, &c.

SUB-SECTION 2.—SOILS, &c., OF THE CARBONIFEROUS LIMESTONE SERIES. The decomposition of these limestones, with their intercalated sandstones and aluminous shales, gives rise, as might be anticipated, to a favorable admixture for most agricultural products. To these are added, in some of the northern counties embraced in this sub-division, many feet of quaternary deposits, still contributing to the variety. The general outline of the country is that of undulating land, neither so broken nor so sandy as the knobs, nor yet so level as in those counties wholly quaternary.

The soil generally seems admirably adapted for small grain and grasses, consequently we see, particularly in Harrison, Lawrence, Monroe, Putnam and Montgomery, fine cereals, luxuriant meadows and picturesque pastures. In the western part of Crawford and Orange, as well as the eastern portions of Martin and Green, where the Ferruginous Sandstones constitute the higher general levels, the surface is more uneven and the soil often less calcareous and productive. These, however, would probably prove fine high and dry pastures for sheep, a species of stock for which there will probably be a gradually increasing demand. The staple agricultural products are corn, wheat and stock, including, besides the flocks of sheep, hogs, cattle, horses and mules.

SUB-SECTION 3.—ROCK QUARRIES, &c.—Perhaps no other formation in Indiana is so replete as this with quarries affording fine building materials, millstones for some purposes, good grindstones, an excellent quality of whetstones, and locally a fair quality of lithographic stone, as well as hydraulic limestone.

This seems to have resulted partly from the quiet waters in which portions of the materials were deposited, partly from the succession of calcareous, silicious and aluminous materials, which, especially near their junction, are thus often modified.

The various quarries for building rock will be found fully described in giving the details of counties in this formation, particularly Monroe, Putnam, Lawrence and Crawford; good localities for grindstones and whetstones are enumerated in the description of Orange county; the lithographic stone is discussed in detailing the resources of Harrison county.

SUB-SECTION 4.—METALLIC ORES AND OTHER MINERAL WEALTH.—Although thin seams of coal show themselves in Indiana, in the sub-carboniferous limestone series, amounting to perhaps eighteen inches or two feet in thickness, which sub-conglomerate coals have in a few other States thickened to workable beds, yet the probabilities are that these seams will not prove profitable in our State; the true coal measures furnishing beds of so much greater thickness.

As already mentioned the chert, occurring in this formation, is often highly charged with hydrous peroxide of iron, giving a deep red color to the adjoining aluminous materials. These being the strata in which the heavy deposits of a similar character furnish abundant iron ore for furnaces, in the counties of Trigg, Lyon, Caldwell, Livingston and Crittenden, Kentucky, it is not improbable that a detailed survey may bring to light similar mineral wealth in this portion of Indiana, particularly as considerable deposits exist in section 35, township 2 north, range 2 west, and have already been seen at several other localities.

The same reasoning applies, although perhaps with less probability, to the discovery of lead ore, as the galena (sulphuret of lead) found in Derbyshire, England, associated with fluor spar, as well as the sulphuret of lead and sulphuret of zinc in Yorkshire, ramify their most productive veins through the rocks under the Millstone Grit series of those counties in Great Britain. Details on this subject can be found in Prof. Phillips' work on the Mountain Limestone, and in the first volume of the Kentucky report.

Nitre has been manufactured somewhat extensively in the caves, and Epsom Salts could also be obtained in considerable quantities.

SUB-SECTION 5.—TIMBER AND PREDOMINANT VEGETATION.—In portions of Kentucky parts of the sub-carboniferous limestone series gives rise to "barrens" with White Oak, Red Oak, and Black Jack Oak; but in Indiana, although a similar character prevails to a small extent, where the upper ferruginous sandstones furnish the main soil, and Cedar occasionally in Crawford county exhibits indigenous luxuriance in the rocky clefts, yet usually the sub-carboniferous limestone region is well timbered, and where the aluminous ingredients are abundant, we have Beech here, as in most parts of southern Indiana, constituting the predominant forest growth, associated, however, with Tulip Tree, Sugar Tree, Black and White Walnut, and Ash.

Some ferns were collected from several counties in this formation, but they do not exhibit the profusion and variety noticed in the sub-carboniferous sandstone detritus, and in soils resulting from the disintegration of aluminous shales of Devonian age. Mr. Larrabee, of Greencastle, informs us that peat has been dug in the south-west portion of Putnam county. The cereals and grasses luxuriate in Indiana on the soils resulting from sub-carboniferous limestones and their intervening argillo-silicious shales.

In some of the moist, rich bottoms, north of White river, in Lawrence county, an extensive growth of our American genus of the Mezereum family, the Leatherwood, Moose-wood or Wicopy, (Dirca palustris, L.,) the fibrous bark of which the Indians used for thongs, has given its name to a creek which empties south of Bedford into White river.

SUB-SECTION 6.—MINERAL AND OTHER SPRINGS.—The celebrated sulphur springs of Orange county flow from the base of the grindstone grits over a bed of limestone, as described hereafter. Several chalybeate springs will be found noticed in the details of these sub-carboniferous limestone counties, and copious springs of sparkling water, often rather hard from filtering through the limestone cavities, are quite abundant. Frequently the water pours out from above the limestone strata, being arrested after filtration through the shaly sandstones.

One remarkable geographical feature of this formation consists in the "sink holes" of some regions, where a limestone stratum, usually the sub-member "B," has caved in, leaving a depression through which the water filters, until perhaps clay washings close the cracks and fissures sufficiently for ponds to form, often convenient for stock.

A yet more striking phenomenon consists in the entire disappearance of good-sized streams for miles, and their reappearance, or the outbursts of other subterranean water in the form of fathomless springs, capable of furnishing fine water power. Lawrence, Orange and Crawford furnish some of these remarkable localities, the details of which will be given under those heads. No artesian wells or borings, so far as we know, have yet been attempted in this formation, and the numerous cavities for the escape of water might interfere with success.

SUB-SECTION 7.—MISCELLANEOUS FACTS, AS TO THE PREVALENCE OF DISEASE, &c.— The greater area of this region is very healthy, and it has fortunately been selected for the site of many of our collegiate institutions. Wabash College, at Crawfordsville, is close to its lowest members. The State University, at Bloomington, is chiefly on the lower limestone. The Asbury University, at Greencastle, is a few feet above the uppermost limestone; and Hanover College, Jefferson county, is in a region probably equally salubrious, but of Silurian age.

Some portions of the country are, it is true, liable to intermittents, as in the bottoms of White river; in others typhoid fever and pneumonia may visit as elsewhere; or again, milk-sickness, in Spice and Brushy valleys, and a few other localities, admonishes to caution in permitting cattle to range out of cultivated fields, yet on the whole it may be considered as having established its claims to decided general healthfulness.

The numerous caves of this cavernous limestone are chiefly in a continuous, gently curved line, nearly equidistant from the margin of the coal field; and the Mammoth cave of Kentucky is nearly a continuation of the same curve, about 130 miles, in a direct line, from its twin sister, the Wyandot.

Sub-Section 8.—Characteristic Fossils of this Formation.—

Corals: Lithostrotion Canadense,
L. harmodites,
L. Stokesi,
Zaphrentis centralis,
Z. Cliffordana,
[Z. spinulosa, Z. Dalii,]
Cyathoxonia cynodon,
[Amplexus coralloides,]
Trochophyllum Verneuilanum.

Crinoids: Actinocrinus proboscidialis,
A. tuberosus,
Agaricocrinus Wortheni,
Synbathocrinus Swallovi,
Platycrinus Wortheni,
Forbesiocrinus Wortheni?
Pentremites sp.? (perhaps the oblongus of Phillips.)

Echinites: Archæocidaris Wortheni.

Bryozoa: Archimedipora Archimedes,
Retepora laxa, (Phil.)
R. irregularis, (Phil.)
Polypora flustriformis, (McL.)
Fenestella membranacea, D'Orb.)
And two new species of Ceriopora, hereafter described.

Conchifers: Edmondia sulcata, Ph., (Sanguinolites of McCoy, Allorisma of King.)

Brachiopods: Spirifer striatus,
S. attenuatus,
S. Forbesi,
S. Sowerbyi,
Orthis crinistria,
O. Keokuk, (Hall,)
Terebratula hastata; also, var. sacculus,
T. lamellosa,
Productus semirecticulatus,
P. punctatus,
P. tenuistriatus,
P. elegans.

Pteropods: Conularia Crawfordsvillensis.

Gasteropods: Pileopsis (Capulus) pabulocrinus,
Euomphalus catillus.

SUB-SECTION 9.—A DETAILED DESCRIPTION OF EACH COUNTY IN THE FORMATION.—

MONTGOMERY COUNTY.

The eastern portion of this county is underlaid by the sub-carboniferous sandstone; but, as we approach the center, going west, we observe the junction, and have a fine opportunity to collect, near Crawfordsville, the fossils of the limestone above, many of which, however, have become detached, and are found in the decomposing aluminous and silicious shales beneath. High portions of the county are said to present evidences of the Millstone Grit; but all the underlying rocks, of whatever character, are covered by a heavy deposit of Quaternary, amounting frequently to over a hundred feet, and it is therefore only at deep cuts, made by the water courses, that we have access to the rocky substratum.

The county is generally somewhat level or undulating, being only broken near the water courses. The soil is rich, and agricultural products abundant. Corn, wheat and stock are exported: pork being a staple article, and mules selling for $50 00 at weaning time. Some of the sandy loam portions have furnished excellent Red-top crops. Other portions, somewhat wet, and requiring drainage, have been successfully underdrained, at an average cost of one dollar per acre. The materials employed were White Oak rails and staves, which, after twelve years, are still sound, and the drains unobstructed.

Several flocks of fine-wooled sheep were observed, and, as water power is good, and some woolen factories are in profitable operation, this branch of agricultural and manufacturing wealth may be expected to extend.

In the neighborhood of Darlington, we observed considerable fields of sorghum and broom corn, of promising growth.

Although rock quarries are not abundant, yet, on Raccoon creek, calcareo-silicious slabs for tombstones and other purposes, have been taken out; and layers of the limestone, by selection, are sufficiently free from arenaceous adulteration to burn into lime. Some of these are the white encrinital, immediately over the knob sandstone; others are bluish limestone, solid and almost destitute of fossils, underlying a silicious limestone or calcareous sandstone, such as is usually termed by quarrymen "bastard limestone."

There are abundant indications of iron ore, chiefly in the Drift, and numerous chalybeate springs exude where these deposits rest on a stratum of quaternary clay. Several of those springs were observed on Cornstalk creek. While Mr. Stephen Field's farm, (Sec. 4, T. 18 N., R. 4 W.,) near Crawfordsville, was in timber, that region was frequently struck by lightning, which induced some to suppose that there might be large bodies of iron near the surface. But the fact that, since the timber has been removed, this seldom occurs, would rather indicate that the tall trees on an elevated point acted as conductors, especially when wet, rather than deposits of hydrous oxide of iron, the only ore seen near there; one whose conducting power can not be great.

The Quaternary bowlders, gravel, &c., seem to have been deposited so unconformably in the depressions of the underlying strata, that, although Prof. Hovey, of Wabash College, in having his well dug, encountered bowlders at seven feet from the surface, yet his immediate neighbor, on the adjoining town lot, passed through 80 to 90 feet of sand, gravel and blue hard-pan, and found at that depth fragments of wood, in dark mud and gravel.

Silurian fossils are near here abundant in the Drift, derived probably from the Upper Mississippi region; and abundant fragments were reported as resembling charcoal or peat. Lights were also stated as being visible at times, apparently emanating from the soil: not having an opportunity to visit the locality, it is difficult to decide whether or not these derived their origin from bubbles of phosphuretted hydrogen, generated usually in swampy places.

A bed of marl was observed at several places near the prairies of western Montgomery, underlying the black muck, and some bowlders were noticed of half a ton weight, the prevalent varieties being granite, gneiss, greenstone and quartz rock.

About twenty years since, when Major Ellston dug a well near the bank of Sugar creek, at Stover's mill, some 60 feet below the surface it was supposed that a four foot bed of coal was struck, after boring through sandstone. Judging from all evidence, it seems reasonable to conclude that these were the black bituminous shales, of Devonian age, which theoretically would be expected to crop out a few miles east, from under the Knob sandstone, as they do on the Wabash, at Americus.

Timber is sufficiently abundant, Beech, Sugar-Tree, Hickory, Ash, some Chestnut, Hackberry and Honey Locust, (Gleditschia triacanthos) undergrowth in places, Elder, Polk-weed, some ferns and stickseed, (Echinospermum lappula.)

In the creeks were observed shells of the genera unio, cyclas, melania, physa and planorbis.

No hog cholera has troubled this county latterly, and milk-sickness does not occur within 25 or 30 miles of Crawfordsville, consequently not in the county.

The Court House at Crawfordsville is 744 feet above the sea; at the College the collection of fossils and minerals is well worthy a visit.

PUTNAM COUNTY.

The prevailing character in this county results from the proximity of the upper limestone members, with occasional admixture of some overlying ferruginous sandstone, and a considerable top dressing of quaternary materials, giving a favorable variety of soil, well adapted for small grain and grasses.

Near Greencastle, the capital, the limestone rock comes sometimes to the surface, but more frequently is a few feet below, sometimes overlaid by sandstones, and a thin seam of sub-conglomerate coal. The limestone furnishes a durable and handsome building material, seldom, however, exceeding, at this locality, over 22 inches in thickness.

Some interesting fossils were obtained here, under the polite guidance of Mr. Larrabee: numerous Bryozoa and remains of actinocrinites, probably A. longirostris? spines of archæocidaris, (A. Wortheni,) Retzia Verneuilana, and Productus tenuicostus, besides fine samples of Stigmaria, in the overlying sandstone, showing beautifully the ramifications of the rootlet or fibres, transversely from the sub-central core or axis, terminating externally so as to form the stigmata, from which it derives its name.*

At Havilah Findley's, 2¼ miles south-east of Putnamville, a coal, somewhat slaty, but otherwise good, is obtained for use.

From four to six miles northerly from Greencastle, the ferruginous sandstone is stated to furnish whetstones. We saw a sample of good quality, but were unable to visit the locality.

Peat is reported as occurring in the south-west part of the county.

As the particulars regarding the sub-conglomerate coal of this county will be found in the report of Mr. Lesquereux, it is unnecessary here to enlarge further on these details.

---

* For this cut, see Appendix.

Greencastle is pleasantly and healthily situated, and its flourishing institution of learning, the Indiana Asbury University, is well patronized. The Court House is 830 feet above the level of the sea.

MONROE COUNTY.

As we find, near the eastern limits of this county, on Bean Blossom, the junction of the Knob-sandstone with the overlying Lower Cavernous limestone, while at the extreme western limit we have the upper oolitic members of the sub-carboniferous limestone, and soon reach the conglomerate in the adjoining county west, Monroe county may be said to embrace nearly the entire range of the sub-carboniferous limestone, with comparatively much less Drift than Montgomery or Putnam counties.

Bloomington, the capital, and the seat of our State University, is surrounded by a fine undulating region, luxuriating in Beech and other fine timber. The Court House is 771 feet above the sea, consequently is nearly 30 feet higher than Crawfordsville, and about 60 feet lower than Greencastle, which accords with their relative stratigraphical geology; they being all in the line of strike, but Crawfordsville being chiefly near the base of the series, Monroe the middle, and Greencastle in the upper members.

Judge Hughes, in digging his well, on an elevated portion of town, passed through 6 feet of clay and 54 feet of solid limestone, beneath which he obtained water.

In the eastern part of the county, geodes are remarkably abundant in the natural low cuts, which reach the chert.

In the north-west portion of the county, there is an interesting and important stone quarry, on the development of which Capt. Love, of Indianapolis, formerly of the U. S. Army, and others, have expended about $20,000. Columns can be obtained here sometimes 18 feet in height, without a crack. The upper portion, for about 6 feet, is usually a hard, close-grained, white limestone, often oolitic, then succeed, in descending order, 12 to 18 feet of a fair building stone, in which are some shales of Enomphalus, Murchisonia, and an occasional Retzia. The substrata are usually somewhat coarser and more friable, until the evaporation of the quarry water hardens them. For superstructures, especially, this "Monroe Marble Quarry" stone is deserving of extensive use, on account of its beauty, ease of working, durability, and fair average strength of material.

The subjoined official letter, written March 9, 1860, to Capt. Love, in reply to his queries, furnishes additional details on this subject:

N. HARMONY, March 9, 1860.

DEAR SIR:—In accordance with your wish that I should state officially the result of my examinations at Stineville, I am pleased to be able to say that, of the many good quarries in Indiana, yours is among the best for a variety of purposes. The rock is geologically in the sub-carboniferous limestone, somewhat oolitic in structure, and having occasionally a few fossils imbedded, such as Retzia Verneuilana, Enomphalus Spurgenensis, and Murchisonia turritella, already described by Prof. Hall as being found in the oolitic rock of another Indiana locality, and altogether resembling considerably the Portland stone of Great Britain, although that is from the more recent, true Oolitic Formation.

An examination of the spec. gravity, and consequent weight per cubic foot, of your Stineville rock, confirms this view, as one sample gave sp. gr. 2.14, weight of a cubic ft. 133.3 lbs.; the other 2.47, weight 153.88 lbs.

According to Mahan's Civil Engineering, used at the U. S. Military Academy, Portland stone gave, in one sample, sp. gr. 2.428, weight of a cubic foot, 148.08; while in another the sp. gr. was 2.145. Judging from analogy, as we have not yet obtained the hydraulic press, which we design using to test the strength of materials, this would give a resistance on each superficial inch at the moment of crushing, equal to from $1\frac{3}{4}$ to 2 tons, about half as much as that of granite; while the average weight producing fractures is on each square inch nearly one, or about one-third that of granite.

The estimate of a transverse strain on prisms of 4 inches long, the cross section being a square of 2 inches on a side, distance between the points of support 3 inches, averages 2682 lbs., being nearly 3 times the strain which well-burned brick will sustain, and nearly equal to Cornish granite, which averages 2808 lbs.

The resistance to abrasion is more than double that of good brick, and about four-fifths that of statuary marble. Undoubtedly, for sustaining great vertical pressure, as in bridge abutments, foundations to large buildings, and the like, more compact rock might be obtained; but for beauty of structure and color, durability, ease of working, and thickness without a crack, say 18 feet, which yours possess, it seems all that need be desired for superstructures.

The following analysis proves that there are no materials, in any quantity, calculated to impair its durability by disintegration, a small amount of iron, equally diffused, acting rather as a cement to the calcareous particles.

One hundred parts of the building rock gave:

Of moisture, expelled at 250° F..	0.05
Of insoluble residuum (silica)..	0.90
Of iron and alumina..	3.00
Of bicarbonate of lime...	95.00
Of bicarbonate of magnesia...	0.22
Of loss and alkalies...	0.83
	100.00

The rock at your quarry is very much of the same character as some obtained in Harrison, Lawrence, Putnam and other counties, while several localities in Wabash, Decatur, Jennings, Jefferson, &c., furnish compact rock for foundations, proving to the citizens of Indiana that it is quite unnecessary to import building materials from other States.

Very respectfully,

RICHARD OWEN,

Assistant State Geologist.

About two miles north-west of Bloomington, on the farm and near the sulphur spring of Mr. Orchard, a very shaly limestone dips slightly to the south-west. At this locality a fine palæoniscus was once found, which can now be seen in the cabinet of the University, under charge of Prof. Wylie. During the short time we could spend at this place, we were unsuccessful in seeing any further remains of fossil fishes.

LAWRENCE COUNTY.

In the north-eastern portion of this county, near Heltonsville, on the head waters of Leatherwood creek, the junction of the sub-carboniferous or knob sandstone with the overlying limestones, can be very satisfactorily studied. One of the tunnels of the Ohio and Mississippi railroad, in this county, also cuts through a portion of both.

The Bedford rock has long been celebrated for its excellent qualities as a building stone, and is extensively shipped; additional localities are being opened, and only require the liberality of railroad directors to furnish switches and other facilities for still more extended sales.

From Mr. Glover's quarry, near town, we obtained specimens of a grit suitable for millstones to grind corn, although not so well adapted for wheat. It occurs chiefly in one limited space, a layer among the building-rock strata, silicified probably by infiltration.

At a quarry one mile south of Bedford, the upper layer, for about 18 inches, is oolitic, beneath which six or eight feet occur of a limestone, that, on account of its cracking on exposure, is rejected by the quarry-men. Next below succeed nine feet of excellent building stone, which can be obtained in slabs of almost any desired size. The same limestone continues for many feet beneath, rendered visible as we go south on the railroad, by a descending grade of 80 feet to the mile; but becomes somewhat bituminous, and has hard lumps which prevent it from being worked. Some of these facts were communicated by Mr. Needham, formerly of the "Dean Marble Quarry," during our examinations of the various Bedford quarries, under the guidance of Judge Duncan, of the State Board, Mr. Stilson, and other gentlemen, who politely accompanied us from town.

In a cut a mile and a half south of town we observed a vertical space, about six feet in width, filled with beautiful calc-spar, in botryoidal pendent masses, between the darker limestones. Beneath this stratum the now argillo-silicious limestone rock is broken into thousands of small fragments, the layers being contorted and folded, as if the deposit had been made in unquiet waters and submitted while plastic to lateral pressure. Still beneath these strata are seen, near the abrupt termination of the rock, White river having here channeled her bed through the limestone, strata richly productal and bryozoic, with a layer of hydraulic limestone imbedding gypsum and selenite in cavities.

This being nearly two hundred feet below Bedford must bring us very near the Knob-sandstone, which shows itself not far distant in the tunnel of the Ohio and Mississippi railroad. Judge Duncan's farm, nearly on the same geological horizon, has abundant fine geodes and afforded a magnificent slab, now in the State collection, about two feet long by twelve to fifteen inches wide, on which can be counted about 140 Spirifers, (of the species striatus,) nearly every one of which is hinge up, and chiefly with the valve open, very much in the position assumed by our fresh-water unios, when exercising their power of locomotion in wet sand or mud. This would appear to confirm the supposition of Woodward and others, that this genus of Brachiopods was free in the adult state. Several specimens on this slab measure two inches along the hinge line. On the same farm are Productus tenuistriatus and Orthis crinistria.

About five miles from Bedford fine stalactites are found in the Pitman cave. Some of these we saw in Dr. Blackwell's cabinet at Bedford. Mr. Herscher of that place has a fine collection of birds, chiefly from our own State.

In addition to the above items, Judge Duncan, member of the State Board for this agricultural district, remarks: "The Bedford rock is shipped on the railroads and sold in Louisville, New Albany, LaFayette, Indianapolis, &c., and is unsurpassed as a building stone." * * "Stoneware is manufactured in various parts of the District." * * * "The staple agricultural products are corn, wheat, oats, hay and tobacco. Our corn and hay are fed to hogs, cattle, horses, sheep and mules. Our wheat and tobacco are partly manufactured before sending to market." * * "Nearly all the varieties of soil and timber in the State can be seen in this District." * * "Hog cholera has prevailed in some localities. Potato rot, I believe, has been universal. The ravages of insects were formerly restricted to stone fruit; but latterly have extended to the apple. Our corn crop is sometimes affected by the grubworm," (Melolonthians or perhaps Agrotidæ.) "In 1858 a great many meadows were injured by the same pest. I have heard complaints latterly of the Chinch-bug" (Lygæus leucopteris) "and also of a small wire-worm," (Elateridæ). * * "Fevers of the various type are the most prevalent diseases."

We walked in the evening to Hamer's well-known mills, two-and-a-half miles south-east from Mitchell's Crossing, and found a large stream of water gushing from under heavy beds of rock, with force sufficient not only for Mr. Hamer's extensive mills, but also for many other works. By partially damming this stream boats have been rowed some distance into the cave, disclosing the usual subterranean wonders. Unfortunately it was too late and dark to make a thorough examination, and the opportunity we confidently expected to have again to visit this locality never presented itself.

The section on Salt creek, at the bridge four miles northerly from Bedford, has been spoken of as finely illustrating the junction of the Knob-sandstone with the Cavernous limestone. It is subjoined in descending order:

	Above the Sea.			
Level of Bedford Court House.	680 ft.	Sub-carboniferous limestone.	Soil and sub-soil, (yellowish.)	25-30
			Limestone, (close-grained.)	2-4
			Red clay and chert.	30-35
			Ringing compact limestone.	5-6
			Red clay and chert.	15-25
			Gray pentremital limestone.	40-45
			Oolite, with Euomphalus.	20-25
			Bryozoic and pentremital limestone.	50-55
			Shaly rock and red clay.	10 ft.
			Gray pentremital limestone, with Terebratula lamellosa and Orthis crinistria.	25-30
			White eucrinital limestone.	4-6
		Sub-carboniferous sandstone	Gray shales and solid sandstone, with a few fossils.	20 ft.

Bed of Salt Creek.

Some cases of milk-sickness are said to have occurred in "Spice Valley," about ten miles south-west from Bedford.

ORANGE COUNTY.

This county comprises the upper limestones of the sub-carboniferous series and the superincumbent ferruginous sandstones assigned by some to the same sub-division, and by other writers considered a portion of the Millstone Grit.

The principal localities of geological interest in this county are the places at which Lost River disappears and flows several miles under ground; the reappearance of a large stream, supposed to be the same, near Orangeville; the quarries from which grindstone and large quantities of the Hindostan whetstones are obtained, also the noted sulphur springs, besides several points at which the sub-conglomerate coal seams are exposed.

Leaving Orleans, after examining some specimens collected by Mr. Elrod, Mr. Braun and others, we reached the farm of Mr. Owen Lindley, (some four miles south-west from town,) who politely furnished us some particulars. This disappearance of Lost River is in section 11, township 2 north, range 1 west; and the bed of the river, here 40 to 50 feet wide, was dry at our visit, although sometimes it is over its banks eight or ten feet, as shown on the trees. A sketch and section of this locality is subjoined.

LOST RIVER, ORANGE COUNTY.—(Sub-Carboniferous Formation.)

R. Owen, Del.

SEC. 10, AT LOST RIVER GULF, ORANGE COUNTY.

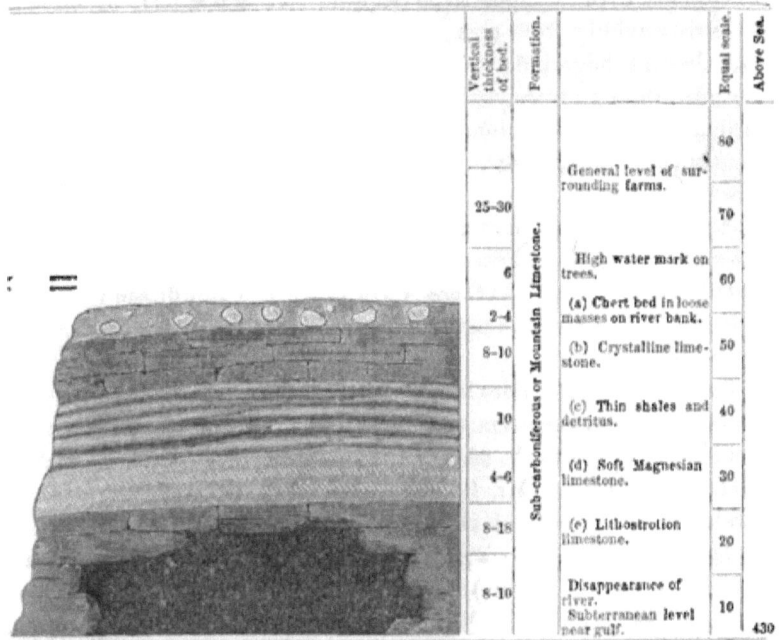

Mr. Lindley informs us there is a still deeper "Gulf," with some water constantly running and disappearing in a cavity, the depth of which has never been ascertained; this is in section 9 of the same township and range as above.

On close examination of the bed of the stream, near Mr. Lindley's, it seems to be formed on the middle limestone and has washed at the entrance eight to ten feet deep, leaving overhead the lithostrotion bed, surmounted by Magnesian limestones, shales, &c., to the amount of from 30 to 40 feet, which materials have consequently been washed away or have caved in to form the bed of the river for some miles back; but which here seemed firm enough to sustain the roof of the natural tunnel.

The chert scattered abundantly on the bank is highly bryozoic, and contains a fair share of mollusks. Some of the lower portions of the limestone are almost lithographic. We were informed that sightless fishes had been found in the subterranean waters of these localities.

The adjoining timber was chiefly Oak, Beech, Sugar-Tree, Black and White Walnut, Tulip Tree and Horse Chestnut.

At Orangeville Mr. Stackhouse kindly conducted us to the spot where Lost River is supposed to reappear. A stream forty or fifty feet wide rises quietly from great depths, apparently from under a Bird's-eye, almost lithographic limestone, which we found surmounted, as we ascended through town, by three other limestone beds, with their intermediate layers of sandstone. A short distance below its emergence from underground, a dam is constructed and a very valuable water power secured.

Although by the barometer, the stream here seems considerably lower, as might be expected, than at Lost River "Gulf," and there are three limestones above, as there is a dip here of about fifteen feet to the mile, this is probably the same middle limestone, two of the upper forming a twin layer.

The whetstone localities in this county are numerous and important. The oldest quarries are in section 5, township 1 north, range 2 west, owned by Mr. Pinnick, and in section 32, towhship 2 north, range 2 west, the property of Mr. Charles. Mr. Wm. H. Cowherd manufactures whetstones extensively near the West Baden Springs; Mr. E. D. Moore (who has also a grindstone quarry, worked by Mr. Pitman, three miles east of Huron,) manufactures whetstones from the following quarries near him: Mr. John F. Carter's whetstone quarry, on section 23, township 3 north, range 2 west; Mr. Thomas Powell's quarry, worked also by Mr. A. Freeman, of Orleans, in the same section, township and range, nearly three-fourths of a mile further north. Mr. Voorhees also quarries whetstones a few miles from Mr. Moore's.

All of these whetstone quarries are in the ferruginous sandstones, at from 40 to 100 feet above the uppermost limestone, (a layer here usually only a foot or two thick,) with Millstone Grit, full of stigmarciæ, capping the ferruginous sandstones, about 110 to 150 feet above this thin limestone bed.

At Mr. Pinnick's there are three whetstone layers, each about two feet thick, separated by aluminous shales, the uppermost layer being the finest grit.

At Mr. Carter's a good many fossil ferns can be collected, and we obtained for the State collection about three feet of a fossil tree, exposed for some ten or twelve feet, and extending, nearly horizontally, apparently much further under the superincumbent mass, which was too refractory even for our large crow-bar. The tree is a Lepidodendron (modulatum?) two feet across its longer diameter, and nine inches through the short diameter, with the subcentral axis (or pith-like cyclindrical bundle of elongated cellular and vascular tissue assigned by Brongniart to the stems of Lycopodiaceæ or clubmosses, which he considers the Lepidodendron closely to have resembled,) near the convex exterior, and occupying the under side, as the tree lay in its bed, somewhat as represented in the following cross section of this and a smaller one, 17½ by 7 inches. On the large tree the scars are two inches long from one acute angle of the rhombus to the other, three-fourths full from one obtuse angle to the other.*

The tree reposes in a south-east and north-west direction, with a slight dip of the larger or supposed root end in a north-east direction.

The following gives the section at the quarry, with Millstone-Grit hills, 50 to 80 feet higher, at no great distance:

	FEET.
Soil and subsoil	6–8
Shaly sandstone	2–4
Upper solid whetstone grit	2
Shaly sandstone	3
Middle whetstone grit	1½
Shaly sandstone	1
Lower whetstone grit	2½
Fire-clay	⅛
Lepidodendron layer	¾

At Mr. T. Powell's quarry the same three layers of whetstone are found, somewhat closer together, with two feet or more of coal† in the position of the above Lepidodendron layer, and 1½ feet shales over the coal. The upper and middle (?) members of pentremital limestone are found respectively about 100 and 160 feet below the whetstone quarry. The Hindostan whetstones sell, wholesale, at $6.00 per 100 pounds.

There is said to be a nitre cave near here, about 50 feet lower than these quarries; also a locality rich in iron, in section 35, township 2

---

*For this wood cut see appendix, and for the original see the State collection.

†The analysis of one gramme coal from Mr. Powell's quarry gave—

Volatile matter	42.6	gas	35.6
		water	7
Coke		carbon	50.4
		ash	7
			100.0

Coke swelled somewhat; ashes whitish gray.

north, range 2 west, which is called the "Iron Mountain." Beech timber is very abundant on the hills around the quarries.

The grindstone quarry of Dr. Bowles is in the sandstone below the upper pentrinital layer of limestone which is here a few feet thick. The best layer for limestones is about 30 feet above the middle limestone, with another calcareous stratum still 30 or 40 feet lower. From the region of the grindstone grit and above this middle limestone flows the long-celebrated sulphur spring known by the name of "French Lick." Dr. Bowles has fitted up a Watering Place and devotes much attention to the causes and cure of milk-sickness. As he thinks of publishing his view on the subject, only a few of his remarks will be given in the chapter touching on this fearful malady.

Near the spring there is another of the same sulphurated hydrogen character, with suitable buildings, &c., kept by Dr. Davis, under the name of West Baden Springs. It is stated that there are in all twelve sources from which sulphur water flows around this immediate neighborhood.

A section near the grindstone quarry gave the following:

	FEET.
Ferruginous sandstone, with whetstone grit............................	110
Upper limestone...	2–4
Space of shaly sandstone and grindstone grit, about................	90
Middle limestone...	20–30
Space of red clay, &c..	40
Lower limestone at least 50, and probably more, say..............	60

Paoli, the capital, is situated on the lower and middle limestones; the Court House is 599 feet above the level of the sea, according to Col. Stansbury and Mr. Williams.

HARRISON COUNTY.

The eastern part of this county is sub-carboniferous sandstone, the junction with the limestone being about five or six miles east of Corydon, the capital of the county. On the western border the higher hills are ferruginous sandstones, consequently Harrison embraces nearly all the sub-carboniferous formation.

Portions of the lower members we find so modified on Indian Creek, four or five miles south-west of Corydon, as to form a good lithographic stone. Mr. Brinkman has opened a quarry, and, by rejecting the outer and more shaly rock, has succeeded in sending for inspection good sized slabs entirely free from any inequalities, and of a smooth, even texture. The stone was tested at Louisville, and a good sample with the design of the Louisville canal and locks still upon it can be seen in the State collection, along with unpolished slabs from the same quarry. It certainly promises so well as to be worthy of a full development and competitive trial in the markets.

At Corydon we found stone steps apparently solid and durable, which when first quarried at Salisbury were so soft that the rock could be readily cut with a broad-axe.

This county affords numerous localities of sub-conglomerate, or as it might here be called, sub-pentrimital coal; here, as almost invariably throughout nature's work, the evidence being given that changes were gradual and that in advance of the period developing so enormously the heavy beds of vegetable matter converted afterwards into fossil fuel, we have thin deposits anticipating and preparing the way for the true coal period.

At Mr. Martin Smith's, on section 22, township 2 south, range 3 east, there is another seam of coal under the second limestone in descending order, as exhibited in the following section:

	FEET.
Sandstone	50
Limestone	2–3
Sandstone	110
Limestone	2–4
Thin seam of coal and fire-clay	½
Sandstone	40
Sandstone and red clay, at least	60

An analysis of this sub-conglomerate coal furnished the following result:

One gramme gave of

Volatile matter	41.5	gas	31.5
		water	10.0
Coke	58.5	carbon	54.5
		ash	4.0
			100.0

The coke did not swell at all; ashes a light, yellowish grey.

On Mr. Asa Rosenberger's farm, in Spencer township, on section 25, township 2 south, range 2 east, we found, as at Mr. M. Smith's, about

2½ inches of coal under the second limestone. It approaches in quality to cannel, and the analysis of one gramme gave:

Volatile matter	48.85	gas	46.85
		water	2·00
Coke	51.15	carbon	41.15
		ash	10.00
			100.00

The coke swelled considerably; ashes grey with a tinge of red.

This coal is underlaid by a few inches of fire-clay, with about two feet of dark roof shales, and some sulphuret of iron; then five to six feet of encrinital limestone, surmounted by shaly sandstone and the upper bed of limestone.

This locality is in Brushy Valley, noted for some severe cases of milk-sickness. Buckeye, which, if eaten by the cattle, produces somewhat similar symptoms, is very abundant here. Mr. Rosenberger's experince with this disease will be given hereafter.

A thin seam of coal occurs under similar circumstances on Mr. Eli Stewart's land, section 31, township 1 south, range 2 east, in Orange county, or near the line; as well as two miles south of that locality on section 18, township 2 south, range 2 east, in Harrison county.

This county has numerous fine springs and water privileges; from one of which a boy who was fishing had just drawn a Menopoma (water puppy) twenty-two inches long. Near a mill, owned by Mr. Hiram Babcock, we obtained a specimen of hydraulic limestone, which we have not yet had time to analyze. Over the layer from which we obtained the sample is blue clay, and above that, chert with Lithostrotion Canadense.

Several localities in this county furnish iron ore in considerable quantity. A sample from the farm of widow Hoagland, section 30, township 2 south, range 3 east, not far from the Martin Smith coal seam, afforded the subjoined result on analysis:

One-tenth of a gramme lost by drying at 350 F., (and then became red,)	0.009
Gave of insoluble silicates	0.052
Gave of sesqui-oxide of iron, (equal to 26.6 per cent. of pure iron,)	0.038
Gave of loss, alkalies and magnesia	0.001
	0.100

CRAWFORD COUNTY.

This county embraces chiefly the middle and upper members of the sub-carboniferous limestone series, giving rise to a country somewhat mountainous and rocky in places, adapting it more for sheep-pastures or vine-hills, than arable fields, although some of the valleys and plateaus afford good farms.

The chief object of attraction in this county is the

WYANDOT CAVE,

Owned by Mr. H. P. Rothrock. He had a survey made, by Dr. D. L. Talbot, from Jeffersonville, of all the ramifications known in 1853; and the later discoveries were laid down by Mr. George I. Langsdale. From the map thus jointly constructed, Mr. Rothrock politely permitted a copy to be taken, which is subjoined on a reduced scale, exhibiting on the west the entrance and main passage, with its various names, of the Old Cave; on the east, the New Cave, with its intercommunicating passages, some of which are dotted to show that they pass underneath the main cave. Thus, the branch connecting the "Wild Cat Avenue" with "The Little Giant Avenue," passes under "Calypso's Island," part of the grand trunk in the New Cave.*

Some years since I had the pleasure of exploring the Mammoth Cave, in Kentucky, and, without desiring for a moment to detract from that justly celebrated and admired subterranean wonder, I can truly state that the Wyandot Cave is almost, if not quite, equally worthy of a visit from the admirers of fine natural scenery, although not explored yet to the same extent as the Mammoth.

To do justice in description to the splendid masses of long, pendent stalactites, uniting sometimes fantastically with the stalagmites below, which burst upon the view perhaps after worming our bodies through an aperture too small for overgrown travelers, or after safely passing the "Dead Fall," whose disturbance and displacement might forever cut off all return to light and life, furnishing a sepulchral catacomb infinitely greater, in the extent of its ramifications, than the wonderful and massive structures of art, the vaunted mausolean pyramids of Egyptian despots; to describe fully the brilliance reflected, even by torch light, from fluted columns of satin-spar, (carbonate of lime) 35 feet high and 72 in circumference, forming the "Pillar of the Constitution," and simi-

---

*See Appendix.

lar scenes, would require a power of language, which at best would feebly shadow forth the reality. To place on canvas the full grandeur of "Monument Mountain," enshrining on its summit a semblance of "Lot's wife," the whole vaulted, by the crumbling of the Magnesian limestone, into an arch 245 feet from the proper floor of the cave, and studded on its oolitic summit with calcareous icicles, which seemed to form the gothic architectural pendants of this "Wallace's Grand Dome," to paint all this might furnish subject for a Rembrandt; but a few rapid outline sketches were all we could hope to carry away as remembrancers.

The numerous Indian relics, in the shape of charred remnants of fires, part of the wood yet unconsumed, portions of bark, which had evidently served as torches, sticks broken and *never* cut, skeletons of several wild animals, and the like, would furnish materials, if the facts were carefully collected, valuable to our archæologists, or to the historian, who desires to preserve all evidence bearing on the manners and customs of the Aborigenes.

To the entomologist, or investigator of specific modifications produced by external causes, the sightless crickets here, in connection with the blind fish and crawfish of the Mammoth Cave, might furnish speculation and argument.

Leaving, however, the scenic and historical description to others, our aim was directed to obtaining the barometrical measurements, at the important points, noting the lithological character of roof, floor and side walls, and to the securing of occasionanal palæontological or mineralogical specimens for the State collection.

The results of the observations made inside the old cave, then in the new cave, and afterwards on the hill which surmounts both outside, are briefly subjoined, referring the heights to low water in Big Blue river.

Mr. Rothrock's house is 30 feet above low water in Big Blue river, and at about 120 feet above the river we entered the old cave, by the only external opening yet discovered to these subterranean wonders. Descending in the old cave to Pigmy Dome, the floor of which is ten feet lower than the cave entrance, we found an abundant efflorescence of Epsom Salts, sometimes quarter of an inch thick, and calcareous tufa in botryoidal form. The filtration of water, and the washing out of the more soluble ingredients from the rock, had here riddled the dolomite roof until it resembled honey comb, and hollowed out side-apertures, which might have passed for a dove-cote.

At "Odd Fellows' Hall," after passing "Lucifer's Gorge," the "Nat-

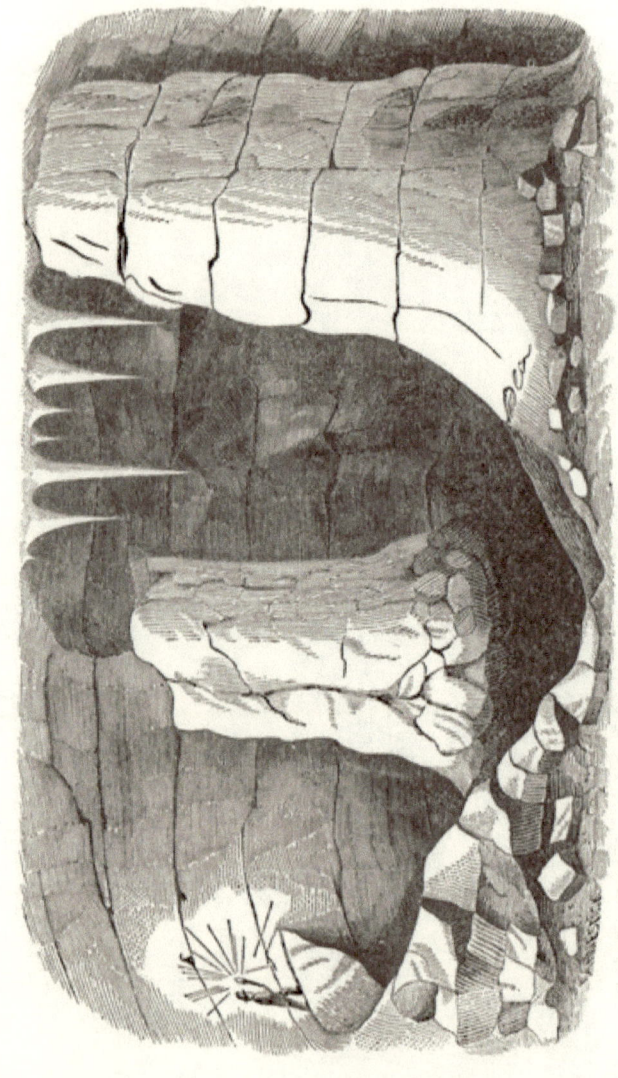

SENATE CHAMBER, CHAIR OF STATE, AND PILLAR OF THE CONSTITUTION.

ural Bridge," and Rothrock's straits, which lead to the New Cave, the roof, 20 feet higher than the Old Cave entrance, is silico-magnesian limestone, with fibrous gypsum, underlaid by more crystalline limestone.

"Jolter's Hole" afforded fine specimens of alabaster and selenite, besides some calc-spar. Ascending to "Spade's Cliffs," we found bastard limestone overhead, and abundant remnants of encrinital stems, as well as corals of the family Cyathophyllidæ.

Descending to "Talbott's Pit," 30 feet below the cave entrance, magnificent stalactites and stalagmites greeted the view, which, on ascending 50 feet, to the further end of "Spade's Cliffs," was gloomed by the myriads of bats, clustering on each other like bees, and hanging head downwards from the ceiling.

On reaching the "Dead Fall," we secured samples of oolitic limestone; and, after passing through the narrow aperture denominate "The Screw Hole," were rewarded by emerging into the very capacious amphitheatre to which very appropriately the name of the "Senate Chamber" has been given, while a somewhat central stalacto-stalagmitic union forms a natural "Chair of State." Facing the "Senate Chamber," or in fact forming pillars which a slight stretch of the imagination might consider the columns of galleries, common in public buildings for deliberative purposes, we find a structure which, from a fresh fracture, reflects light with the splendor of satin, and which effervesces freely with acids. Although breaking usually into prismatic specimens, the longitudinal section thus obtained exhibits numerous and delicate horizontal layers of successive deposition, sometimes slightly tinged with grey, but more generally of a dazzling pearly whiteness. Although generally the cave is dry, here sufficient water trickles into a natural excavation of the pillar, to refresh the weary traveler.

Of this locality we endeavored to give some idea by the foregoing sketch.

It is within about ten feet of being on a level with the entrance to the cave, and terminates the "Old Cave" avenue, in "Pluto's Ravine," three miles from the mouth.

Retracing our steps as far as "Banditti Hall," only 50 feet above the river, and consequently at least 100 feet lower than the Old Cave entrance, the secret door was unlocked, and we glided on our backs, feet foremost, down an inclined plane, over earth and rubbish, at the imminent risk of breaking the Aneroid Barometer; and, passing "Bats' Lodge," stood again erect in the Counterfeiter's Trench, which had been artificially excavated to prevent the necessity of constant stooping in this passage to the main avenue of the New Cave. Here, when it was first explored, were found the remains of Indian fires, supposed to have been kindled when the cave was the resort of the Wyandot tribe, hence the name given to it. Perhaps, when at war with other tribes, they may have resorted to these subterranean hiding places for safety or strategy. The charred remains exhibited White Oak, Hickory, Sassafras and Papaw, with numerous detached pieces of hickory bark, charred at one end, as if used for torches. Scores of dead bats were strewed around; and the skeleton of an Opossum and of a "Wild Cat," to each of which portions of hair and skin adhered, were among the relics.

Near the "Rotunda," we found large quantities of Epsom salts, often as an efflorescence from the Magnesian limestone, and in "Coon's Council Chamber," fine samples of black flinty rock, usually in bands 4 to 5 inches thick, but sometimes in concentric layers of filtration and deposition, that gave the appearance of knots in pine wood. This rock seems to partake of the character of Lydian stone, or flinty Jasper, while the intermediate layers are silico-calcareous, overlying the yellow Magnesian limestone that furnishes the sulphate of Magnesia.

The "Dining Room," upon measurement, proved nearly a hundred feet long by 45 wide, and afforded good samples of Selenite. In the "Sandy Plains," formed by the disintegration of the silico-magnesian limestone, acicular crystals of Epsom salts are abundantly diffused. Here also a Papaw pole was found broken off; no evidences of cutting visible on any of the wood found; but the bark on some was gnawed by animals, apparently rodents. From this point, which appeared to be only sixty feet above Big Blue, we passed over the "Hill of Difficulty," formed chiefly of decomposing dolomitic rock, to "Mammoth Hall," which has a roof stratum of Oolite. This great natural excavation contains the "Monument Mountain," of which we subjoin a sketch designed to show "Lot's Wife," a pyramidal mass of gradually aspiring stalagmite, not, however, so darkly tinged as the noted "Gibraltar Rock," of similar origin, from Spain.

Descending to the "augur hole," we found clear sulphur water, showing the yellowish white deposit beneath in a small natural rock-basin.

Although, much beyond this place, objects of undoubted interest tempted exploration, and some avenues have never yet been traced out, more immediate geological interests having already been subserved, and time passing rapidly, we returned from this point, in order to examine the hill outside.

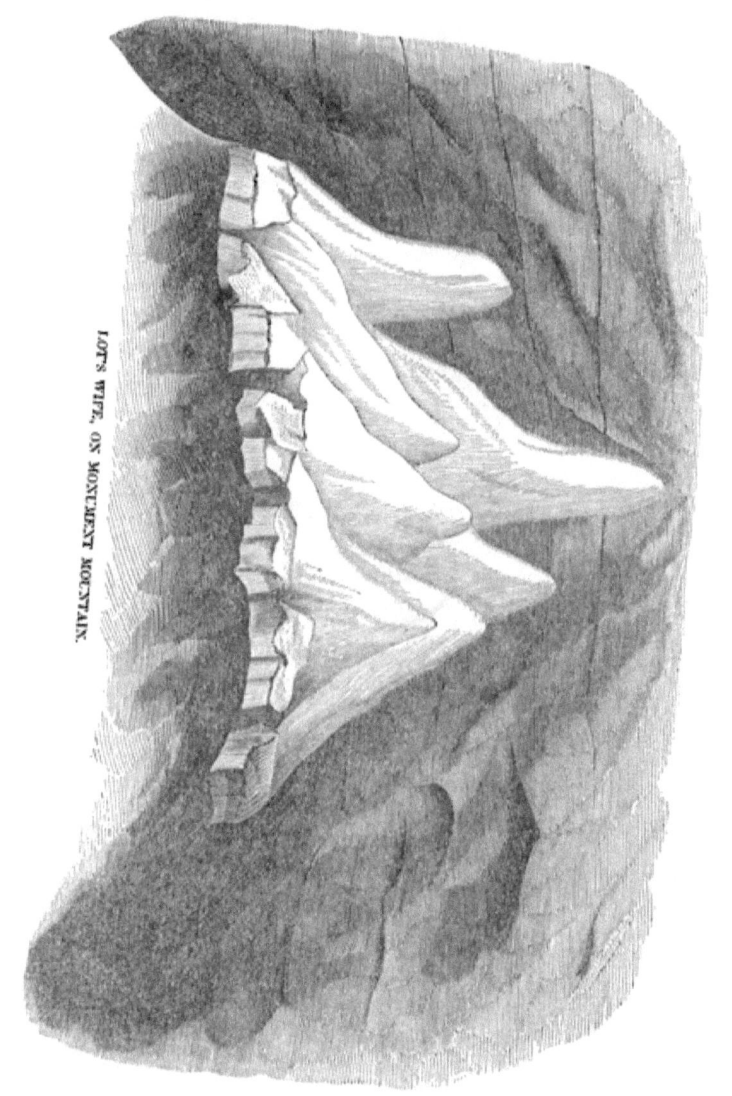

LOT'S WIFE, ON MONUMENT MOUNTAIN.

The upper hundred feet were found composed of ferruginous sandstone, namely, from about 280 to 380 feet above "Big Blue." Then descending, we found a few feet of Bastard limestone, then 50 feet of crystalline, 40 feet of flinty, and finally a few feet of compact limestone; talus covering nearly all below this from view, a space of about 180 feet above the river. Beds of Cherty Limestone were exceedingly abundant, with numerous Bryozoa, near our camp, which stood on a plateau about 40 feet above the river, and fragments of chert showed themselves often between this point and the mouth of the Old Cave. In the bed of Big Blue, and up to nearly the level of Mr. Rothrock's house, magnificent specimens of Lithostrotion Canadense are scattered about, some weighing over fifty pounds.

The Sibert Cave, a short distance from the Wyandot, although not extensive comparatively, is yet more replete with splendid stalactites and stalagmites, often uniting to form pillars, along galleries, extending for several hundred yards, and not yet fully explored. It is not so dry as the Wyandot, but some of the more slippery chasms have already been bridged. It is well worthy a visit from the traveler fond of adventure and remarkable scenery.

For convenience of reference and comparison with the map, the most important distances, and heights, widths, &c., in the Wyandot Cave, are here recapitulated in tabular form:

DISTANCES.	MILES.
Length of "Old Cave,"	3
To Monument Mountain	1¼
From Augur Hole to Junction	1½
Thence to Crawfish Spring	1½
To end of Wabash avenue	1¼
From Sandy Plain to the Throne	1½
Thence to the end of Southern Avenue	1½
From Amphitheatre, south	¾
From Mound to Junction room	¾
All other avenues, about	6
Total as far as explored in 1853	19

From the south-western to the extreme north-eastern limits, about 9 miles. The exact distances in the New Cave were not furnished, but can readily be approximately obtained from the map.

WIDTH AND HEIGHT.	FEET.
Greatest width at any point in "Old Cave," about	180
Greatest height, (varying from 2½ to 100) about	100
Average height, about	20
In "New Cave," greatest breadth	300
Height in "New Cave," from 3 to	245

Which is the height of Wallace's Grand Dome above the proper floor of the Cave.

TEMPERATURE.—The thermometer indicates usually, at different points, a variation of from 54° F. to 57°; but at the mouth of the New Cave it was noticed by us as low as 52° F. on May 28, 1860. A strong, cool current of air rushes out of the Cave in summer, as at the Mammoth Cave, Ky.; and the same capability of continuous exercise without fatigue, so frequently observed by visitors in the latter, is here also remarked.

The Wyandot Cave is in Sec. 17, T. 3 S., R. 2 E., and can be reached by a few miles travel from the Ohio, or by way of Corydon, Harrison county.

The growth of timber around the Cave is Buckeye, Sugar Tree, Beech, Cedar, Oak, Tulip Tree, Hickory and Sassafras.

This county is also noted for the beautiful oolitic limestone, furnished at the capital, Levenworth, and other localities, both for building purposes, and for the purest quality of white lime; an intermediate layer of sandstone at the Fredonia bluff is used extensively in Loussville, at the gas works.

Sub-conglomerate coal shows itself at several places, particularly at Mr. Houghton's, on Sec. 32, T. 3 S., R. 2 E., and near the Levenworth graveyard, where 5 inches, with fire clay and sandstone shales, show themselves below the 2d Archimedes limestone, or member "B" of the upper sub-carboniferous limestone, which here exhibited abundance of Productus shells and Pentremites.

The analysis of a gramme of Mr. Houghton's sub-conglomerate coal gave:

Volatile matter	39.3	gas	30.3
		water	9.0
Coke	60.7	carbon	48.7
		ash	12.0
			100.0

Coke scarcely altered in appearance; ashes red.

The same quantity of coal from near the Levenworth graveyard, afforded, on analysis, the following results:

Volatile matter	40.0	gas	29.0
		water	11.0
Coke	60.0	carbon	40.0
		ash	20.0
			100.0

Coke did not alter its appearance by burning; ashes reddish grey.

Between the Levenworth and Houghton localities for sub-conglomerate coal, we obtained, at "Dry Run," a section extending through nearly two hundred feet of sub-carboniferous limestone, thus:

	FEET.
Red soil, subsoil and shaly limestone	40
Limestone	2
Yellow, grey and blue shales	15
Sandstone	20
Solid limestone	30
Shaly limestone	10
Buff limestone	25
Impure sandstone and shales	10
Limestone	10
Calcareo-aluminous sandstone	6
Shaly limestone	2
Solid limestone	10
Aluminous sandstone	5
Limestone	5
Sandstone	4
Level of Dry Run	0

Several remarkable springs issue from under the cavernous limestone in this county, one of which is said to be yet unfathomed, and samples from some limestone layers near there were furnished us for analysis, as hydraulic.

Iron ore is found at Mr. Lambdin's, near Mt. Prospect, in considerable quantities, and one sample of lead ore was given us from this county. Milk-sickness is prevalent in portions hereafter alluded to.

SEC. VI.—COUNTIES IN THE COAL MEASURES.

The valuable and interesting report of Prof. Leo Lesquereux, giving the details of most of our coal counties and identifying the stratigraphy of the beds, renders it unnecessary for me to do more, under this head, than to give the analysis of the coals, as far as examined, and some details regarding subjects in the same counties not strictly connected with coal examinations, as also a few observations in counties of the Coal Measures, which time did not permit the Professor to reach.

SUB-SECTION 1.—GENERAL DESCRIPTION.—Although, as noticed in the preceding section, sub-conglomerate coals, which in some instances elsewhere have proved valuable workable seams, are found in Harrison, Crawford, Orange and Putnam, yet in the Coal Measures proper, we include only the deposits found above the Millstone Grit, in the counties of Warren, Fountain, Parke, Vermillion, Clay, Vigo, Owen, Green, Sullivan, Martin, Daviess, Knox, Dubois, Pike, Gibson, Perry, Spencer, Warrick, Vanderburgh and Posey. In the eastern parts of a few of these, the sub-carboniferous limestone is found; but in all of them coal is worked more or less, and in some quite extensively.

SUB-SECTION 2.—THE RESULTING SOIL, &c.—The disintegration of sandstones in the Coal Measures is not calculated to produce the best soils; but as the deposition of limestones, of various thicknesses, during this period was frequent, and as many parts of the formation are in the river bottoms or on prairies, there is thus often a modification, which furnishes very fair or even highly superior farming lands, as will be observed in the several details. Indeed part of these counties furnish the vast quantities of Indian corn, which emanate in flat-boats by fleets from the Wabash, and for which a somewhat large proportion of arenaceous materials is by no means objectionable. The Wabash and White river bottoms, as well as some of the adjoining prairies, are of the above sandy-loam character, with occasionally large amounts of organic matter as proved by reference to the analysis of soils from the farms of Hon. G. D. Wagner and of Hon. J. D. Williams.

SUB SECTION 3.—QUARRIES, &c.—Many of the sandstones of the Coal Measures furnish an excellent freestone for building materials, and the intercalations of limestone are sufficient to furnish, in most of the counties of this formation, materials suitable to burn into lime. In the entire 1,234 feet of Coal Measures belonging to the Eastern or Apalachian field, as given in the excellent Manual of Coal, by Dr. J. P. Lesley, there are nine limestones, varying from eight to seventy feet in thickness, and in the connected section from the Kentucky report, somewhat modified by Mr. Lesquereux, for Indiana, and subjoined with this report, an equal number of limestones present themselves, besides two or three thinner beds of calcareous deposition. They are more prevalent with the higher than the lower coals.

Quarries of one or the other materials are opened extensively in several counties, which will be noticed in giving their details. Excellent potter's and fire-clays will also be found enumerated in describing the resources of the separate counties.

Sub-Section 4.—Metallic Ores, &c.—A few localities in the Indiana Coal Measures afford indications of zinc ore which may on further examination prove sufficiently abundant to be workable. At least one county has, associated with the zinc ore, an important admixture of cobalt ore, well worthy of detailed examination, and accurate qualitative analysis.

The edge of the coal field is our great dependence for workable iron ore, and as might be expected we find it in several counties as mentioned in the detailed descriptions.

Sub-Section 5.—Prevalent Timber and other Vegetation.—The growth of timber in this formation is very various; but perhaps there is in the uplands rather greater predominance of Oak and proportionately less Beech than in the other systems. Our Indiana prairies are chiefly in the Coal Measures, especially those of Warren, Fountain, Vigo, Sullivan and Knox.

Sub-Section 6.—Springs and Artesian Wells.—Much of the water is hard from the presence of the limestones above noticed; sulphur springs not unfrequently occur from filtration through coal charged with sulphurous combinations, and chalybeates also are found. Some wells and springs, analyzed, afforded a most unusual amount of alumina, particularly in regions of milk-sickness.

The favorable positions for successful salt-boring, are alluded to in Mr. Lesquereux's report, and the well-known Artesian boring at La-Fayette, not far from the edge of the field, added to the fact of the strata all inclining to the central coal field, render the theoretical probabilities encouraging for similar attempts.

Sub-Section 7.—Miscellaneous Facts, &c.—In regions of the Coal Measures, where aluminous shales are abundant, milk-sickness is apt to be found, as well as when those prevail in other geological formations; an important fact, frequently alluded to by my late brother, and which he intended to fully elucidate in this report. In a subsequent chapter the subject will be again brought up.

SUB-SECTION 8.—CHARACTERISTIC FOSSILS.—Among the numerous coal plants, so characteristic of this period, as to furnish by specific difference, data for the determination of successive beds of the coal deposit, only some of the most common will be here enumerated, reserving a more extended list for subsequent remarks; this remark applies also to the other fossils, chiefly mollusks and fishes.

Fossils of Coal Measures:

Ferns: Neuropteris,
Pecopteris arborescens,
Lepidodendron modulatum,
L. vetustum?
Psaronius,
Sigillaria reniformis,
Stigmaria,
Syringodendron pachyderma,
Calamites,
Asterophyllites,
Sphenophyllum Schlotheimii,
Besides the trunks of trees formerly described by Lyell and others, as Palm trees.

Foramenifera: Fusuliuia cylindrica.
Corals: Chætetes milleporaceus, (Edwards and Haime,) Newburg.
Crinoids:
Mollusks: Spirifer attenuatus,
Patella,
Mytilus,
Ambonychia Grayvillensis,
Productus Providensis,
Nautilus ferratus,
Sp. cameratus?

Several species of the genera,
Pleurotomaria,
Bellerophon,
Dentalium,
Macrocheilus,
Pecten, Pinna.

Also a Palæoniscus,
Sharks teeth very abundant, and Ichthyodorulites.

SUB-SECTION 9.—SOME DETAILS OF EACH COUNTY IN THIS FORMATION:

WARREN COUNTY.

Mr. Lesquereux, as stated in his report, having only the opportunity to examine, on his way to Illinois, to a small extent, the formation around Williamsport, a few observations made by our corps in Warren county, under the valuable guidance of Mr. Wagner, then President of the State Board, are here subjoined.

At Independence the sub-carboniferous limestone, with Bryozoa and Productus punctatus, is represented, by its upper member of three feet in thickness, near high-water mark in the Wabash, thickening to five feet, and becoming cherty a short distance below town, with overlying sandstone. The same calcareous bed still shows itself at Williamsport, overlaid by a sandstone which rises gradually twenty-five feet to form the plateau on which the hotel and central portion of the town are built. Main street passes, by a moderate grade, over seventy to seventy-five feet of ferruginous sandstone and cuts, ten feet through the hill, exposing about six feet of sub-conglomerate coal shales. A few rods west, in sinking a well, these coal shales furnished several inches of coal. The adjoining hills are, by quaternary deposits on the above strata, elevated upon an average 150 feet.

The Millstone Grit is finely developed at various places in the county, forming fine bluffs of from 30 to 75 feet on Kickapoo, on "Little Pine," and forming the picturesque falls of Fall creek, near the railroad, back of Williamsport.

Coal shows itself at many places on "Big Pine" and elsewhere in the county. Politely conducted by Mr. Knaur, we saw, at two localities, a two foot seam on "Mud-Pine Creek," a branch of Big Pine, with fire-clay underneath, several feet of aluminous shales over, and 30 to 60 feet of quaternary superstratification. These are on sections 19 and 20, township 23 north, range 8 west. A mile and a half further south is a

---

*From this stratum of limestone were obtained the casts of a Productus and Spirifer, apparently Productus punctatus, and perhaps Spirifer incressatus of Eichwald, although from the prolonged hinge-line more like the Spirifer increbescens of Hall, from the Kaskaskia limestone, yet apparently differing in having the folds obliterated or lacking on the mesial lobe, which is very strongly marked with lines of increase.

coal bed three feet thick, overlaid by shales of sandstone, with a coarse sandstone thirty feet thick, showing itself somewhat further down the creek, which appeared to be Millstone Grit. The edge of the coal basin exhibits in its coarse sandstone a dip S. and S. S. W., sometimes amounting to 45°, with twenty feet of sandstone unconformably over it, having scarcely any preceptible inclination to the horizon.

Near Burr's Mill, on Big Pine, some coal shows itself, overlaid by hard shales and a thin limestone, which coal seam a mile from there augments to three feet in thickness, and is worked by stripping, through the enterprise of Mr. Butts and others.

One gramme of this coal afforded, on coking, &c., the following result:

Volatile matter	45.0	gas	40.0
		water	5.0
Coke	55.0	carbon	40.0
		ashes	15.0
			100.0

Swelled but little in coking; ashes dark grey.

Half a mile south of this, Mr. Kiester is drifting into the bluff and propping the ten to fifteen feet of aluminous shales and shaly sandstones, so as to take out in wheelbarrows, or trunks, the product of the three-foot coal deposit. No limestone showed itself here.

The analysis gave:

Volatile matter	42.0	gas	40.0
		water	2.0
Coke	58.0	carbon	51.5
		ashes	6.5
			100.0

Coke swelled a little; ashes light colored.

These proprietors sell their coal chiefly to blacksmiths for eight cents per bushel at the bank.

We heard of other coal openings six miles north of the above, also one eight miles south-west of Williamsport, said to be better than those we saw.

The rock quarried abundantly in this county and Fountain, extensively used and proved to be durable in bridge piers, foundations of warehouses, exposed to alternate wet and dry conditions, can be had at Mr. Hayne's quarry; also half-way between Williamsport and Attica are several fine bluffs near the Wabash, to which the distance is so short and the descent so gradual, as to promise a good result from the laying of a track to facilitate delivery in increased quantities. This close-grained sandstone has sufficient admixture of calcareous ingredients to effervesce slightly with acids. These quarries seem to occupy the place of the grindstone and whetstone grits in Orange county.

The agricultural prospects of the country can be judged of, by examining the analysis furnished in Dr. Peter's report of soils taken from Hon. G. D. Wagner's farm, adjoining a grove of Hickory, Bur Oak, Walnut, Grey Ash, Buckeye, Red Elm, Cherry, Sassafras and some Black Jack Oak, with an undergrowth of Hazel, Elder-bushes and Red Bud; and also from seeing the splendid cattle that graze throughout the country.

From the adjoining farm of Mr. Wagner, Sen., we obtained a sample of Bog Iron ore for analysis. Other iron ores are reported at several points, especially near Pine Village; chalybeate springs are very common.

A remarkable belt of bowlders and other quaternary Drift, passing from Parrish's Grove, in a south-easterly direction through the county, will be noticed hereafter.

At a digging made by Mr. Robert Pierson for lead, we found a considerable amount of sulphuret of zinc, which may have cobalt associated with it, as in Fountain county.

Some milk-sickness is reported as existing in timbered portions of the county.

FOUNTAIN AND PARKE COUNTIES.

Dr. Bigelow kindly piloted me to a coal opening two and a half miles south of Attica, not now worked. Among the materials thrown out were pieces of silicious limestone containing sulphuret of zinc, associated with which Mr. E. T. Cox, of the Arkansas Survey, to whom its qualitative analysis had been assigned, detected notable quantities of cobalt, which probably, upon further examination and development, may prove of considerable commercial importance.

The building rock, described in speaking of the quarries of Warren county, is worked extensively at the prosperous town of Attica, and shipped usually under the name of the Attica stone. The beautiful bluffs or cliffs near town we regretted not having time to visit.

Mr. Lesquereux has so fully discussed the Coal Measures of Fountain county, that I have only to add a few localities seen at another time.

Near Mr. Scott's old diggings, (which, as nearly as could be ascertained in the absence of the proprietors, is either on section 6, township 18 north, range 8 west, or on section 1, township 18 north, range 9 west,) Mr. Ripple and Mr. Mesner are working a four and a half to five feet seam tolerably extensively, on land belonging to Mr. Woods. The roof of this coal bank is a bluish "soapstone,"* the floor a light colored fire-clay.

East of Mr. Thomas', on Coal creek, are other openings not yet visited.

The coal of Mr. Thomas, near Lodiville, furnished on analysis of one gramme:

$$\text{Volatile matter} \ldots \ldots 45.0 \begin{cases} \text{gas} & 37.0 \\ \text{water} & 8.0 \end{cases}$$

$$\text{Coke} \ldots \ldots 55.0 \begin{cases} \text{carbon} & 51.5 \\ \text{ashes} & 4.5 \end{cases}$$

$$100.0$$

Sandstone is quarried abundantly near Covington, the capital of Fountain; and the town displays a handsome Court House and Odd-Fellows' Hall.

Fine barns, good orchards, fields of sorghum, buckwheat, &c., besides other agricultural indications, denoted prosperity and enterprise.

In Parke county, the coal near Clinton Lock, so fully described by Mr. Lesquereux, is leased and worked on the land of Messrs. T. Jones & Co. by Mr. Griffith; another opening is owned by Mr. Walter G. Crabb, a third by Mr. Joseph Blake, and a fourth by Mr. G. M. Griffith.

Besides Mr. John W. Campbell, the Abdallah Company, Mr. Beattie Harrison and Mr. Fagan Boyd, own openings at or near section 34, township 15 north, range 7 west, in the region of Little Raccoon Creek, to which Dr. Dare, of Rockville, also a proprietor of coal land, was good enough to pilot us.

Mr. Lesquereux has also described the excellent coal of Hon. W. G. Coffin, whose absence East, prevented our having the pleasure of his company, which as a member of the State Board he would otherwise have given in the explorations of his county; but under his son's directions and those of Dr. Hubbs, who courteously conducted us, we had a good opportunity to examine the Sugar Creek Coal.

---

*So denominated by the miners, but not so in strict mineralolgical language.

Mr. Coffin's coal gave, on analysis of one gramme:

Volatile matter	48.0	{ gas	42.0
		water	6.0
Coke	52.0	{ carbon	49.0
		ashes	3.0
			100.0

Coke swelled somewhat; ashes light grey.

A gramme of Mr. Campbell's coal afforded:

Volatile matter	49.0	{ gas	42.0
		water	7.0
Coke	51.0	{ carbon	49.0
		ashes	2.0
			100.0

Coke swelled slightly; ashes reddish brown.

A considerable amount of milk-sickness was reported in this county, and some examinations were made with reference to the subject, which will be reported under that general head.

VERMILLION AND CLAY COUNTIES.

As our corps had an opportunity of visiting some coal openings, during an exploration subsequent to the one made in company with Mr. Lesquereux, a few additional localities are here subjoined: One occurs a mile and a half south-east from Newport, and several others are found on the creek close to that town. The coal of one bank owned by Messrs. Bell and Groves gave the following components:

Volatile matter	47.0	{ gas	33.0
		water	10.0
Coke	53.0	{ carbon	54.0
		ashes	3.0
			100.0

Coke swelled very little; ashes reddish grey.

Mr. John W. Thomas owns a bank one mile from Newburg, on Little Vermillion; Mr. W. A. Henderson another two miles west of town, and Mr. Bennett a third, six miles north-west from Newburg, worked by stripping. The two former are No. 9, with No. 11 above not worked.

Mr. John Wright on section 13, township 14 north, range 9 west, also Mr. Samuel Davidson and Mr. Van Ness own coal banks about one mile west of Clinton, reported as having a seam over fourteen feet thick.

Around Eugene coal is extensively worked on Big Vermillion, where three beds frequently show themselves, sometimes only separated by a clay parting of two to three feet; the upper coal from two to three feet thick, being selected as the best. The roof of this consists here of about five feet of black shales and several feet of sandstone, usually overlaid by twenty-five to thirty feet of gravel. Underneath is fire-clay; large Septaria are sometimes found immediately over the coal and under the black shales. At Mr. Collet's bank there is a rich deposit of iron ore.

The openings chiefly worked, are Mr. John Heapburn's, nearest town, Mr. Miller Jones' bank, leased to Mr. Dunlavy, and Mr. Joseph Collett's, 2½ miles from town, on the north side of Big Vermillion. Mr. John Groendyke's, and Mr. Samuel Groendyke's banks are on the south side, near town. Mr. Harrison Elsby's is below "Hanging Rock," 4 miles from Eugene, and Mr. Wm. Hughes' bank, mentioned by the late State Geologist, in his Report of 1857, is 4½ miles from town; thickness of coal, 2½ feet.

On Little Vermillion, there are numerous coal openings, the seam being represented as thicker than those above described. Mr. Martin Patrick's name was the only one ascertained. Sandstone quarries are also found on the same stream. On Big Vermillion, one stratum in the subjoined section, where the bank is from 75 to 100 feet high, affords solid building material:

	FEET.
Quaternary	25–35
Sandstone and shales	8–10
Solid bed of sandstone	6–8
Sandstone, somewhat shaly	10–15
Shales	2
Thin shales	5
Black slaty shales	5
Coal, dipping slightly under water	2–2½
Level of water in Big Vermillion	0

Coal sells in Eugene at 7 cents, delivered, in Perrysville at 10 cents, At the latter town, in descending to the ferry, a bed of 3 to 4 feet of solid dark limestone shows itself over more shaly beds, with the black slates underneath.

The well-known "Indiana Furnace" is in this county, on Sec. 23, T. 14 N., R. 10 W., and has been in operation 23 years. It is owned by Messrs. E. B. Sparks & Co., who employ 75 hands, using the hot blast, and obtaining heat from the gases given off by the combustion of metal and the charcoal. They pay $1.50 per ton for ore delivered. It is found abundantly, of several varieties, in all the hills around, as well as close by their furnace, over a five foot vein of coal. By mixing several ores, previously roasted to expel the sulphur, they often avoid the necessity of fluxing with limestone, although when necessary it can be obtained near there. They can run ten tons of metal per day, using twenty-five tons of ore and drawing twice in twenty-four hours; they ship the iron on the Terre Haute Railroad, at Sandford, seven miles distant. Frequently they manufacture also their own fire-brick.

Clay county is equally favored as regards coal. A sample from the Brazil shaft of the Splint or Boghead coal afforded:

Volatile matter	53.0	gas	48.0
		water	5.0
Coke	47.0	carbon	44.0
		ash	3.0
			100.0

Swelled in cooking; ashes light grey.

The coal is used and much liked by the proprietors of the Rolling-Mill at Indianapolis. Mr. Campbell, who, with four other Scotchmen, works it, conducted me, (28th November, 1859,) after a descent in the shaft of nearly 100 feet deep, along the three foot three inch drift for one hundred and eleven yards, nearly north-east, since however greatly extended. About thirty feet from the shaft is a "horseback" of no great detriment; but at the extremity of one chamber the coal entirely disappears through a similar cause. They furnished the following shafting:

	FEET.	INCHES.
Soil and subsoil	10	
Limestone	4	
Aluminous shales	28	
Coal		10
Sandstone	28	
Shales	2	
Coal	3	3
Fire-clay		6
Blue shales, indefinitely down		

The Highland coal of the Staunton Company gave on analysis:

Volatile matter..	44.0	{ gas	39.0
		water	5.0
Coke..	56.0	{ carbon	55.0
		ash	1.0
			100.0

Swelled in cooking; ashes grey.

The other coals of Clay county will be found enumerated in the list of coals tabulated in the appendix.

Mr. Talbott has a thriving pottery in this county, and there may be others which we did not see.

VIGO AND OWEN COUNTY.

Besides the coal mentioned by the late State Geologist as found on Honey creek, and the banks, fully described by Mr. Lesquereux, which Mr. McQuilken works by stripping in his bottom fields and by a shaft near the railroad, on sections 7 and 8, township 12 north, range 9 west, there are several coal proprietors near him along Sugar creek, of Vigo county, who take coal to Terre Haute, a few miles distant; also, Mr. Jonas Seely, eight miles east of town, at Woods' Mill. Mr. Frederick Miller, on Coal creek, of Vigo county, has a four foot and a half seam on section 30, township 13 north, range 9 west; and Mr. Ferrin and Mr. Ross, have banks near Middletown, with a thinner bed of coal; the former is on section 16, township 10 north, range 10 west.

Mr. C. R. Clarke, who lives on the edge of a prairie near Prairie-town, to which the Wabash sometimes extends itself, although five miles distant, accompanied us to these localities chiefly with a view to the examination of a region in that neighborhood troubled with milk-sickness, in places which have been fenced in for thirty years,—cattle dying if permitted to browse before the dew is off. No metallic poisons were detected in the springs by the sulphureted hydrogen gas test, although they reported that lead had been found near there. It may have been left in spots by the Indians, as they had six mounds near here, used as burial places, either of natural quaternary deposit, or raised artificially from those materials.

Doubtless there are also other coal proprietors and banks in this county, whose names did not reach us.

Above Fairbanks, near Middletown, there is a flourishing stone-ware pottery, and Mr. S. W. Gapen has one also one mile south of the latter town. The ware is sold chiefly in Carlisle or taken into Illinois. The potter's clay is a quaternary deposit obtained just under the soil two miles south-west of his place; the limestone employed is quarried two miles east, the sandstone two miles west of Middletown.

Mr. McQuilken has quarried both limestone and sandstone, a few miles N. N. E. of his farm, which find a ready market and are easily boated on the Wabash. Mr. Peter Hulse furnishes Terre Haute with fire-clay from his place, nine miles east of town, near the edge of Clay county. The details will be found fully given in Mr. Lesquereux's report regarding the coals of Owen county, and the probability of obtaining profitable results in the oily products from distillation of a cannel coal from a bed discovered by him on the land of Mr. Henry Jackson.

The particulars regarding the Mammalian bones found near Gosport, have been so well detailed by Rev. Theophilus Wylie, of the State University, as to render it only necessary here to call attention to that fact.

GREENE AND SULLIVAN COUNTIES.

The coals of the former will be found fully described in the report of Prof. Lesquereux.

The subjoined is the only analysis which time permitted of the Greene county coal. It is from the sub-conglomerate, two feet and a half seam, of Mr. Thornton Hays, on section 16, township 6 north, range 4 west.

One gramme gave:

$$
\begin{array}{lr}
\text{Volatile matter} \dotfill 44.5 \begin{cases} \text{gas} & 36.00 \\ \text{water} & 8.50 \end{cases} \\
\text{Coke} \dotfill 55.5 \begin{cases} \text{carbon} & 53.50 \\ \text{ashes} & 2.00 \end{cases} \\
\hline
& 100.00
\end{array}
$$

Coke swelled somewhat; ashes steel grey.

The Richland furnace, which has been carried on for a number of years, was unfortunately temporarily suspended in its operations while we were there, so that we had not a favorable opportunity for inspecting or ascertaining its facilities. The ore is said to average from the furnace forty per cent. of pure iron.

In Sullivan county, the coal described by Mr. Lesquereux as belonging to Messrs. Elliott and Sharpe, was formerly owned by Mr. Isaacs, and is now raised from a depth of fifty feet by Messrs. James Elliott, Ralph Elliott and David Sharpe. They convey it on a switch about half a mile to the Evansville and Crawfordsville Railroad, at Farmersburg, having undertaken to supply the Terre Haute gas works. According to their calculation they could furnish annually for the market 60,000 bushels.

These gentlemen presented for the State collection some coal plants obtained from their mine, and decided by Mr. Lesquereux to be Sigillaria reniformis, Sphenophyllum Schlotheimii, Pecopteris arborescens, a Neuropteris and Syringodendron pachyderma. Good limonite was observed by Mr. Lesquereux.

On a previous exploration, I was informed by the landlord at Sullivan's Station; that Mr. Thomas Grant, of Evansville, had made a boring for a company near the Station, which is nine miles from the Wabash river, reaching a three foot vein of coal at 150 feet below the surface, a seven foot vein at 500 feet in depth, and no other seam in the remaining 80 feet which terminated the work. A three-foot bed of bastard limestone, which crops out about three-quarters of a mile east of the town of Sullivan, he thinks was reached at a depth of about 200 feet.

At Princeton, Gibson county, there is an agency for the coals of Sullivan; which a reference to the appendix will show is mined at many points, near Currysville and Busseron creek.

From the Merom heights a most magnificent view is obtained over the Wabash into Illinois, and the following interesting section can be studied, at an escarpment 160 feet in perpendicular descent, formed by the denudation of the river through a long succession of ages:

	FEET.
Quaternary	30
Shaly sandstone	15
Solid sandstone	25
Limestone, partly covered by ferruginous tufa	2–5
Coal	1–2
Black shales	5
Productal and encrinital limestone	1
Brashy coal	1
Fire-clay	8–10
Shales	30
Talus, probably shales	40–50
Level of Wabash river	00

The upper limestone has almost the character of a conglomerate rock and has seams of coal one to two inches thick, deposited and enclosed in its substance. Of this we secured specimens. The coal is not adapted for blacksmithing, but will burn in grates.

Prairies are numerous in this county, and Birch trees are not uncommon along the streams. The soil is generally sandy, especially near the river, and even for several miles thence, where the recent quaternary constitutes the superficial deposit.

MARTIN AND DAVIESS COUNTIES.

In addition to the coal of Martin county described in the report of Prof. Lesquereux, attention deserves to be called to the excellent marble and oolitic limestone quarries of Mr. Ralph Delamater, on section 13, township 4 north, range 3 west. We found extensive beds of each, about three feet in thickness, susceptible of receiving a fine polish, and easily shipped, as it is close to White river. The marble is a mottled grey.

Good iron ore can be found at various places. A sample furnished by Dr. W. F. Delamater, of Dover Hill, from McCameron township, gave evidence, on analysis, of containing about 44 per cent. of iron.

An iron ore from the farm of Mr. Moses C. Edwards, also obtained through the politeness of Dr. Delamater, afforded on quantitative analysis of a tenth of a gramme:

Loss by drying	0.004
Protoxide of iron	a trace
Insoluble silicates	0.027
Peroxide of iron	0.058
Alumina, Magnesia, alkalies and loss, not separately estimated	0.011
	0.100

Consequently this ore contains over 40 per cent. of iron, according to the rule employed for estimating.

Near the edge of Lawrence county, in the region of Willow Valley Station, the Messrs. Elliott own a large deposit of iron ore, which contains a good per centage of iron, but, from the quantity of associated silica, would be refractory to work. There is also iron ore on the land of Mr. O'Brian and of Mr. Hanna, living in Petersburg, Pike county.

Three miles south-east of the White River Shoals, Mr. Sullivan, of New Albany, and others, own a coal bank, into which we were informed they had already, in Nov., 1859, drifted about 400 yards, working a four foot seam.

At Owensburg, the blacksmiths obtain a jet-black, sub-conglomerate coal, almost destitute of sulphur, from an opening four miles south-west of that town.

The bank of natural points, mentioned in Mr. Lesquereux's report as being over the sub-conglomerate coal near Dover Hill, consists of fine aluminous materials mingled with the various iron ores, so as to form several different tints, such as yellow ochre, red ochre and umber. This locality has been purchased from Dr. Delamater by a Cincinnati company, who have extensive arrangements for washing off all impurities, drying, barreling, shipping, &c. Mr. Munson, one of the firm residing in Cincinnati, politely furnished a wood-cut exhibiting a section and back-ground of the locality, which illustration is herewith subjoined.

We found similar red paint on the farm of Mr. Henry Inman, section 28, township 5 north, range 2 west.

Between Indian creek and White river, on section 22, township 4 north, range 3 west, a lithographic stone is obtained, which, although it contains occasionally Productus tenuistriatus, &c., could probably be quarried in blocks of moderate size, sufficiently uniform in texture for the use of the artist.

On the same section a deposit of fibrous gypsum and selenite, and a strong chalybeate spring are owned by Dr. Delamater. Fragments of zinc-blende were also shown us from Mr. Phillip Baker's land, three miles from Indian springs.

Some of the fire-clays of Martin county are manufactured into stoneware by Mr. Stookey.

The timber adjoining Dover Hill is Chestnut Oak, Tulip-tree, Hickory, White and Black Walnut, Sycamore, with an undergrowth of Persimmon, Papaw and Redbud. There are also some Sugar and Beech trees, the latter exhibiting on their bark frequent cicatrices of the scratched furrows made in the ascent by bears, which used to be very numerous in this county.

The well-known and valuable watering places distinguished as the "Indian Springs," owned by Mr. Donahue, and the "Trinity Springs," by Dr. Dunn, are situated in Martin county.

At the former locality there is a white-sulphur and a stronger black-

IRON, COAL AND PAINT BANKS, NEAR DOVER HILL, MARTIN COUNTY

sulphur spring; with heavy deposits of sulphur and some of the characteristic odor of sulphureted hydrogen gas. However there does not seem to be enough of this to blacken the paint of white lead, on the wood-work around, as always occurs at Blue-Lick Springs and Drennon Springs, Ky. When transported, in closely corked bottles, to the Laboratory, the water of the white sulphur appeared, on treatment with solution of arsenic, to give:

Of sulphureted hydrogen, only... 0.00031
There was also a small amount of carbonic acid.
The specific gravity, water being 1,000, is............................. 1.00014
200 grammes dried at 300 F., gave of sold matter................. 0.65
And, after ignition, gave of solid matter............................ 0.60

A quantitative analysis, which was made, showed some alumina and magnesia, and notable quantities of lime, potash and soda, particularly the latter, some of them as chlorides and sulphates; but, before recording the exact figures, it is thought best to repeat the analysis of this and to compare with the black sulphur, as also with the water of Trinity Springs, when time permits.

The latter, as indicated by the name, issue in *three* fine streams just below the town.

This mineral water seems to be the congenial element for peculiar confervæ, which in their turn furnish a suitable habitat for numerous infusorial animals, whose specific forms it would be highly interesting to examine microscopically.

Near Indian Springs a remarkable white Magnesian mineral, which cuts readily with a knife, and resembles the meerschaum used for pipes, deserves an accurate quantitative analysis.

Altogether this county, although somewhat rugged and broken for easy farming, is replete with mineral wealth of various kinds, well worthy of being developed.

Daviess county ships large quantities of coal and has even sent it to the St. Louis market. Besides the enterprising firm of Messrs. J. B. Legg & Co., Messrs. Church, Raymond and Tranter are working their respective banks somewhat extensively; and, with increased demand, can readily produce greater supplies.

A sample of coal from Mr. Church's bank gave:

Volatile matter 44.0	gas	39.0
	water	5.0
Coke 56.0	carbon	53.0
	ashes	3.0

100.0

While examining Mr. Nelson Jackson's coal, on Teal creek, we observed a good many water snipes (Totanus Semipalmatus?); and Birch trees (Betula nigra) are not uncommon; but White Oak prevails on the eastern sub-division.

There is a considerable amount of prairie in this county, particularly in its north-western portion.

KNOX AND DUBOIS COUNTIES.

Hon. J. D. Williams, Vice-President of the State Board, remarks on his District, that wheat and corn are the staple products, clay predominating in part of the soil; sand near the river; Oak, Poplar and Walnut are the principal timber; chills and fever and pneumonia the prevailing diseases. Hog cholera visited portions of the District in the spring and summer of 1858.

The sandstone of his District was used in the construction of piers for the railroad bridge, where the Ohio and Mississippi road crosses the west branch of White river, and in similar constructions. Lime is brought in flat-boats out of Lost river.

The good coal near Wheatland, described by Mr. Lesquereux as No. 4, is owned by Mr. Ashcraft, and is conveniently situated on White river, besides being reached by a switch from the Ohio and Mississippi railroad. It makes excellent coke and is extensively shipped.

Near Edwardsport, on a different exploration, we had an opportunity to examine a coal bank belonging to Mr. Thomas Curry, the bed three feet thick, parts being of clear, black bituminous coal, a portion rather earthy, with some sulphuret of iron, fire-clay underneath, and ferruginous sandstone forming the roof shales. Near this place Messrs. J. and B. Hargas, the steam-mill company, have, on section 1, township 4 north, range 8 west, the same coal, and a third opening below town on Mr. J. R. Haddam's place, although forty feet lower, seems from the quality and roof shales to be the same seam. These three openings supply the town, and as yet no coal is exported.

Two miles below Edwardsport the coal shows itself in the bed of White river; and is also worked at Apraw.

Through the kindness of Mr. Peck, of Vincennês, we obtained a sample of lead ore of rich quality from this county; but as yet there seems no certainty as to its extent. Mr. James Dick found it on his farm near Dicksburg, two miles below the railroad bridge across White river, probably in a quaternary deposit.

In the south-west portion of this county, much of which is prairie, there is, between the Wabash and White river, a Cypress swamp, embracing a township, or over 17,000 acres, probably the most northerly latitude to which the American Bald Cypress (Toxodium distichum) extends itself. This valuable timber is cut for cask-staves, shingles, &c. The swamp land commissioners have cut a wide ditch, about four miles long, from Deshee creek to White river, with a view of draing these lands.

Of the Dubois county coal several analyses were made and are subjoined.

One gramme of the Portersville coal gave:

Volatile matter	45.0	gas	39.0
		water	6.0
Coke	55.0	carbon	50.0
		ashes	5.0
			100.0

Coke swelled considerable; ashes light grey.

Two sub-conglomerate coals which I had an opportunity to examine in this county, near Celestine, were analyzed. That of Mr. Hawhee, from a seam two feet thick, worked two miles east of Celestine, gave, from a gramme:

Volatile matter	44.0	gas	39.0
		water	5.0
Coke	56.0	carbon	53.0
		ashes	3.0
			100.0

Coke swelled a little; ashes grey.

One gramme from the bank of Mr. Samuel H. Jacobs, worked by stripping, one mile east of Celestine, coal two feet thick, gave:

Volatile matter	42.0	gas	34.0
		water	8.0
Coke	58.0	carbon	55.0
		ash	3.0
			100.0

Coke swelled somewhat; ashes light grey.

The coal described by Mr. Lesquereux as being one and a half miles east of Portersville is worked by Mr. Ralph Atkinson. Milk-sickness has prevailed in portions of this county, especially near White river, to a considerable extent. The information obtained will be detailed under that head. Mr. Wade, living near Jasper, (3½ miles south) has indications of iron ore on his farm; also Mr. Edmondson, 14 miles N. N. E. of town.

In this county they have limestone enough to supply kilns, and the eastern portion furnishes sandstone for building purposes from the Millstone Grit.

PIKE AND GIBSON COUNTIES.

Of the coals, extensively worked in the former county, we had only an opportunity to analyze two; that from Mr. Hughes' mine, near Parkersburg, gave, from a gramme:

Volatile matter	44.0	gas	38.0
		water	6.0
Coke	56.0	carbon	51.0
		ash	5.0
			100.0

Swelled somewhat in coking; ashes dark steel grey.

Besides the banks of Dr. Posey, seven feet thick, and of Mr. Fechlin, near Kinderhook, which ship large quantities, the mine owned by the Rhodes family, close to the canal, has furnished flat-boat loads of coal pretty regularly out of White river, for sale, to the Wabash towns. I can testify that New Harmony, Posey county, purchased many thousand bushels, as much as twenty-five years since, for steam-mill purposes, from that source, and the same bank still supplies this town with most of that fuel; a few blacksmiths using the Pittsburg in part.

The coal from Mr. Rhodes bank gave, on analysis of one gramme:

Volatile matter	44.4	gas	30.1
		water	14.3
Coke	55.6	carbon	54.9
		ashes	0.7
			100.0

It puffed out slightly in coking; ashes reddish grey.

Gibson county is not yet working any openings extensively, except perhaps at Dongola on the Potoka, although a few families and blacksmiths supply themselves at several points. About a mile below the town of Potoka, at the Flouring Mill, I obtained the following section:

	FEET.
Quaternary soils, marls and clays	20–30
Aluminous shales	40–50
Impure limestone	2
Black shales	5
Coal	1
Fire-clay	8
Talus	5
Bed of Potoka river	0

The talus was supposed to cover up a shaly sandstone, as that rock was found at the same level a few miles distance on White river, overlaid by a thin stratum of coal.

Beech timber, Eupatorium perfoliatum, and milk-sickness can be found, not unfrequently, in the White river flats near the junction of Knox and Gibson. The southern part of the county has numerous fine farms, generally with substantial barns and houses; the Fair-Grounds are commodious, fairs well attended and the county noted for good stock.

PERRY AND SPENCER COUNTIES.

The chief locality in Perry, from which coal is extensively shipped, is Cannelton, on the Ohio river, noted also for its large and flourishing Cotton Factory.

Some of the hills back of town have been tunneled in various directions, and a good article obtained, often having the properties of cannel coal. In these mines they have encountered a few horsebacks, but have had no serious difficulty. The dip in a south-west direction is usually 17 or 18 feet in the mile. The supplies furnished by these coal mines sell rapidly at the wharf to passing steamboats. An analysis of one specimen from this place gave:

Volatile matter................................ 47.0	{ gas	39.0
	{ water	8.0
Coke... 53.0	{ carbon	50.0
	{ ashes	3.0
		100.0

Coke swelled somewhat; ashes light grey.

From a box full, sent through the attention of Hon. Hamilton Smith, ten pounds of this coal, tested for oily products, furnished these results:

POUNDS.
Ammoniacal liquor... 0.600
Crude oil 2.400
Coke.. 6.138
   Or, estimated by per centage:
Ammoniacal liquor... 6.00
Crude oil 24.00
Coke.. 71.38
Volatile matter and loss... 8.62
                                                           100.00

This is equal to sixty gallons of crude oil in a ton of 2,000 pounds. The Breckenridge coal averages about seventy-five gallons to the ton. Other coals of less importance, because further from market, are found in Perry county, on the land of widow Alvis, that of Mr. Mauk, of Mr. Van Winkle, &c.

Probably as heavy a deposit of iron ore as any in Indiana is to be found near Leopold, in Perry county, as it occupies a bed from four to eight feet or more in thickness, (judging from partial excavations,) extending through the greater portion of two hills near the town, and forty to sixty feet above it, besides being found in beds of about the same level in other hills north-west and south-east, from the first, at from one to two miles distance, as well as on section 11, township 4 south, range 2 west; also at Mr. Jackson Williams' place, one mile west of Troy, on Mr. Abraham Lusher's farm, and other localities in the county.

The one is, at Leopold, associated with the higher ferruginous sandstones, about 290 feet above the upper member of the sub-carboniferous limestone found at no great distance on Oil creek,* the bed of which, at that crossing is about 330 feet below Leopold. As this stream during the winter is navigable, if it were found profitable to work the ore, the metal could be readily transported on the creek six miles to the Ohio river.

The ore, on analysis of one-tenth of a gramme,

Lost by drying	0.0080
Gave of protoxide of iron (Fe. O.) only	a trace
Of insoluble silicates	0.0160
Of lesquioxide or peroxide of iron, ($Fe_2 O_3$)	0.0695
Alumina	0.0030
Lime	a trace
Magnesia, alkalies and loss, not separately determined	0.0035
	0.1000

This ore therefore contains 48.6 per cent. of iron.

Near Cannelton, as well as in other parts of the coal field, sandstone is quarried for building purposes. Hon. Hamilton Smith, wishing to introduce it for extended sale, asked an official opinion, which I furnished, as fully as a general examination, without applying special tests, would permit, in these words:

NEW HARMONY, IND., March 10, 1861.

Hon. Hamilton Smith, Cannelton, Perry County:

DEAR SIR:—In reply to your favor of the 4th inst., I offer the following remarks on the strength and durability of sandstones generally as a

---

*This stream derives its name from the oily scum frequently found floating on its surface, in a manner similar to that which first attracted attention in Ohio and elsewhere, and led to boring for oil, now so profitably practised. As these borings are usually successful, near the margin of the coal basin reaching the reservoirs of oil commonly in the carboniferous conglomerate, it is highly probably that borings might be remunerative in portions of Perry county.

building material, also some observations on the particular variety you propose introducing into market.

Durability.—According to Prof. Mahan, in his excellent work on Engineering, used at the United States Military Academy, those stones which are fine-grained, absorb least water, are of greatest specific gravity, and most free from potash, clay, iron, and similar chemical combinations; "are also most durable under ordinary exposure."

The absorption of water by some fine-grained sandstones is almost unappreciable after one day's immersion, and the annual wear by some experiments, is about one-tenth of an inch. The weight of a cubic foot of sandstone varies from 144 to 158 pounds, Cornish granite 172 pounds; the specific gravity of sandstones is from about 2.23 to 2.53, while granite is 2.6.

Some of the oldest buildings and bridges of Europe are constructed of sandstone, especially in Scotland.

Strength of Materials.—Although stone generally will sustain much less tensile and transverse force than good timber, yet it will bear from three to six times as much as brick, where the object is to resist a great *crushing force*, as in the walls of buildings or the piers of bridges, during a long period; it is certainly a most valuable material, *if well selected*. Each superficial inch of granite will sustain a crushing weight of from 2.8 to 4.7 tons; sandstone from 1.40 to 3.94; the latter being a sandstone of the Coal Measures.

Varieties of Sandstone.—It is evident from the above that there is considerable difference in the quality of sandstones. The freestone of Edinburgh, Scotland, which has stood for centuries unimpaired in buildings and bridges, is from the Coal Measures. And I may add a large granary, erected at New Harmony forty-five years since by the Germans, is from the higher series of the Coal Measures. It seems as substantial as the first day, except at one place where some salted meat by being piled against it, caused some scaling and crumbling. A sandstone from near Fredonia, employed in the Louisville gas works, is from just beneath the Coal Measures; and the carboniferous sandstone forming the foundation of a warehouse, which I observed at Williamsport, Warren county, Indiana, exposed alternately to high and low water of the Wabash, as well as to freezing and thawing, exhibited scarcely any perceptible scaling or abrasion, after about thirty years. It may be proper to remark that sandstones are not so suitable for roads, pavements, or steps much used, because their resistance to abrasion from friction is one-fifth that of marble and about one-sixteenth that of granite; but the comparative ease with which freestones are dressed will recommend them even for some of these purposes, if the liability to friction is not very great.

Although I have not examined the sandstones of Cannelton with special reference to these points, the theoretical indications are decidedly in its favor, as it is from near the base of the Coal Measures.

Yours respectfully,

[Signed,]   RICHARD OWEN.

The late State Geologist engaged the valuable services of Mr. J. Lesley to make a lithographical survey of part of Indiana, in order to exhibit the advantages of such surveys for those parts of an entire State, which from the presence of coal beds, or other valuable minerals, demand a more detailed acquaintance with their relative levels. Perry county being of this character was selected, and the beautiful map executed by that accomplished Topographical Geologist is now framed and suspended in the Geological Room of our State Capitol. On a somewhat reduced scale, these could be photolithographed, when several thousand impressions are desired for about twenty dollars per thousand. Mr. Lesley's report, and accompanying estimate, shows the nature and approximate average cost of such work, when applied to any desired region.

Perry county has been noted during many years for the Troy Pottery, mentioned by the late State Geologist as having been established by Mr. McClure and others, from Staffordshire, England. It is now owned by Messrs. Sanders' Bros., who manufacture an earthenware, usually known as the Rockingham and Yellow Ware, into such articles of kitchen use, as fruit jars, pitchers of tasteful patterns, besides spittoons, &c. The ware is stated by them not to be affected by acids, the glazing being composed chiefly of white lead, borax, sand and common clay; still it is always best not to leave canned fruit, or any other acid article of food, in tinware or earthenware, except with the salt glaze or felspathic enamel, exposed for a considerable time to the action of the atmosphere. At first they made a white ware; but found it would not justify them under the circumstances. They also manufacture good fire-brick at five dollars per thousand, grinding up a portion of the old saggers with the fire-clay, and sometimes portions of sandstone.

If the demand for roofing and draining tiles, or for paving brick, justified them, they could at this locality, from the same potters' materials, with a vitrifying glaze, make a hard tile brick quite impervious to water; at a cost for the roofing tile of twelve dollars a thousand.

The coal seams show themselves here, one exhibiting only its roof shales at low stage of water in the Ohio river, the second, from the fire-clay of which they manufacture their saggers, is below high water mark, and the third or highest having only a few inches of coal, but whose fire-clay here often occupies ten feet, furnishes the potters' main material for earthenware.

In this thriving county, a city, with several thousand inhabitants, and important manufactures, but which we seek in vain even on late maps, has been raised in a few years from the forest wilderness, through the industry and enterprise of a Swiss population, who have named it "Tell-City."

In some portions of this county milk-sickness still occasionally visits the inhabitants, unless they keep up their cattle on tame pastures.

The Spencer county coals have been minutely described by Mr. Lesquereux. The analysis of a gramme from Mr. Robert Woods' coal bank, one mile west of Elizabeth, on section 19, township 4 south, range 5 west, gave:

Volatile matter	48.5	gas	45.00
		water	3.50
Coke	51.5	carbon	48.00
		ashes	3.50
			100.00

Coal is probably most extensively worked at present by the Messrs. McGrail, on Mr. B. Shrodes land, near Rockport, and by Mr. R. Woods; Mr. Lewelyn Jones sends from his bank, on section 25, township 5 south, range 7 west, coal for blacksmiths' use in Gentryville. The section given by Mr. Lesquereux as furnished by me, was obtained through the politeness of Mr. Lewis G. Smith, formerly interested with the company who undertook the boring in search of coal.

The Taylorsville coal could not be visited, at the time the corps was near there, in consequence of the Ohio being out of its banks.

At Maxwell, on another occasion, near the line of Perry, a bank was seen which had been opened close to the river.

The county is also abundantly supplied with sandstone, furnished chiefly from bluffs similar to that seen at Rockport, and which has re-

R. D. Owen, Del.

LADY WASHINGTON ROCK, AT ROCKPORT.—(COAL MEASURES.)

ceived the name of "Lady Washington Rock." The subjoined engraving of that scene is from a sketch made by the late State Geologist, some years since.

WARRICK AND VANDERBUGH COUNTIES.

Besides the coal banks of Messrs. Bethel Bros., under the charge of Mr. Hutchison, as mentioned in Mr. Lesquereux's report, there are also coal mines belonging to Messrs. Bethel Bros., near Newburg, furnishing coal for sale on the Ohio river; the names of the other proprietors around there we regret not to have obtained. Limestone is also quarried about Newburg.

Near Millersburg, Mr. Isaac C. Miller, on section 11, township 5 south, range 9 west, has reached a twin vein, now worked by Messrs. Young and Whetstone, from which they sell extensively and could readily ship annually 100,000 bushels.

Mr. George C. Hart strips a bank on his fine farm, in section 25, township 5 south, range 8 west, from a four foot two inch seam, from which he sells abundantly in Boonville, and could furnish the same annual product above stated. So also could Mr. John A. Reynolds, who from beds four feet six inches to five feet ten inches, on section 6, township 6 south, range 7 west, and on section 1, township 6 south, range 8 west, wagons coal to the same town, selling at ten cents per bushel. At some of these openings coal No. 9 is worked; at others No. 9 seems almost united with No. 11.

Some milk-sickness is reported in this county.

In Vanderburgh county the chief coal, mined for sale, (although some shows itself near Judge Silas Stevens', on "Pigeon,") is that raised from the Bodiam shaft, one hundred and seventy feet, by the enterprise of a company at Evansville, under the able superintendence of Mr. Wm. Kesterman, to whose politeness we are indebted for the section credited by Mr. Lesquereux to myself.

In the vicinity of this city, the industrious Germans have in many cases planted vineyards and manufactured considerable quantities of wine, thus often in a few years quadrupling the value of hill-land, too broken for the most profitable arable farming.

Orchards are also abundant in this county, and although little or no Beech and Sugar-Tree grow, they have abundance of good timber, such as Oak, (white, shingle, red and black,) Elm, Hickory, Sassafras, Catalpa and Sweet Gum.

This county also furnishes lime for use and exportation, and quarries a considerable amount of sandstone. Occasional cases of milk-sickness occur in portions.

POSEY COUNTY.

Although, as remarked by Mr. Lesquereux, but little coal is worked in Posey, except annually a few bushels for blacksmiths' use, from the bank of Hon. J. C. Pitts, near Springfield, and one or two others, yet, when the demand justifies a shaft, doubtless coal can be readily reached.

Already, near West Franklin, Mr. Priest has reached coal of which he sent a sample for analysis. No doubt earthy matter from other portions of the boring have mingled with the coal, or the per centage of ashes in this, No. 9, coal would not be so high. One gramme gave:

Volatile matter	39.0	gas	36.0
		water	3.0
Coke	61.0	carbon	42.0
		ashes	19.0
			100.0

Beneath the sandstone of the cut-off, near New Harmony, two thin beds of coal have been observed recently, one only having been formerly noted.

On this sandstone, which at places is shaly, reposes the vast bed of quaternary marl, newer than the Drift or Erratic quaternary, older than the alluvial gravel and other quaternary of the Ohio and Wabash.

This bed, in some places twelve to fifteen feet or more thick, and in which land fresh water shells of the genera Helix, Helicina, Paludina, Planorbis and others are found, and which will be specifically enumerated in the general list, furnishes a valuable fertilizer, the analysis of which was given by the late State Geologist in his first report. It seems of lacustrine origin, or at least extends for miles back from the Wabash into Illinois, and into Posey county, covered with the later quaternary clays and subsoils. At its junction with the second river bottom of gravel and alluvial deposit it sometimes seems to rest upon it; but this is deceptive, existing only at the junction, where it has washed on to the later deposit. The marl bed is probably of the same age as the "loess" of the Rhine, and the circumstances of its formation will be hereafter discussed.

In the "second bottom," near New Harmony, in digging a well for a steam-mill, branches of dicotyledonous wood resembling cotton wood were taken out at about fifty feet under the gravel level. Still a few feet beneath this "second bottom" we find high water mark, and a "first bottom" enriched almost annually by the deposits of mud left after the subsidence of the Wabash in its Nile-like overflows.

A section at the cut-off Ferry, where the river has cut through part of the rock, exhibits the following:

	FEET.
Quaternary soil, subsoil and clay	50–100
Bed of marl	6–8
Shaly sandstone	20–25
A few inches of upper coal	½
A few inches of the lower coal, (No. 13,) with fire-clay under	½
Penciled aluminous shales of Dr. D. D. Owen, section	20
Low water in Wabash	0

In digging a well a short distance back from the cut-off, the workman, an intelligent observer who had been in California, reported the following: After passing through two or three feet of soil and subsoil, he encountered clay and marl, until at twenty-one feet below the surface, he found eight to twelve inches of a coal seam, (samples from which were sent to the laboratory,) with two feet of fire-clay under the coal, and five feet of common clay to the bottom of the well, viz.: twenty-eight feet from the surface. This, from concomitant evidence, would appear to be in place, and may be coal No. 12 covered, after denudation of the materials of secondary age, by twenty-eight feet of quaternary alluvium.

In this county there is some milk-sickness, on Big creek, the details of which will be given under that special head. On the same creek, limestone is found which is burned for lime; and at West Franklin Mr. Febre has a kiln, holding 400 barrels, which he fills every two weeks, shipping the lime on the Ohio river.

Mr. Charles Fitch owns a quarry one mile and a half below Mt. Vernon, the county seat, from which building materials and lime can be obtained advantageously. The sandstone of the Cut-off is used extensively for foundations, and an associated limestone, showing itself near the dam, has interesting fossils, which have been described by Dr. Norwood, Messrs. Cox, Pratten, &c.

*SEC. VII.—COUNTIES IN THE DRIFT OR ERRATIC QUATERNARY.

Sub-Section 1.—General Description.—Most of the beautiful and productive counties north of the Wabash are so covered by the great northern Drift as to be included in this category; and indeed many counties south of the Wabash have from 50 to 60 feet of quaternary deposits overlying the rocks. Still, as in the latter counties, the larger streams cut through the drift and permit an examination of the underlying rocks, we rank them accordingly. In some of these northern counties rocks have been rendered visible at various points; but as these are few and detached, although they will be noted and their stratigraphy assigned, yet as a whole it was thought best to embrace under the head of Quaternary counties the following: Steuben, LaGrange, Elkhart, St. Joseph, LaPorte, Porter, Lake, DeKalb, Noble, Kosciusko, Marshall, Starke, Jasper, Newton, Allen, Whitley, Fulton, Pulaski, White and Benton.

Sub-Section 2.—Resulting Soils, &c.—From the diversity of material brought down, during this period, and scattered as it were broadcast over the secondary rocks or the detrital remains, the soil in the Drift counties is usually remarkably fertile. As a general rule, from the quantity of argillaceous shales and disintegrating bowlders in which alumina is prominent, as well as from the latitude in Indiana suiting those crops, wheat, rye, timothy, clover and potatoes form staple articles. Corn is likewise grown, but not so extensively as in the more arenaceous Wabash bottoms, also oats and Sorghum. A superior kind of sugar-kettle has been patented in St. Joseph county, by Mr. Miller, and success attends their granulation of the saccharine juices from this variety of corn. The tops also, from the same variety, are reported as making better brooms than those manufactured from the broom-corn, while the refuse is more acceptable to cattle. Some animals of the finest stock I have seen in the State were raised on the prairies of these northern counties.

---

*This should properly be Sec. 5, as it was originally designed that the five geological systems should constitute five sections, which would have made the Carboniferous, as in the index, the 4th section, and the Sub-Carboniferous Sandstone Series was designed as a sub-division, and marked Sec. 4^1; so also the Sub-Carboniferous Limestone Series should have been Sec. 4^2, and the Coal Measures Sec. 4^3. But as this was inadvertently overlooked in reading the proof, the reader will please make the correction to correspond with the index.

Portions of the ridges are sandy, while into the low prairies has been washing for centuries, decomposing vegetable matter, muck or humus, sometimes extending many feet below the surface, and furnishing the righest soil when sufficiently drained.

SUB-SECTION 3.—ROCK QUARRIES, &c.—Rock quarries are not abundant throughout this formation, still in several counties they will be found noted, particularly in Allen, White and Jasper.

The immense deposits of marl, sometimes replete with shells chiefly of the genera physa, planorbis, cyclas and unio, sometimes a clay marl, particularly in St. Joseph, LaPorte, Porter and Lake, are of great commercial and agricultural value, as well as for burning into lime, as for the fertilizing of the soil; but more particularly for the manufacture of artificial stone and brick; provided that enterprise, so successfully commenced, should extend itself as it promises. On this subject I take the liberty of making the following extracts from a private communication obligingly addressed to me by Dr. John W. Young, in reply to enquiries which I made on the subject, not doubting that this will prove interesting to many :

BLOOMINGTON, Jan. 9th, 1860.

"*Dr. R. Owen, New Harmony:*

"DEAR SIR: * * * You wish to be informed of the address of the company about organizing for the manufacture of artificial marble-blocks, brick, &c. The gentleman at the head of the movement is Mr. Clark D. Page, Rochester, New York. The patentee lives at Charleston, South Carolina, a physician, and quite a scientific man. I have written to Mr. Page informing him, as far as I could, from what you were kind enough to tell me of the fine *marl bed* at South Bend. I think it quite likely that he will visit the place, and possibly, in connection with others, may organize a company to manufacture the marl into brick, marble-blocks, ornamental window tops and sills, mantel-pieces, tombstones, &c., &c., whatever marble is used for. South Bend would be a good point from which to reach Chicago and Detroit, and with a little effort a mammoth concern might be built up in a few years. I have requested Mr. Page to send you a brick to Indianapolis,[*] care of Governor Willard.

For an ordinary sized brick, the manner of preparing is to take one

---

[*] Unfortunately this specimen for the State collection never reached its destination, so far as I have been able to ascertain.—Note by R. Owen.

part good roached lime and nine of marl. Slack the lime and add sufficient to make a very thick liquid; mix this with the marl, and mould in the ordinary moulds; let it dry apparently well on the surface, and press in steel moulds with a pressure of about eight tons.

For ornamental work use about one-sixth lime and press, using a plunger that will give any pressure you may desire.

When the lime and marl are good and white the effect is beautiful beyond description. The article pressed, be it brick or ornamental work, will present the same finish the mould and plunger possess; in some cases looking very much like finely gloosed marble. This, I think, could easily be clouded as might be desired with the proper coloring matter imperfectly worked into the mortar.

Brick made in this way are extremely hard, requiring nearly as much force to break them as would be necessary were they made of "pig metal." Besides they are entirely impervious to water or moisture, and can be made and sold as low as the Milwaukee brick, (or even lower,) which, I believe, cost $25.00 per thousand.

Mr. Page has invented a lime-kiln, which he has patented for burning *marl*. The best lime used with the marl is that made from the marl itself. * * Mr. Page has also patented a lime-kiln for burning lime and cement. * * The kilns are a great improvement over the old method of burning, in fact they are mammoth crucibles and will disengage in the most perfect manner the carbonic acid, thereby increasing the value of the lime or cement twenty per cent., so says the Chief Engineer of the United States. * * * *

Respectfully,

(Signed,)   JOHN W. YOUNG.

The deposits of clay for the manufacture of various articles, as well as of that peculiar variety of the buff-brick, sold abundantly at Milwaukee, will be described in the details of counties.

SUB-SECTION 4.—METALLIC ORES, &c.—With the progress of railroads and consequent demand for railroad iron, as well as the increased development of our domestic manufactures, especially in Rolling-Mills, Foundries, and the like, there will doubtless arise an increase in the demand for iron, after its first extraction from the ore, and of a concomitant desire to find large bodies of easily-worked and productive iron ore. Under these circumstances, the immense value of the great northern iron deposits, which actually grow, in the low grounds, by accretion or external accession of parts, can scarcely be overestimated.

It is not asserted that foreign materials are converted and assimilated as nourishment, but that year by year a sensible growth or addition is made by accretion, or by deposition of similar solid particles around the solid nucleus of hydrated brown oxide of iron, from the highly charged chalybeate waters of many of those regions, filtering through fragmentary, conglomerated, ferruginous materials, brought, in the erratic period, perhaps from a distant matrix, and arrested in their porous percolation, by an impervious substratum, to reconsolidate into the most important metallic ore yet offered to the skill and industry of man.

The numerous localities at which this so-called Bog-Iron ore are found, will be given in detail in describing the counties; but it may be well here to state that with a suitable iron bar, should the survey be continued, several acres of these marshy lands could be probed by an industrious laborer in one day; the records of which could well repay the time and means expended thus to examine the whole region supposed to have profitable deposits of this ore.

Occasionally other metallic depoits are found in the Drift, as copper, lead, silver and gold, but as these are usually detached fragments or at best have accumulated in small pockets formed by the finer materials sifting into rock fissures, &c., we can scarcely rely upon finding deposits except accidentally, not remunerative as a steady object of search, as in the case of iron ore.

SUB-SECTION 5.—TIMBER AND PREDOMINANT VEGETATION.—A considsiderable portion of these northern counties is treeless, and, especially south of Lake Michigan, nearly through our State, but at least to the upper Wabash, we find extensive prairies, connecting with those of Illinois, generally low and sometimes wet; while higher and drier tracts, commonly destitute of timber, with, however, occasional groves, occupy a position to the east and west of the low prairies.

In this boundless expanse, this ocean-like land, level sometimes as a floor, with perhaps no path to guide the traveler and scarcely any two objects which by comparison can enable him to estimate distances, nature has provided for the brave denizen of these American "Steppes" a diurnal polar star, a directive sign, like the moss on the north side of trees to the backwoodsman, or almost like the compass to the wanderer on the trackless sea. A plant of the composite family grows abundantly in the prairies, with its thick, dry, resinuous leaves, all flattened to one plane, as if fresh from the pressure of a herbarium, surmounted by a gay, yellow, asteriod flower; and this plant, Silphium lacineatum, or rosin weed, even at its earliest exit from the soil, and ever afterwards, in its developments, ranges this broad foliacious plane due north and south, thus presenting one face of the leaf east, the other west. Instead of an upper side covered with nature's varnish for protection, an under side presenting the breathing stomata of most leaf-bearing vegetation, these leaves are nearly the same on both sides, rough and resinous. To this peculiarity of ranging its leaf-plane north and south it owes the name of compass-plant, and to its highly resinous composition the name of rosin-weed. Another plant common on these prairies of the same genus, S. terebinthinaceum or Prairie-dock, which has also an inclination to range its leaves north and south, but not so uniformly to be depended upon, contributes with the compass-plant, artemisia and others, to give to the prairie fires the remarkably dense volumes of black smoke, which are matters of astonishment until we trace them to these vast reservoirs of inflammable resins, furnishing the carbon more rapidly than it can combine with the oxygen of the atmosphere.

Other peculiar vegetation of the prairies will be alluded to in the details of counties, such as the Cattail-flag, various grasses, usually rather coarse; wild indigo and ferns often skirting the edges, with sometimes an undergrowth attempt at timber in the form of willows, aspens, and the like.

In some northern counties there is abundance of fine timber, especially White Oak, with some Beech and Sugar-Tree; and, towards the lakes, Cedars, Pines and Tameracks, (Larch or Hackmatack,) and Alders, the interspaces dotted beneath by such quantities of a genus from the Heath family, as to require a special train at the gathering season, under the names of Huckleberry (or Whortleberry) train; while another genus of the same family, the Cranberry, furnishes, from otherwise useless swamps, the palatable relish to heighten the savory flesh of the native buffalo, deer or pinnated grouse, (prairie-hen,) which formerly enlivened these vast plains or still rush and whir through the prairie.

Sub-Section 6.—Springs, &c.—At numerous localities chalybeate springs were seen, and at more than one place encrusting springs. One of these, so highly charged with carbonate of lime, held for a time in solution probably by an excess of carbonic acid, and under certain circumstances parting with the gas, and depositing the calcareous incrustation in such quantity as even to encase human bodies buried within its influences, will be found described among the details of Porter county.

At most places on the Indiana prairies water can be obtained by digging a moderate distance, as, although sometimes clay with little intermission extends from one to two hundred feet deep, usually beds of sand are found, after penetrating which, a supply of water is often obtained, it being arrested by an impervious substratum.

SUB-SECTION 7.—MISCELLANEOUS FACTS.—Among other interesting facts, it was gratifying to learn that this northern region, whether prairie or timber, so far as we could hear, is not visited by milk-sickness.

The formation and drainage of prairies, the phenomenon of acres of dead shells being sometimes found on the surface of these natural meadows, the examination how far the craw-fish and the ants may contribute to elevate some prairies, the improvement of the swamp-muck "swales" and the dry-sand ridges, with various similar descriptions will be found alluded to under the physical geography of those regions.

SUB-SECTION 8.—CHARACTERISTIC FOSSILS OF THE QUATERNARY.—In the Drift proper, or old quaternary accumulation, resulting from the erratic bowlders, gravel, clay, &c., brought from the north by whatever agency may have been employed, we seldom find fossils which have been evidently derived from the older strata in our State, but chiefly petrifactions of the secondary period; though from the more local middle and later Quaternary we obtain in this country abundant specimens of the Mammoth and Mastodon, species of which are found in the European Tertiary strata, also in beds probably contemporaneous with the loess deposits along the Rhine, we find at least one megatheroid animal resembling the giant half armadillo, half sloth of the South American Pampas-plains. These fossil remains of the Ohio river loess have thus far shown themselves on the Kentucky side, particularly near Henderson; but as an exactly similar deposit, characterized by the same shells, is found on the opposite shore in Vanderburgh and Posey counties, it is quite probable further search may reveal them in Indiana; where, as already remarked, mammalian remains of the proboscidean family, comprising the Mammoth (Elephas primigenius,) and Mastodon have been found.

These and the shells, &c., will be enumerated in the tabulated list of Indiana fossils.

SUB-SECTION 9.—COUNTIES IN THE QUATERNARY FORMATION:—

STEUBEN, LA GRANGE AND ELKHART COUNTIES.

Steuben is one of the counties which we failed to reach, and which remains to be examined; it is represented as resembling in many respects LaGrange and Elkhart; well watered by streams and lakes, and having good timber; in other parts there is prairie with occasionally higher and more sandy ridges.

Most of the north part of LaGrange county comprises Oak openings, having some Bur Oak timber, with no underbrush, and a yellow hard-pan at from two to six inches under the surface, beneath which they reach water in the so-called "water-gravel." The surface soil is a whitish sand and gravel, with a mixture of reddish loam. The southern portion of the county, especially about Hawpatch, is a black sandy soil and had originally an extensive growth of wild plums and haws, extending into the north part of Noble. Of LaGrange county we did not see as much as we desired, but would indicate the south-west part as having probably considerable deposits of bog iron ore, as well as the country about Lima.

In Elkhart county there are to be found on the river and at the town of the same name, flourishing flouring mills, also a paper mill. At Goshen, the capital, the water privilege is also fine and good mills abundant. One in the process of erection had its substantial cellar walls carried up of bowlders broken so as to give one flat face. These are seen at various places piled up for sale, as we serve wood, and selling for eight dollars a cord.

We were informed that some years since they obtained a coarse and somewhat soft sandstone grit, twelve miles west of Goshen, from which they manufactured grindstones. Five or six miles north-east of town, bog iron ore is obtained from the Cornell-Marsh; also from Mr. Storm's place, about two miles north of Goshen.

The drift here and elsewhere affords good gravel and sand for ballast, mortar, &c. They burn some of their marl into lime. The bridge across the Elkhart, at Goshen, is 600 feet long, on account of high water.

Along the valley of the head waters of the Elkhart we noticed some Beech, Sugar-Tree, Black and White Walnut, Cherry, Oak, (white, black and red,) of vigorous growth and ample dimensions.

Poplar (Tulip-tree) lumber, where we were in Goshen, was one dollar a hundred for inch stuff; firewood one dollar and fifty cents per cord. In order to compare produce in Northern Indiana with those in the Southern part of the State, we may here add that in October, 1859, the time of our visit, wheat was selling at a dollar, corn at twenty-five cents, potatoes the same, prairie grass five to six dollars per ton ; labor seventy-five cents to one dollar a day; brick four dollars per thousand.

At Millersburg, near the line of Elkhart and Noble, we observed piles of staves and lumber ready for the freight train.

ST. JOSEPH AND La PORTE COUNTIES.

Mr. Miller, of South Bend, a member of the State Board of Agriculture, after furnishing hospitable entertainment, conducted us to examine the resources of St. Joseph county.

They distinguished, besides the low or swamp-muck lands, three soils ; the first and best is the bur-oak and prairie land, producing as high as thirty bushels of wheat to the acre, also fine crops of potatoes. Secondly, the soil of the thick woods, not quite so rich; and thirdly, the black or scrub-oak and sandy woods' land, of rather inferior quality.

The low lands of the Kankakee, around South Bend, sell, for agricultural purposes, at from $30 to $50.

The muck or marsh is from three to twelve feet deep in organic matter, which in many places continues to increase by water, heat, rank vegetation, &c., with an occasional admixture of sand from below by the labors of craw-fish, frogs and ants.

In these swamps, frequently when ditching and draining, the great deposits of bog-iron ore are encountered, and their position often indicated by the appearance of the water around, also by bends in streams. The lumps, sometimes eighteen inches through, are chiefly at one or two feet below the surface. At Mishawaka, Mr. Hunt many years since established iron-works, which are now carried on by a highly experienced iron master, Mr. Niles. He considers this variety of ore among the best and easiest to work, next to the Vermont, superior even to that of the "Hanging Rock," Ohio, and the hematite of Tennessee. When prices justified they manufactured about 800 tons per annum, by using the marl for flux, instead of limestone, and charcoal for fuel. The average yield of the ore was about $33\frac{1}{2}$ per cent., although some gives as high as 50 to 60 per cent. They usually hauled their ore five or six miles, and as railroad companies increased the rate of freights, and at one time iron fell to $17.00 a ton, whereas now it commands, according to quality, $27.00 to $50.00, they discontinued the use of the bog ore.

Beneath the swamp-muck beds, in which these valuable deposits are found, a shell marl, three to ten feet thick, is obtained, in which are large and abundant specimens, some well preserved, of shells belonging to the genera physa, planorbis, cyclas and unio. At many places this is dug and moulded into brick-shaped masses of considerable size, so as to be readily piled in a kiln, burnt and used for all purposes to which lime is usually applied, being of an excellent quality and white color.

An extensive manufacture of this kind is carried on near the fine Catholic College of Notre Dame, beautifully situated a mile or two north of South Bend.

In the bed of the St. Joseph and in the banks is found the valuable blue clay, from which the celebrated buff brick are manufactured. The same clay is moulded into articles of pottery ware, including crocks for stove-pipes, &c. In places this is a hundred feet deep, according to Mr. Dahoff of this town, and, as will be seen, nearly two hundred in other localities. The same authority informed me that water is found usually at from thirty to fifty feet deep, after passing a few feet of dark, sandy loam, about four of yellowish or red sand, then grey sand and some gravel, in which frequently they penetrate from one to five thin beds (seven to eight inches thick) of "lime cakes" or "hard-pan," probably an indurated marl.

Chalybeate springs issue abundantly about high-water mark from the St. Joseph river, in the vicinity of South Bend.

In LaPorte county, especially around the Kankakee region, we saw abundance of bog iron ore, one average specimen of which afforded us on analysis of one-tenth of a gramme, in the laboratory, 63 per cent. of pure iron.

Regarding the Agricultural District embracing these two counties, as also Marshall and Elkhart, under the presiding care of Mr. Wm. Miller, that gentleman writes: "There is bog ore in great abundance in all the above counties; it has been worked at Mishawaka, in St. Joseph county, also in LaPorte county, and near Plymouth, in Marshall county. Bar-iron has been made in considerable quantities in St. Joseph and Marshall counties, and in Laporte castings have been made from the bog-ore." * "There is abundance of marl in all the northern counties of the State, which makes a superior quality of lime. The beds are frequently ten feet deep, composed of shells of various kinds." * "A grindstone quarry was worked in Elkhart county; but abandoned because the grit was found to be of inferior quality." * "Products—wheat, corn, oats, potatoes and beef." * "Soil, sandy loam." * * "Timber—Oak, Ash, Hickory, Maple, Poplar, Pine and all timber common to Indiana." * "No milk-sickness nor hog cholera. The potato rot and Hessian fly prevail to some extent. Fruit is much injured by the curculio and other insects." * "Iron, sulphur and other springs are abundant. Diseases mostly bilious."

While in this county, we visited, by desire of His Excellency, the late Gov. Willard, the site of the Northern State Prison, and examined pretty thoroughly the region around Michigan City.

We traveled, near Gosport Mills and Calumet, in the southern part of Porter, over alternations of low prairies with a black subsoil, and sand ridges; the former covered sometimes for acres with nothing but a waving level of yellow helianthoid flowers, again diversified by willows, aspens, flags, indigo, rag weed, Eupatorium, golden rod and hazel bushes, interspersing the predominant yellow with purple and pink, blue, white, and even black, (the indigo pod,) on a ground of green, enlived probably by a lake or pond bearing the large leaves of aquatic growth all over its surface; the sand ridges, sometimes a whitish loam, sometimes a drifting sand, bearing abundant ferns along the marginal junction with the prairie, dotted in its interior with Whortleberries, Alder-bushes, Sumach, Sassafras and Hickories, also large White Oak, Spanish Oak, and small Black Jack and Black Oak; most of which gradually gave way to a growth of tall Pines and Tameracks, as we neared the Lake, with still an abundant undergrowth of ferns, especially a variety with neuropteroid venation; having a nervures closely resembling that of its great carboniferous prototypes in the genus Neuropteris.

Passing, at the southern outskirts of the city, the Penitentiary, a large saw mill, and stave and shingle mill, we entered Michigan City and proceeded to Trail creek, which cuts through the great sand ridge; we found a small river, with fifteen to twenty feet of water near its mouth and wide enough for a moderate sized vessel to turn in, cutting through a sand drift which has blown up to form a ridge from 100 to 175 feet high. The surveyor's level made it 176 at one point, and we found it in some places only twenty feet wide on the top. It extends west, we were informed, to Indiana City, and some asserted to Chicago, so closely washed by the waves that the sand lately rolled down in an arenaceous avalanche, denominated "the Hoosier slide." Yet in the early settlement of the country between the lake-waters and this sand ridge the mail stage and other carriages were driven undisturbed by the lake waters, along the beach, from Michigan City to Chicago. The light house occupies part of the elevation and exhibits sixty feet above its summit a steady warning to vessels on the Lake. A beautiful "lookout" has also been erected by the liberality of Mr. Blair, a citizen of wealth and enterprise, who has also successfully drained several thousand acres of swamps, on plans hereafter described.

This drifting sand ridge, formerly covered with pines, is now chiefly overgrown with stunted Oaks, Alders, and the like.

In company with Col. Seely, the Warden, and Major Dunn, one of the Commissioners, we examined the site, plan, and surroundings of the Northern State Prison. The State owns one hundred acres of land, seventy of which is a bottom of rich, alluvial sandy loam, cultivated in corn, potatoes and the like; the thirty acres on more elevated ground being reserved for buildings, &c. The Prison walls were already, on Sept. 10, 1860, when we were there, laid up to grade, and enclosed about eight acres. The foundation for a cell-house to contain 300 prisoners was already laid, as also for a workshop 200 feet long. In a brick kiln, adjoining, half a million of brick were nearly ready for use, and, for three weeks previous to our visit, twenty car loads of substantial building rock had been arriving daily from Joliet, Illinois.

For the advantages to them of trade, the Michigan Central Railroad Company donated to our State, with the right of way for all time to come, a track costing a thousand dollars and connecting with their own. From this facility the blue clay for buff brick, as well as superposed clay for red brick, are brought on the track twelve miles from near Gosserts Mills, at a less cost, it is estimated, than would suffice to cart them from points much nearer at which they are also found. The Warden's buildings are to be on the east of the Prison, and a handsome pine grove skirts it on the west. One hundred and eighty prisoners were at that time employed in performing the principal labor, with the above large amount of constructive materials. A steam engine and fixtures were already on the ground.

With these gentlemen and others who were so obliging as to accompany us, we examined a saline sulphureted hydrogen spring, about three miles from town, belonging to Mr. James Walker; it readily blackens silver and deposits a thick sediment of sulphur. The large hollow gum sunk by the early settlers is much gnawed all around by the deer and other wild animals resorting to the attractive saline "Lick." A chalybeate spring flows out at a short distance from the sulphur water. Some Beech timber was noticed around this spring.

On returning towards town we passed a fine cool spring filtering through sand and flowing out over the blue clay. The water is so highly prized as to sell in town for one cent per bucket. About a quarter of a mile before reaching the Lake we pass a drifting sand bank, about 600 feet long, 200 feet wide, and averaging from 50 to 60 feet above the Lake level, already giving growth to dwarf oaks, asclepias and other vegetation. About 100 bushels of Huckleberries are shipped daily from here during the fruit season, requiring an extra freight train.

The city contains about 4,000 inhabitants, and if, without too much expense, the harbor could be kept free from sand bars, and the extensive piers or projecting wharf, for loading vessels of somewhat heavy draught and tonnage, were maintained at all times in thorough repair, the facilities here for cheap transportation would be of great value to the northern counties. As a matter of home interest it seems highly desirable that Indiana should maintain here or at some other point, if there be a better, along her Lake-coast, a harbor worthy of the State; otherwise her commerce is necessarily diverted to outlets in the adjoining States, the cost of transportation thereby increased to our citizens, and the profits of the carrying trade also lost to them. We were informed that wheat and similar articles could be shipped on the Lake for five cents, when by railroad the cost would be twenty; also, that pine lumber, iron, copper, &c., would be imported at lower rates to furnish raw materials and quicken domestic manufactures into greater energy.

When the winds prevail from the north they bring by their force, and the action of the agitated waters, quantities of sand out of this southern beach of the Lake, which material again sometimes washes out to form bars; but 1,000 feet from shore all the Lake bottom is said to be a hard clay; at points on the shore this clay shows itself nearly as hard as rock; at other places black sand was abundant, and a few rounded fragments of rock, chiefly granitic or arenaceous.

Going south towards the capital, LaPorte, we passed, at Gen. Orr's, on the north edge of Door Prairie, the height of land 306.5 feet by the railroad survey, above Lake Michigan, and 224 feet higher than LaPorte, others consider "Bald Point," near there, as even higher. Judge Lawson informed us that some of the lakes around that county seat are about 200 feet above the surface of Lake Michigan. Our barometrical observations, allowing for the daily meridian rise, made Stony

Lake somewhat over 200 feet above Lake Michigan. The periodical fluctuations, diurnal, secular, &c., which all these northern lakes undergo, have been made the subject of a paper by Col. Whittlesey, pulished under the auspices of the Smithsonian Institution. Some of the smaller lakes have no visible inlet, nor outlet.

The prairies, with Boneset, Indigo, Golden-rod, ferns, &c., interspersed with oak groves and "barrens," continued to Kankakee bridge, where our barometer made the river at least fifty feet higher than Lake Michigan.

PORTER AND LAKE COUNTIES.

In Porter county we had an excellent opportunity, enjoying the hospitality and guidance of Mr. Freeman, member of the State Board and Swamp Land Commissioner, to examine thoroughly the bog-iron deposits of the Kankakee country, and other portions of his District. We took specimens for analysis from section 5, township 33 north, range 6 west. We also took samples of the different soils, prairie, oak openings, &c., on Mr. Cornell's farm, S. E. quarter of section 8, township 33 north, range 6 west, and from that of Mr. Stoddard, N. W. quarter of section 32, township 34 north, range 5 west, also from Mr. Milan Cornell's, on N. E. quater of section 31, township 35 north, range 5 west, which soils, however, were not reached in analysis by the limited number to which Dr. Peter was necessarily at present restricted for want of additional funds.

From these low prairies they cut from two to four tons of a grass which they denominate "Blue Joint," (Calamagrostis Canadensis). The yield of wheat, corn, (40 to 60 bushels,) clover, sorghum, potatoes, pumpkins, &c., is also good. We observed likewise some good apple orchards. Among the natural growth were noticed boneset, ferns, wild indigo, vast quantities of dwarf willows, which, according to the old settlers, would soon overspread these prairies, unless the grass was burned, cocklebur, (Xanthium strumerium, var. echinatum,) May weed, (Maruta cotula,) often called in the west Dogfennel,* some mullein, blackberries and strawberries. Cranberries grow in the north part of Porter county. The quaking aspen (Populus tremuloides) and sassafras are also fast taking possession of the prairie. The groves,

---

*Some, I think, call the wild Chamomile (Matricaria fœtida) Dogfennel; but there is usually great confusion about popular terms.

singularly enough called "Oak Openings," are chiefly here of Bur-Oak, with some Hickory, (usually Carya amera).

About a mile west of Valparaiso, on an elevation, red brick are burnt from the clay, although other portions of the county furnish the material for the buff brick. Thirty or forty feet below the clay of Valparaiso about twenty-five feet of rich, black muck, then marl, sometimes with helix, paludina and other shells, occasionally destitute of any, and below this a hardened calcareous tufa, strongly impregnated with iron. This they occasionally burn into lime, but it does not furnish so good an article as the shell marl, on account of the iron. Beneath this there is another swamp muck bed exposed in this hill-side. Near here was an old grave yard, from which bodies were removed when they desired to form a new Cemetery. When taken up the corpses were found encrusted with the above calcareous tufa, and in tolerable preservation, except parts of the extremities, the broken sections of some exhibiting an internal hollow.

On Mr. Howell's elevated land, about three-quarters of a mile east of Valparaiso, on section 30, township 35 north, range 5 west, we were shown good grey crystalline limestone which had been quarried and burned into lime; but as the layer is only two or three feet thick, and apparently local in extent, it was soon abandoned. Unfortunately no fossils were found, the lithographic or lithological character, however, indicates a rock of Upper Silurian age.

The parallelism of the sand ridges in this and other counties, south of Lake Michigan, to the curve of the lake shore, as well as other points connected with this great expanse of water, will be discussed in treating of the physical geography.

Although portions of the Kankakee marsh are often rather wet in spring for plowing, yet we observed some fine prairie land here drained by Hog creek, which was well fenced, and bearing good crops of wheat, corn, and of the "Blue Joint grass," said to equal our domestic "Timothy" or Herd's grass. The low prairie is in places, white with dead shells, chiefly of the genus paludina, supposed to have been killed by the drying of the marsh.

In Lake county, near the Illinois line, we ascended gradually from the Kankakee region to the Lake. Some of the low prairies, as we were informed by an intelligent farmer, who for many years boated on Kankakee river, have latterly been drained, and where formerly (not over three or four years since,) a man would have mired, he now hauls tons of hay with wagon and team. Bog-ore is abundant here; and land sells at from five to eight dollars per acre. Near here, on an arm of the Grand prairie, an Artesian boring has been made which passed through fifteen feet of clay, then fifty-four of sand, then six feet of blue clay or hard-pan, total seventy-six feet. Some water was obtained at thirty-two feet, and finally a good supply. On other adjacent parts of these prairies, they reach water in quicksand, after passing through about twenty-two feet of clay. In the barrens or groves they commonly obtain water after penetrating thirty to thirty-five feet of sand.

From these low lands of rich, black soil, and some bowlders and gravel, with the wild indigo, ferns, wild red-top, &c., and island-like barrens of Black Jack, Black Oak, Sumach, Hazel-bushes, wild Plums and Haws, we rose gradually to rolling prairies of Flutter-dock and Rosin weed; then to groves and ridges, with Cedar Lake in the distance, and the head waters of West Creek on the west, to a "divortia aquarium," or water shed, a few miles south of Crown Point. This ridge is something more than 100 feet above Kankakee river, and composed chiefly of altered or late Quaternary deposits, not having any bowlders or much rounded gravel; but vast quantities of sandstone detritus, angular and shaly, with a growth of scrubby Oaks and Hickories. Descending, gradually, nearly to our former Kankakee level, we reached Crown Point on a rich prairie, with some bowlders and groves, passing by fields which had evidently borne heavy crops of oats. At Crown Point, the capital, Capt. Smith, formerly of the 16th U. S. Infantry, and now Swamp Land Commissioner, indicated the low grounds between the ridges as being rich in bog-iron ore, but especially the regions between the two Calumets and from Great Calumet to the Lake.

Crossing Deep river and progressing amid rich black muck prairies of abundant ferns, such as Aspidium marginale, of the wild Sunflowers, (H. rigidus occidentalis and mollis) the ubiquitous Iron-weed, (Vernonia fasciculata) Boneset, and a creeper with a cucumber-like seed vessel, (probably Echinocystus lobata,) with more rarely Flutter-dock and Rosin weed, quaking aspen and willows, or farms with Orange hedges, Sorghum crops, Buck-wheat, corn and orchards, we reached Hobert's Station to camp, and resume the route already described in detailing Porter county and its resources.

To the Zoologist it may not be uninteresting to learn that we saw several specimens of the Batrachian reptile, Menobranchus, taken by fishermen from the Lake, while we were at Chicago, Illinois, twelve miles from the State line.

In Lake county we disturbed frequent flocks of cranes (probably

Grus Americana) and sometimes three or four white Herons, which appeared in the distance to be the Ardea leuce.

DEKALB AND NOBLE COUNTIES.

Of DeKalb county we saw but little, except in its south-west corner, on our way from Legineer to Fort Wayne. Judging from what we saw and heard, we should say that the soil is generally productive, the woodless tracts being a sandy loam, the forests more clayey, the land sometimes heavily timbered with little or no underbrush, sometimes comprising low prairie. Being meandered by the St. Joseph of the Maumee and its tributaries, the county is sufficiently undulating for drainage, and well watered for agricultural purposes.

Bog-ore is found in many parts of the marshy prairies, especially near its junction with Noble county.

At Legineer, in Noble county, we observed piles of bowlders corded for sale, also staves and lumber, indicating abundance of timber. At Rochester, about three-quarters of a mile east of Legineer, bog-ore was worked, which they dug a few miles south of that place; this furnace was not visited for want of time. The ore brought from Avilla is used for chimney backs, because it does not burn out. There is also abundance of marl, similar to that in St. Joseph, although it does not calcine quite so freely. At Rome there is a chalybeate spring.

Between Legineer and Kendallville, Beech timber is not uncommon, and where ditches were cut for drainage we observed a very black soil, often underlaid by gravel. Several gravel quarries were open for the transportation of ballast to the railroad; in some we examined, the gravel was in a bed from six to twelve feet thick. Near Kendallville, at Lisbon, is a height of land or summit level, whence waters run south to the Gulf of Mexico, and north-east to the Gulf of St. Lawrence. Somewhat north of Kendallville to the Michigan line are found oak openings, with no undergrowth, and with yellow hard-pan at from two to six inches below the surface. For wheat this is first-rate, for corn second-rate land. South of these are the so-called "bastard openings," with Black Walnut, Poplar and Blue Ash, undergrowth Hazel and Sassafras. This is considered here first quality of soil for all farming purposes, affording 40 to 45 bushels of corn per acre. It is a blackish sandy loam, with gravelly subsoil.

Adjoining the lakes and streams they have also wet prairies, with muck or humus at top and, immediately below, a quicksand and gravel, then marl and blue clay. They sometimes dig wells through the hardpan that require no curbing.

KOSCIUSKO, MARSHALL AND STARKE COUNTIES.

The northern portion of Kosciusko is chiefly prairie, the southern heavily timbered.

Near the junction of Marshall and Kosciusko, a farmer showed us a variety of maize called "Butcher Corn," deep red, heavy grained and long-eared, represented as productive.

In the northern part of Kosciusko down to Leesburgh, which is about 100 feet above the level of "Yellow river," at Plymouth, we observed a heavy drift of sand and bowlders, exhibiting at some sections about 30 feet, over clay; the soil, black, with occasional Beech timber and numerous ferns. Continuing south we found good farms with substantial barns and excellent roads; crops of buckwheat and clover luxuriant.

As we entered the timber we enumerated Beech, Sugar-Tree, Elm, Oak, Gum, Hickory, Poplar, White and Black Walnut, then crossed low ground on corduroy, with ferns, flags, smart weed, Willow, Swamp-Maple, Blue Ash, &c.; succeeded by the silkweed or milkweed (Asclepias cornuti) and Boneset for some miles. We crossed the railroad track at Bourbon, a good-sized thriving place, on a level a few feet higher than the fern thicket or brake. Near here some of the tributaries of Tippecanoe take their origin, and we shortly reach the deep-green shades of the Tamerack. Many of these were noticed to be dead, perhaps from the draining of those lands rendering the soil too dry for its favorable growth. An industrious and enterprising population has settled around here, evinced by the wheat, which the Hon. Mr. Thralls, of Warsaw, was taking in at the rate of 1,200 bushels per day, and which averages considerably over sixty pounds to the bushel. This gentleman informed me that abundance of bog-iron ore had been dug near town. It was first discovered in 1838, and an account of it was published in the "Eaton Register," Ohio. As yet they have no furnace nearer than Rochester. Ore is also dug near Milford and sent to the Rochester and Mishawaka furnaces. Near town, in digging wells, water is reached at sixteen or twenty feet in sand, after passing through gravel and two to four feet of hard-pan, made up of pebbles cemented in clay. On the north sice of Turkey-creek prairie the same materials, which extend but from nine to fourteen feet in depth, occupy in the same prairie, two miles south, a bed forty to forty-five feet deep. Yet the surface of the prairie, tested by the barometer, varied scarcely a foot in level. It is dry enough for an abundant growth of Rosin weed and Flutter dock. Ferns also constitute here a very prevalent vegetation; in this county most of those collected are of the Bracken genus "Pteris," with the spores in a continuous *marginal* line of fructification.

We made Eagle Lake from twenty to twenty-five feet lower than Warsaw, and then ascended sixty feet or more, over sand, gravel and abundant bowlders, with Quaternary hills still twenty to twenty-five feet above us, covered chiefly with Oak timber. Although some Sumach grew at this elevation, the corn looked very well, while orchards and substantial barns gave evidence of a good and productive agriculture. Near Fairview we passed a saw mill and brick-kiln; the clay indicated by the latter soon giving growth to a predominance of Beech timber, as we neared the line of Whitley county.

In Marshall county, which we entered prior to the examination of Kosciusko, at its north-west corner, after crossing the Kankakee bridge in Laporte county, we found near West York sandy barrens with White Oak and Hickory, about thirty feet higher than the river, also some small prairies and bowlders, about twenty feet above the Kankakee level at the bridge; and finally a small lake about a mile long by half a mile wide, which they seemed to be draining. The bowlders were of good size, mostly granitic. The prairie growth was indigo plant, asclepias, Helianthoid, yellow flowers, golden-rod and ferns; in low spots Eupatorium, Willow and Aspen, on the higher sand ridges, trending here chiefly east and west, small pines, oaks and hazel bushes. The prairie subsoil appeared from creek sections to be a blue clay.

As we approached Plymouth, the county seat, the timber became larger, being Hickory, Swamp Maple, Sassafras and White Oak; occasionally we passed, on "corduroy," swales with abundant ferns, iron weed, smart weed and some bowlders. Nine different species of ferns were identified here, chiefly of the Tribe Aspidieæ, represented near Plymouth by the three genera Aspidium, Onoclea and Cystopteris; also the gracefully waving Maiden-hair, Adiantum pedatum, sole representative here of the Pteridean tribe so common in Kosciusko county. Of the True Fern Family only one species, the Beech Polypody, was obtained; of the Adder's-Tongue Family, a Botrychium, and of the Flowering Fern Family an Osmunda, (probably O. cinnamomea,) with its high and separate central fertile fronds, a fern found by us so abundantly in the counties rich in disintegrating aluminous shales.

Plymouth, the thriving county seat, containing about 1,500 inhabitants, and surrounded by an industrious population, some of them Germans, is in its lower part about twenty feet above Yellow river, a tributary of the Kankakee, and by our barometrical observations about the same level here as the Kankakee at the LaPorte county bridge, and which, judging from the Canal Engineer's Surveys, (as by their estimate Yellow river, ten miles from its mouth, is 690 feet above the sea,) can not vary much, at Plymouth, from 700 feet above the level of the ocean.

Bog-ore is abundant near town and works have been carried on, a few miles south of town, which we would have examined had we not been informed that the furnace was not then in operation.

In digging wells around here they pass through from two to three feet of gravel and reach yellow clay.

As we traveled south-east towards Kosciusko county, sand, gravel and bowlders continued, with some rich, black soil, occasionally, and Beech timber, varied by a White Walnut and Tulip-tree ridge of more arenaceous character. Ferns were sometimes still so abundant as to constitute a "fern-brake," if this be not tautology, or perhaps more properly a "Bracken-thicket."

Of Starke county we saw less than we should have done could we have crossed the Kankakee lower down; but as the Upper Silurian rocks make their appearance a few feet below the general surface in the northwest part of Pulaski, it seems probable that the disintegration of the superincumbent black aluminous shales, of Devonian age, may have contributed, to some extent, towards giving this Kankakee region its peculiar character, modified, no doubt, somewhat by the subsequent drift.

In boring near the edge of Marshall county they penetrated about 100 feet of sharp sand.

Although portions of the railroad track are laid on piles, in order to avoid the water during the wet season, yet some of the prairie farms in this county seemed to be productive, having large and promising corn fields.

The cat-tail flag (Typha latifolia and angustifolia) is very abundant, and being used extensively by coopers for tightening the stave-joints might perhaps pay for exportation to places where it does not grow.

The railroad conductor informed us that the bog-ore of this county sometimes extends to a depth of twenty feet.

At San Pierre there is an extensive sand ridge bearing E. N. E. and

W. S. W., forming oak openings ten or fifteen feet higher than the prairies around. From some of the former a pure clear sand is carried out for ballast. A few bowlders show themselves occasionally.

JASPER AND NEWTON COUNTIES.

The great Quaternary ridge of the Grand Prairie culminating in Mounts Gilbo and Nebo, gives origin to several head waters flowing north and south. About three miles north of the Benton county line in Jasper, following Carpenter's creek, one of the above-mentioned sources which flows north into the Iroquois river, we found the bluff banks of the creek composed of eighteen to twenty feet of black, bituminous, aluminous shales of Devonian age, which show themselves on the Wabash at Delphi and Americus, and which has been thinned out so that, at some points on the creek, while tracing the slate a mile and a half, we found the silico-magnesian limestone of Upper Silurian date, the same found abundantly further east in White and Pulaski counties. Mr. Jordan now owns the land formerly the property of an early settler and hunter, Carpenter, whose name the Grove and Creek now bear from his having died and been buried there. Close to Mr. Jordan's house, which is built on the edge of the grove, they quarry sandstone from an upper bed overlying these aluminous shales, being thirty feet higher by the barometer. The strike of this Knob-sandstone extends, according to Mr. Jordan, as much as ten miles in a northerly direction, passing east of Rensselaer, the Devonian disappearing here as the dip is quite perceptible in a westerly direction. This we afterwards verified by a visit to the Phillips' quarry hereafter described, when visiting that town.

When Mr. Jordan first settled here, he could see a hog anywhere in his grove, when they happened to be in there; now he looks in vain as the undergrowth forms a thicket of Hazel, Sassafras and Hickory bushes.

At Rensselaer we crossed the Iroquois river and found several feet in thickness, exposed, of a yellowish, silico-calcareous rock, breaking in irregular masses, having calc-spar and indistinct casts of fossils, stratigraphically fifty feet lower than the base of the sandstone at Carpenter's Grove. We finally recognized Stromatopora concentrica, confirming the previous opinion of its Upper Silurian age. Here it is too cherty to burn into lime; but a mile below town they have lime-kilns, and Mr. Alters has a good quarry six miles north-west from Carpenter's

Grove. Taking an easterly direction, somewhat south, towards Bradford, we passed numerous bowlders and saw some fine Merino sheep, clover and sorghum fields; at about three miles from town we reached the sandstone quarry, which we were seeking. It is chiefly on the property of Mr. Simon Phillips and can be struck anywhere, on about an eight acre tract, at from two to five feet below the level of the prairie, and twenty-five to thirty above the level of the Iroquois river at Rensselaer. It is never struck in digging wells on the north side of the river. The stratum is quarried in slabs from one to two feet thick, sometimes of a coarse grit, at others resembling the freestone overlying the Upper Silurian in the Wabash bank a few miles below Logansport, thus forming a valuable building material. The dip here appeared rather southerly of west, whereas at Rensselaer it appeared to be north of west, and at one point appeared to have an anticlinal axis.

Returning to Rensselaer, we took the Chicago road towards Morocco, passing Osage hedges and bowlders scattered over the prairie, in which flourished the rosin weed, flutter dock, coarse grapes, white clover, golden rod, smart weed, iron weed, rag weed (Ambrosia bidentata) and wild indigo; the oak groves occupying ridges of sand, gravel and bowlders, with sometimes marshy sloughs or swales between the ridges, if the interval happened to be of narrow dimensions.

We were informed that in the western part of the county a thin limestone is occasionally found, which from its position and the character given in describing it, would appear to be sub-carboniferous limestone.

As we traversed Beaver prairie and the southern portion of Newton county, generally, the prairie and bowlders continued; but as we neared Morocco, the soil became whiter looking, being comparatively high and dry. This town, the county seat of the newly organized county, Newton, taken from Jasper, is by our barometer about 100 feet above the bed of the Iroquois, at Rensselaer; but as a thunder storm overtook us as we entered town this level may require some correction. However, the barometer still indicating the same reading next morning, there is probably not much error. This place proved very interesting from the Artesian boring undertaken at the steam mill by Mr. G. W. Clark, who had then penetrated 128 feet, and intends continuing the work. At 72 feet below the surface they encountered a few feet of sand; but with that exception the 128 feet proved a pure clay, becoming bluer, as they bored further. Samples were taken for analysis at different depths. It seems almost destitute of silicious ingredients, and would make a valuable clay for many purposes. With an Oberhaeuser, magnifying 300 diameters, no microscopic-organisms could be detected in this clay, even after moistening.

Another boring was made at Brook, on the Iroquois, over 100 feet deep, through a considerable amount of gravel before reaching the clay, which then continued uninterruptedly.

Few, if any, bowlders were observed after leaving Morocco, on the route towards Beaver Lake; the country here is rather sparsely settled, yet some fine low prairies exhibit an extensive growth of a grass, which we could scarcely distinguish from tame red-top. Ferns and mimosa bushes were common as we approached the sand ridges, with scrubby timber. We disturbed several flocks of cranes and a few fine white specimens of the genus Ardea, probably the A. leuce.

We have been gradually descending as we pass near Beaver Lake, being now at least a hundred feet below the level of Morocco, until finally at the Kankakee crossing in Illinois,* (there being no suitable bridge or ferry short of Momence on the west, or St. Joseph county on the east,) the barometer made the bed of the river 180 feet below the steam-mill at Morocco.

A calcareous tufa is dug in this county just below the humus or swamp muck, which becomes very solid on exposure to the atmosphere.

Beaver Lake has been purchased and partially drained by Major Dunn and Mr. M. Bright, being forty feet above high water in the Kankakee river, at its nearest point, which is five miles distant. Already between 8,000 and 9,000 acres of marsh have been drained and a portion cultivated. One of the renters, in digging his well, went through twenty feet of pure black swamp muck, chiefly decaying or decayed vegetable matter, woody fibre, leaves, ferns, mosses, &c.

Bogus Island, in the centre of the lake, so called from its formerly having been the resort of counterfeiters, is covered with wild Black Cherry Trees: (The Cerasus Virginiana, of Michaux, and the Padus serotina, of Ehrhart.) As the remains of beaver dams are numerous in the lake, it is supposed to have been the resort of the fast-disappearing castor fiber, hence the name of the lake.

---

*In this connection, it may be interesting for some to learn that a company has been recently organized for the straightening of the Kankakee river, which in its windings is three times as long as the direct line; by means of which, and the removal of obstructions, they hope to deepen the channel and form a drain that will run off its high waters and that of its tributaries more rapidly than now, and into which cross ditches can be cut, thereby rendering many thousand acres so much drier than at present, as to bring land up from three and four dollars per acre to thirty and forty.

ALLEN AND WHITLEY COUNTIES.

Allen county is included among those characterized chiefly by the Quaternary, because the drift exists to a very considerable depth over most of the county, forming small prairies, with occasional timber. Deep wells have been sunk, particularly one at Fort Wayne, by Mr. Barry, General Superintendent for the Railroad Company, which has passed chiefly through blue clay, with a stratum of sand at eighteen feet and another at thirty feet.

There are also numerous quarries, pockets as they here term them, having gravel and sand in a bank from ten to fifteen feet deep, out of which they cart large quantities of ballast for the railroad. Near the junction of the St. Mary and St. Joseph to which I was kindly conducted by Mr. McCulloch, stood the old Fort and General Tipton's house, and not far off a hill of pure sand.

The heavy Quaternary deposits seem at some period to have caused a remarkable change in the water courses of this region. St. Joseph of Maumee flows from its source to Fort Wayne in a south-west direction; St. Mary's river has a general steady north-west course to this city; but here meeting apparently a natural barrier, instead of continuing as the upper Wabash does, flowing west with the dip, these two rivers unite, and, under the name of the Maumee, return nearly on their former course, almost due east.

This subject will be more fully discussed in the chapter on the physical geography of Indiana.

Mr. Jesse L. Williams, formerly Chief Engineer of the Wabash and Erie Canal, now residing at Fort Wayne, furnished valuable information regarding the levels run through the State, and Hon. Allen Hamilton, whose generous hospitality we experienced, directed our attention to a deposit of shell marl under the prairie swamp-muck in parts of the county, also to some sulphur springs six miles from Fort Wayne, and gave other details regarding this important county.

During the period intervening between our first and second visit, Mr. James Humphreys, of Fort Wayne, had politely sent a copy of his article, contributed to one of their newspapers, giving some details regarding a rock quarry recently opened about nine miles south-east of city. We accordingly took the Pickway road in order to examine this quarry formerly owned by Mr. Heer. We found it on the east half of the north-east quarter of section 35, township 30, range 13 east, now the property of Mr. James A. Key, a foot or two below the suface of a field, which we made thirty to thirty-five feet higher than the level of St. Mary's river, near the Adams county line, and nearly a hundred feet above the Maumee at Fort Wayne. The excavations were filled to some extent with water and it was difficult to be certain of the dip, as well as to obtain any decidedly characteristic fossils; but, as nearly as could be ascertained, there is an easterly dip of several degrees, perhaps ten or twelve. Lithologically the different beds afford varied characters: some being in layers three to six inches thick, blue and bituminous, with vertical striæ similar to those found in the Upper Silurian, eighteen to twenty miles south, on the Wabash river, in Wells county; other beds grey, crystalline and more fossiliferous. The fossils are not quite satisfactory and conclusive, still the abundance of Atrypa aspera, a considerable portion of a fish, which strongly resembles the tubercular triangular body, either of the Pterichthis or Coccosteus, probably the under side, near the caudal elongation, of Coccosteus cuspidatus, the frequency of a gibbous orthis, found also in Devonian limestone, at Lower Sandusky, as well as of Cyathophyllum helianthoides and a Calymene found at Jeffersonville, would seem to indicate Devonian, which is confirmed by finding, mingled with the Drift, quantities of detritus, that seems derived from the disintegration of black aluminous shales, such as those of New Albany, Delphi and elsewhere. On the other hand Terebratula reticularis and Bryozoa assignable to either, the uncertainty of some casts, the close resemblance of other fossils to Upper Silurian types, as T. Wilsoni, and the proximity of the Upper Silurian all along the upper Wabash to Huntington and even lower, would throw some theoretical probability on these being the higher beds of Upper Silurian age.

As, however, at Huntington there is an axis of dislocation, and the dip at this Allen county quarry seems east, while at Wabash, Peru and Logansport it is west, and as the Devonian rocks rising from under the Michigan coal fields are supposed to reach the northern counties of Indiana, leaving perhaps only one tier of counties between them and Allen, this may be, as at New Albany, the junction of the Devonian with the underlying Silurian rocks, and the Devonian may originally have extended from Lake Erie to Lake Michigan, as will be more fully discussed hereafter.

The question regarding the probability of obtaining water by Artesian borings will also be brought up under that general head.

In traveling from Fort Wayne to the quarry we passed heavier timber than in the north part of the county, where swamp-muck and clay prairies, with willow growth, &c., alternated with higher oak openings, composed of bowlders, gravel and sand ridges. The timber is mostly Beech, Sugar-Tree, Poplar and Black Walnut. In passing we again saw the fine nursery and farm of Mr. Nelson, formerly visited, part of whose orchard was so remarkably improved by plaster. The analysis of the soil and remarks on this subject in Dr. Peter's report will be found highly interesting and useful. They have here some fine fruit, although occasionally troubled with the bark louse (Coccidæ) and the Weevil, (probably Curculio or Rhychænus, Nenuphar). These bottoms raise also good crops of corn, and the county cultivates wheat successfully, as well as potatoes. The flying weevil (Anacampsis cerealella) and potato rot have sometimes shortened the crops. As we went south, Dogwood (Cornus Florida) became added for the first time, in the northern counties, to the above enumerated timber, also large Oaks and Hickories.

In the north portion of Whitley, which we entered near Fairview, and traversed thence towards the Pierson railroad station, as well as near Cherebusco, we have a height of land or water-shed, some springs flowing through Eel river and Tippecanoe, finally to the Gulf of Mexico, other streamlets converging to form the Elkhart and discharging through the Big St. Joseph into Lake Michigan, ultimately to reach the Atlantic through the river and Gulf of St. Lawrence.

This part of the county has numerous small lakes and prairies, having abundant bowlders, cat-tail flags and willows, with intervening quaternary ridges or oak openings, chiefly of gravel and sand.

On nearing Columbia we passed some fine farms and an extensive Tamerack swamp, probably 100 feet lower than the "divide;" also a large fern thicket and thick undergrowth of Hazel bushes.

The corn crops (Sept. 13) were very promising, and the clover fields seemed to have afforded a heavy cut.

Columbia, the county seat, is a thriving place with nearly a thousand inhabitants, situated on Blue creek, one of the tributaries of Eel. Bogore is dug west of town. The wells in Columbia average from fourteen to twenty feet in depth, chiefly through clay. Half a mile north of town a farmer living on more elevated ground has soft water, the only well around known to have this peculiarity, although his digging was also through clay. Probably this elevated quaternary has no calcareous debris near to affect the filtering water.

In the Drift near town there were large quantities of dark aluminous shales, angular as if not transported far, but very friable when handled; some black sand, similar to that of Lakes Superior and Michigan, showed itself in the ravines, yet a preponderance of clay in the soil was evinced by its cracking when dry.

As we progress south the prairies are smaller and the timber more abundant, as well as larger, Oak, Hickory, Elm and Sassafras, with Hazel undergrowth; ferns less abundant, but bowlders yet large sized and numerous.

The land is sufficiently undulating, generally, not to be too wet for cultivation, and besides the above crops, we observed good sorghum and buckwheat; the corn is usually cut and shocked, which is generally considered the most thrfty mode of harvesting it. Hoop-poles, staves and lumber seemed abundant, and a shingling machine was observed in operation near the line of Whitley and Allen, about Huntsville.

FULTON AND PULASKI COUNTIES.

Of Fulton county we did not see as much as we desired, having been informed, correctly or incorrectly, that the Iron Works at Chippewa, near Rochester, were not in operation when we were in the neighborhood. No doubt the general remarks made in connection with the furnace at Mishawaka would apply here, both as to the quality of the bog-ore and the facility with which it is worked; only requiring transportation facilities, cheap fuel, marl or limestone in some quantity, and a fair average price for pig iron to make the work profitable.

Both Fulton and Pulaski have extensive prairies, with oak openings, and portions were heavily timbered; they are well watered, as well as drained to some extent, by the Tippecanoe river winding through their interior. Gravel and bowlders are not so common here as in counties further east, clay and swamp-muck constituting the prairies and swales, sand the oak openings. These are fenced with small oak timber, without splitting, simply "spotting" the round rails in places to prevent their rolling.

On approaching Francisville, the prairie widens with some island-like patches of timber, chiefly oak. Limestone can be struck here any where, and even as far north as Medaryville, at from two and a half to eleven feet below the surface, reaching indefinitely downwards. A well which we examined was made by blasting through seven to eight feet of rock, the layers being from four to six inches thick.

The rock, from the few samples we saw with fragments of fossils appeared to be of the same character as that found, a short distance south-west, in the bed of the Little Menon, hereafter described, in White county, and, if so, of Upper Silurian age. This may seem somewhat singular, considering the fall from the head-waters of the Big Metamonong to the region of the Wabash at Americus or at Delphi, where the Devonian black slate overlies the Upper Silurian pentamerus limestone, did we not reflect that these Upper Silurian rocks extend up the Wabash, (which river crosses the strike, with a bearing not very dissimilar, from the direction of the Tippecanoe river,) as far as Logansport and even Huntington; the dip corresponding in direction with the descent of the river, although greater in degree.

Many good farms were seem in these counties; corn, wheat and potatoes being staple products.

WHITE AND BENTON COUNTIES.

Conducted by Mr. Watson, we examined the rock near the railroad station at New Bedford, in the bed of the Little Menon, from which it is extensively quarried for lime, although in places rather silicious. They also burn lime at West Bradford, and even four or five miles higher up the Big Metamonong.

There seems at New Bradford a dip (probably only locally so much inclined,) of from 4° to 10° in a westerly direction, sometimes a little north, sometimes south of west. The rock is a yellowish silico-calcareous stone, and from the fossils found, few of which were quite perfect or satifactory, it seems an upper bed of Upper Silurian age, much like the coralline limestone of Schoharie, New York, which Prof. Hall considers part of the Niagaria Group. Stromatapora concentrica was the most satisfactory, while immense quantities of crinoidal stems, fragments of Bucania or Bellerophon, of a coral closely allied to Favosites Niagarensis and of Orthoceratites seem to confirm the testimony.

A few miles west of this place sandstone is quarried which probably corresponds to the silicious bed seen below Logansport, on the Mississinewa, above Peru, and at Mr. Irish's quarry, Pendleton, about the junction of the Devonian with the overlying sub-carboniferous sandstone. This stratigraphical level, at New Bedford, is probably seventy-five or eighty feet above the waters of Tippecanoe river, near Monticello, where a fine railroad bridge crosses that stream sixty feet above its bed.

Large portions of the county are a fine, farming, prairie region, with a considerable amount of Aspen. Near Norway, bowlders are abundant, not so much rounded by attrition as in other localities, soil sandy; and the growth, in the extensive prairies, of ferns, chiefly polypods, boneset, wild indigo, mimosa bushes, asclepias, smart-weed, (Polygonum hydropiper,) May-weed, willows, &c. Sand ridges somewhat higher, give growth to Oak, Sumach, Sassafras and Hazel bushes.

Continuing west, through the Grand Prairie, which we had entered in White county, we gradually rose, keeping Mt. Gilbo in view, until we ascended the great Quaternary ridge or gravel bank, which in this, its eastern culmination, is from eighty to a hundred feet above the general level of the country. The ridge extends westward, sometimes as a sand bar, sometimes as a gravel bar, or apparent ancient bank of a lake, at a somewhat lower elevation, until, at eight or nine miles distant, it comes up by its western and more arenaceous culmination, Mt. Nebo, nearly to the same level as at its eastern summit, Mt. Gilbo.

From this ridge of coarse gravel, finely exposed by an excavation, Carpenter's creek flows north, as before indicated, cutting through the sub-carboniferous sandstone and underlying the Devonian black shales; and the head-waters of Big Pine flow south, cutting through the sub-carboniferous limestone, and the Millstone Grit or Carboniferous Conglomerate.

Mt. Gilbo ridge is about a quarter of a mile wide and extends at its high level about half a mile; the land, including the Mount, is owned by Mr. Simon Brown, of Fountain county. Mt. Nebo is the property of Mr. John W. Swan, in section 20, township 26 north, range 7 west. Two miles and three-quarters south-east of Mt. Nebo, a limestone quarry is owned by Mr. J. W. Nutt, from which slabs are obtained eighteen to twenty inches thick, in beds with a considerable dip north of west. Another is on the property of Mr. Evan Stevenson, on Big Pine, eight miles north-east of Oxford; and a third, somewhat further south, on the farm of Messrs. Sample and Seabury. The quarried specimens of stone, which we saw, indicated the same yellowish, silico-calcareous rock of Upper Silurian age seen on the Little Menon; and these beds are probably reached where the Devonian black shales have been swept off by denudation.

Mr. Posey, of Warren county, a relative of General Posey, whose name was selected for our extreme south-west Ohio-Wabash county, called our attention to a belt of bowlders, about two miles wide, extending from Illinois, south of the Nebo-Gilbo ridge, past Parish's Grove, a mile and a half north of Wagner's Grove, in a south-east direction, towards central Indiana, in the region of Indianapolis, perhaps a mile north of that city. This is not very different from the slightly curved strike-line of the Black Slate.

From an average of the observations made in ascending from the Wabash, at LaFayette, to the Grand Prairie and returning by way of the Battle Ground, again to the Wabash, our barometer makes the swales and low places in the prairie, where we sometimes saw black shales, about 285 feet higher than the level of low water in the Wabash, underneath the LaFayette bridge; which places are often characterized by small ponds with water lilies and other aquatic plants, and along side of these a growth of smart-weed, (Polygonum hydropiper,) flags, boneset, galingale, (Cyperus strigosus,) as well as willows and aspens, when not kept down by fire. On somewhat higher portions we saw extensive plains with ferns, wild indigo, mimosa, and a vast representation of helianthoid compositæ, and other flowers with every variety of hue.

The more rolling and drier prairie, twenty to forty feet above the swales, is usually characterized by the compass-plant, flutter dock, and a plant denominated by the residents Quinine-weed. Mounts Gilbo and Nebo being about eighty feet higher than these rolling prairies, must be at least about 390 or 400 feet above the Wabash, at LaFayette, or about 880 feet above the level of the ocean, viz.: the LaFayette Court House being, according to Messrs. Stansbury and Williams, 538 feet above the sea, we must deduct from this 48 or 50 feet, which the Court House stands above the Wabash, leaving 490 feet,* and to this elevation add the 390 feet, or thereby which the culminating point of the Gilbo-Nebo ridge attains over the Wabash, and we have the total as above, 880 feet above the sea.

The swales and ridges in this county generally run in an easterly and westerly curve, the latter sometimes having a growth of scrubby Oaks, Hickories, Sassafras, and an undergrowth of Hazel.

A well, dug on this prairie, at least 150 feet below Mt. Gilbo, is thirty-one feet deep, and water usually stands in it to the height of eleven feet. The owner said he dug through two or three feet of dark soil, about four of yellow subsoil, then hard pan, gravel and bluish clay through the rest of the distance, to a thin bed of quicksand.

We saw some fine flocks of sheep and good osage hedges; but very generally here the fine large corn fields are entirely without fences, as

---

*This agrees very nearly with the altitude given by Lieut. Ellett for low water in the Wabash, at LaFayette, which he states to be 492 feet above high tide.

hogs are not permitted to run at large and the numerous herds of cattle, constituting the staple profit, are kept from straying and from trespassing by the skill of a single herdsman.

Flights of cranes were seen and we frequently shot, for camp use, the pinnated grous, (Tetrao cupido,) or, in the groves, the wild pigeon, (Ectopistes migratoria,) besides startling the meadow lark (Sturnella ludoviciana) and a few smaller birds, from their prairie nests.

For the Geologist and Physical Geographer, the Botanist and Zoologist, as well as the lover of scenery such as the boundless vision of the day and the gorgeous sunset of the evening often afford, this ocean-like prairie region, and these island-like groves are replete with interest and instruction.

CHAPTER III.

PHYSICAL GEOGRAPHY OF INDIANA.

SECTION 1.—OROGRAPHY, STATISTICS OF ALTITUDES, EXAMINATION OF WATER-SHEDS, &c.—A synoptical table of hypsometrical statistics in our State is subjoined in the appendix, as it was thought that it might be interesting to some to have these various heights in Indiana in a condensed form; as well those alluded to in the report, usually obtained by barometrical observation, as also others chiefly derived from the surveys of Col. Stansbury, (U. S. Topographical Engineer,) and Mr. Jesse L. Williams, (Chief State Engineer of the Wabash and Erie Canal,) and from the writings of Mr. Charles Ellet on the Mississippi basin, published by the Smithsonian Institution; likewise that of Mr. Blodget, author of a valuable work on Climatology, and a few similar sources. For convenience of reference they are arranged alphabetically, according to counties, giving also the exact place.

The addition of the altitudes, found by Messrs. Stansbury and Williams, while surveying in Indiana, as obtained by them at two hundred and eight stations, and reported to our Legislature in 1836, gives a total, which divided by the number of stations, averages a fraction over 678 feet for the mean elevation of the land in Indiana over high tide in the ocean. A nearly similar result is obtained by taking the average between the maximum and minimum heights observed in Indiana. Thus on adding the greatest elevation run in our State by the levels of Messrs. Stansbury and Williams, the summit between Sand and Salt creeks, 1,057 feet above the ocean, to the lowest portion of Indiana, 207 feet high, viz: the mouth of the Wabash, as estimated by Mr. Ellet above high tide, and dividing the result by two, we have an average of 677 feet, only one foot less than the average obtained by the other calculation.

According to Mr. Chas. Ellet, in his valuable remarks on the Mississippi Valley and the improvement of navigation in the Ohio river, the altitude of Lake Michigan is 610 feet. By the observations made at Chicago, as a military post, Mr. Blodget gives that city at 591 feet, and Milwaukee, on the MSS. authority of Messrs. Marsh and Lapham, at 600 feet above the ocean. We cannot therefore be far wrong in assuming 600 feet as about the average of the observations, remembering however, as shown in the interesting contributions of Col. Whittlesey to the Smithsonian Institution, that the level of the water fluctuates many times every day a few inches, corresponding apparently with a diurnal rise and fall in the barometer; and that, periodically, usually in several years, perhaps dependent upon a succession of dry summers, there is an ebb, and after several wet years a flow, amounting to several feet.

Thus then, the average of the land being 678 and of the lake 600 feet, if the general level of our State were depressed about 80 or 100 feet, a portion of the waters of Lake Michigan might readily flow almost directly south into the Gulf of Mexico, through the valley of the Mississippi; or a similar result could be obtained by excavating a canal through Lake county, perhaps twenty miles in length, and not even at the deepest cut 100 feet below the general level of the county, by which the waters of the lake could be made to flow into the Kankakee, (which however, at the mouth of West Creek, is probably but a few feet lower than the lake,) and thence through the Illinois river into the Mississippi.

It seems probable, as we shall see hereafter in treating of the physical geography of this basin, that at some former period the waters of the lakes may have so discharged themselves; but as the present observations are designed first to bear upon the dividing ridges or heights of land in Indiana, we now proceed to discuss those points.

As there exist in Indiana many dividing ridges, or water-sheds, or divortia aquarum, a term used by Humboldt and others, we must be careful to avoid the error into which I found a citizen of one of our northern counties had fallen, who contended he occupied the highest portion of Indiana, indeed, he added, he might say of North America, because, from his farm, water flowed south ultimately into the Gulf of Mexico, as well as north-east into the Gulf of St. Lawrence.

We must bear in mind that the great valley of the Mississippi is a central basin rising towards Lake Itasca, the source of the mighty river which drains this valley, acccording to some authors, from an elevation of about 1,800 feet above its lower termination at the Gulf of

Mexico. On this height of land, after great rains, canoes can be paddled, from south-flowing streams, into waters that discharge into Hudson's Bay; and, not very far from the source of the Mississippi, head waters reach rivers emptying into the Arctic Ocean, wherever the drainage to the Gulf of St. Lawrence is, from near the same region, an eastern side slope from the head or higher plateau of this great valley; the diverging waters being separated often only by insignificant ridges or even swamps, in Wisconsin, Illinois and Indiana.

From the west or right bank of the Mississippi we have a gradual rise, estimated at about six feet to the mile, until within two to four hundred miles of the Rocky Mountains, so that the emigrant, when he leaves the great carboniferous basin and reaches, among the adjacent Permian, Cretaceous and Tertiary strata, the higher lands extending from Council Bluffs nearly due south, is at Fort Riley, Kansas, or at Fort Arbuckle, Indian Territory, already from 1,000 to 1,200 feet above the ocean; reaching Fort Kearney, Nebraska, or, by a more southern route, Fort Belknap, Texas, he is at an altitude of from 1600 to 2,300 feet; and at Fort Laramie over 4,500 feet. If he pursues his journey by the banks of Sweet water and crosses the South Pass, three hundred and twenty miles west of Fort Laramie, he is about 7,000 feet above the sea, and is in sight of culminating points over 13,000 feet high, which further north, in the Cascade Range that connects the Sierra Nevada with the Rocky Mountains, attain an elevation of 16,000 feet or more. From this Pass, reached by a gradual ascent, an almost equally imperceptible descent brings him off these heights to a table land, with saline lakes, &c., still on a general level of more than 4,000 feet high, whence some waters flow west to the Pacific.*

A similar gradual rise may be traced from the east or left bank of the Mississippi, towards the Allegheny range, although not to an elevation equaling the western culmination. If we examine the hypsometrical details of that region determined by the talent and energy of a Guyot, or consult some of the transvere sections given in the admirable work of the great climatologist, Blodget, we find in those corresponding somewhat to the latitude of Indiana, the following ascent:

---

*Fort Hall is 4,500 feet above the sea; Great Salt Lake, Capt. Stansbury places at 4,351 feet.

	FEET ABOVE SEA.
Mississippi river, at St. Louis, (low water)	381
Lake Erie	565
Pittsburg	700
Plain near Pittsburg	1,100
Blue Ridge of Pennsylvania, general average	1,100
Alleghenies at latitude $37\frac{1}{2}°$	2,650
Mt. Marcy, highest point in the State of New York	5,344
High Pinnacle of Blue Ridge	5,701
White Mountains, average of the eight highest peaks	5,836
Mt. Washington, (culminating point of northern section)	6,288
Black Dome of the Black Mountain, main chain	6,707

From a region, which is almost the geographic center of this continent, (equidistant from the Pacific near Vancouver's Island, the Atlantic at the Gulf of St. Lawrence, the Gulf of Mexico near the mouth of the Mississippi, and the Arctic Ocean, at Dolphin and Union Straits, or at Simpson's Straits,) the summit of a marshy plateau, as Sir J. Richardson, in his Arctic Expedition, styles the dividing ridge of Arctic and Atlantic waters, in about latitude 48° north, and latitude 17° to 18° west of Washington, all the drainage received from the great eastern slopes of the Rocky Mountains, or the western slopes of the Appalachian range, is carried from an elevation scarcely exceeding 800 feet,* northward to the Arctic Ocean, through a dreary and scarcely habitable expanse of lakes and marshes; southward and eastward, through valleys as fertile as any in the world, embracing, with the St. Lawrence Basin and Texas Slope included, an area of considerably over a million and a half of square miles.

Collecting in its onward course at least one Giant Tributary that had meandered several thousand miles before its confluence with the mighty Father of Waters, (this Michi-Sibi or Great River of the Ogibawa or Chippawa Indians,) besides many other affluents that in their thousand miles might themselves drain a moderate-sized continent, this deep and quiet stream bears on its bosom the countless steamers and flat-boats, which are the medium of ultimately diffusing to all parts of the globe the agricultural wealth of our rich valleys; the former bringing in their return trips the comforts and luxuries sent by the mechanical skill of regions less extensively adapted to an agriculture, which from the extent of Ter-

---
*See Blodget's Chimalogy of the United States, page 102.

ritory and consequent variety of climate can excel equally in the hardy tubers, graminæ and fruits of northern growth, in the valuable cereals and fruits, flocks and herds of a temperate latitude, and the luxurious products of an almost tropical vegetation, such as sugar, rice, figs, oranges, bananas, &c., as well as the commercial staple, cotton.

Of the Mississippi Valley, Indiana forms no unimportant part, being situated centrally as regards latitude,* and occupying the productive region which, in conjunction with Illinois, forms the last terrace of the Atlantic Slope in its western convergence to the east bank of the Mississippi. As we have already seen, no high mountains exist in our State; the greatest hyper-oceanic elevation in Indiana, of between 1,000 and 1,100 feet, is attained about midway on the eastern border of Indiana among the rocks geologically the lowest. The height gradually diminishes going south-west to the coal basin on our western borders; the uplands of which average from 400 to 600 feet above the sea, and north of the Wabash the country is much more level, at a medium average of about 700 feet above high tide in the ocean.

The great backbone or Rocky Mountain Range runs through our continent somewhat east of south and west of north, preserving nearly the same trend in the Andes of South America, while the Allegheny Range constitutes, like the mountains of Brazil, a subordinate elevation forming in its linear extension an angle of about 60° with the backbone; so that a distance, equal to 45°, estimated on the equator, set off from the northern head of the Rocky Mountains near the mouth of McKenzie river, to the coast of New Foundland, and thence again along the Alleghany strike to the southern base of the Rocky Mountains among the table lands of Chihuahua, and once more back to the place of beginning, froms an *equilateral* triangle. A rather abrupt descent, sloping west from the Rocky Mountains, forms the Pacific coast, somewhat parallel to the backbone, and a rapid south-east slant, carries the Allegheny Atlantic slopes to that ocean, in a coast line nearly parallel to the elevation. The descent to the *interior* central depression is much more gradual, but these exhibit also some parallelism to the ranges of elevation. Thus a line, connecting Great Bear Lake, Great Slave Lake, Ithabasca and Winnepeg lakes with the general direction of the Mississippi, would run nearly in the strike of the Rocky Mountains. Lakes Erie and Ontario, the river and Gulf of St. Lawrence, all have a trend somewhat parallel to the Apalachian chain, while Lakes Supe-

---

*From the Ohio in about $37\frac{1}{2}°$ north latitude, to Lake Michigan at nearly 42°.

rior and Michigan form a more central north and south drainage for part of the Mississippi Valley.

Sec. 2.—Physical Geography of Indiana as part of the Hydrographic Basin of the Mississippi Valley, including her Prairies, Sand Plains, Lakes, &c.—From the preceding section it seems probable that the physical character of the Mississippi Valley is due partly to the transported and deposited materials in the gradually subsiding waters which existed in by-gone ages between the Rocky Mountains and Allegheny Range, sometimes as a salt-water ocean receding and again ebbing and overflowing, afterwards probably as a great fresh-water expanse or lacustrine basin, and finally as a main stream, with tributaries of running water gradually narrowing and deepening its channel, leaving first great sand-plains, the shores and sand-bars of the ocean, on the present elevated plateaus; then prairies rolling and gravelly, or arenaceous; then lower swamp-muck prairies, savannas and the like; finally successive river terraces forming often second and third river-bottoms, with marl and fluviatile shells, the latter of species almost invariably identical with recent.

If we cast our eyes on a good map exhibiting the physical geography of the globe we are struck by the fact that nearly all great valleys constitute in their lowest portions prairies, heaths, steppes, selvas, llanos, savannas, pampas, everglades, &c., with their peculiar flora and fauna, modified chiefly by latitude, while more elevated parts of the same valleys constitute sandy plains, or rise, most generally in the south, to plateaus often arid, sometimes with saline lakes in their interior; and not unfrequently the surmounting of a subordinate ridge leads us by this table land to the main range. These seas or salt lakes are found usually diminishing in their hydrographical extent.*

Thus the great marshy and mossy tundra of Siberia becomes gradu-

---

*Lake Titicaca is rather more than half the size of Lake Erie, comprising an area of 4,000 square miles. Lieut. Gibbon reports that it is gradually filling up, "that the water is getting shallower every year." "Finally," he says, "there will be a single stream flowing through what in future ages may be called Titicaca valley." This region includes the plain of Cuzco, which is in itself three times the extent of Switzerland. "Warren's Phy. Geo., p. 17. Regarding the Great Salt Lake, Capt. Stansbury remarks in his account of the expedition to Utah, p. 105: "Upon the slope of a ridge connected with this plain thirteen distinct successive benches or water-marks were counted, which had evidently at one time been washed by the lake, and must have been the result of its action continued for some time at each level. The highest of these is now about two hundred feet above the valley.

ally higher and drier in the Kirghis Steppes* of Central Asia, and is bounded on the south by arid table lands, as those of Gobi and Iran, which lead to the subordinate ridges and ultimately to the highest mountain range, with the not far-distant sea of Aral every year gradually diminishing in extent. Central Europe has its northern marshes, its extensive heaths, and its more elevated southern lands and adjoining Asiatic Caspian Sea, of salt water and depressed basin, eighty-three feet below the ocean. Africa and Australia, as far as known, do not deviate materially from this type.

In South America we have the Llanos of Carraccas, extending to the Orinoco, and occupying over 16,000 square miles. Interrupted only by the forests of the Amazon, this valley extends through the grassy plains of the Apure, and pampas of the LaPlata rivers, to the Straits of Magellan. South-west of the Llanos, the table lands and Despoblado desert, one-third the size of the African Sahara, and Lake Titicaca rapidly filling up, all conform to the general rule, while south of the pampas the Desert of Patagonia exhibits the same principle in physical geography.

In North America, the Mississippi Valley has centrally low prairies of coarse grasses, cat-flags, &c., and rises to higher rolling prairies, (luxuriating in a composite vegetation, chiefly of the helianthus genus,) similar to the grassy Steppes of Europe and Asia; finally on the south-west forming an Artemesia table-land, part of which is an arid desert, with the Great Salt Lake, and others, leading to a culminating ridge.

An examination and comparison of these and many similar facts would seem to point to the conclusion that the vast bodies of waters which originally filled these valleys had gradually altered their level, at long intervals, either through a gentle elevation of the land, probably from volcanic action, or sometimes through the denuding effect of the running water cutting deeper channels.†

To account for the various phenomena geognostic and geographic which have given to Indiana its peculiar orographic and potamographic

---

*These are characterized by the immortal Humboldt as grassy plains, having a growth of Rosacea, Fritillarias, Cipripedias and of high herbaceous plants, such as Astragalus, Saussurias and Papilionaceæ. The salt and soda deserts are those of Kerman, Seistan, Beloochistan, Mekran and Moulton.

†These effects I have noticed at several places in our county and elsewhere as amounting, in twenty-five years, to nearly that number of feet, or an average excavating power of rivulets, in our sandy loam, of one foot per annum; after reaching a stiff argillaceous soil, the rate was diminished at least a half.

configuration, or in other words its elevations, depressions and river drainage, we are almost forced to conclude that during the coal period proper, excluding the sub-carboniferous, as that is chiefly marine, the western part of our State, and nearly the whole of Illinois, were long slightly submerged, or at least so depressed as to constitute fresh-water marshes, a fact indicated by much of the coal flora; but that occasionally, for periods of considerable duration, long enough to form limestone beds two to twelve feet thick, containing marine organisms, the waters of the ocean must have covered parts of this valley. By an increased depression, at a subsequent period, over portions of the great valley which now form plains from one to five thousand feet above the ocean, and which were possibly near the shore, the sea must have remained for ages to permit the life, death and subsequent fossilization of a Permian Flora and Fauna, as seen in Kansas, a Triassic found in Connecticut and other eastern States; patches of Oolite,* both in our Atlantic and Pacific ridges; a cretaceous deposit found occupying a belt from Nebraska, through Arkansas to Tennessee, and portions of the Atlantic States, as well as the superincumbent Tertiary with lignite beds, and locally abundant marine shells, Echinoderms, corals, &c.

After the lapse of these long periods a vast accumulation of materials rounded by attrition has evidently been brought by some agency, now usually supposed glacial, from the north, depositing in the higher latitude enormous rounded bowlders, weighing hundreds of tons. In the latitudes south of 50° north, in the United States, these bowlders usually diminish to the weight of at most a few tons, then a few hundred pounds, and finally, south of the Ohio river, we have rarely evidence of erratics and find mostly a transported gravel, sand or quaternary clay, reposing on the rocky substratum. On the eastern shores of our pre-adamite ocean, Prof. Hitchcock saw White-Mountain bowlders at over 5,000 feet of elevation; and on the western Rocky Mountain border of the same supposed ancient sea, Col. Fremont detected erratics at about the same hyperoceanic level. In the central portion of the valley, except about Vicksburg and other parts of the State of Mississippi, either the waters were too deep for the formation of Permian Oolitic, Cretaceous and Tertiary deposits, or they have been swept away, leaving no traces, down to the Coal Measures.

During the Drift or Erratic or Older Quaternary period, whether or

---

*See Col. Fremont's expedition to the Rocky Mountains, pages 131 and 277; also, Hitchcock's Geology of the Globe, p. 99.

not an estuary in our valley aided the glaciers and icebergs of the north to disperse these erratics in a southerly direction,* and groove the rocks over which the angular fragments were transported, it seems at least evident that a great depression finally remained, possibly in part the crater of a subterranean volcano,† in part an estuary, converted as it dried up into a later Quaternary valley of deundation, ultimately retaining in places bodies of fresh water, then constituting a great chain of lakes and river courses, imbedding particularly in the valley of the Ohio, with recent shells,‡ the skeleton of the Megalonyx, Mastodon, and similar mammals, in Alabama the Zeuglodon, &c. Humboldt, Mrs. Somerville and other eminent writers on physical geography consider all great plains, steppes, llanos, pampas and prairies as the ancient bed of a water basin, ocean, estuary, salt or fresh water lake; and the evidence offered throughout the entire prairie region, visited during the Indiana survey, goes to prove that, although prairies are of very varied elevation, the lower ones now rapidly draining,§ are usually the most level, with a swamp-muck surface and marl or a clayey sub-soil, and are almost invariably surrounded, wholly or partially, and sometimes also traversed, by sandy or gravelly ridges. More elevated prairies are usually more sandy or gravelly, or undulating from deundation, as they become higher and drier. The clay sometimes extends to hundreds of feet in depth, with only an occasional thin bed of sand intervening, and is so fine, like the impalpable mud deposited in the beds of some rivers and lakes, and well exhibited on their drying up, as to offer no chance for the necessary air and moisture to penetrate far enough for the growth of forest-tree seeds, whose roots require a subsoil-permeating-space almost equal to their aerial ramification. That this necessary aeration

---

*As Prof. Hitchcock remarks, the Drift may not always have the north to south direction observed in the erratics of North America and Scandinavia.

†The evidences towards this conclusion are very strong in the miles of basaltic columns and dikes, amagdaloid and other volcanic rocks on the north shore of Lake Superior.

‡The deposits imbedding the shells and bones are often associated with a rich marl bed, extending occasionally some miles each way from the great rivers, and evidently older than the modified later Quaternary of the second bottoms, or last terrace but one. Sir Chas. Lyell considered these marl beds the equivalent of the loess of the Rhine. Forty-two feet below the surface of the modified Drift in the Wabash valley we found coarse-grained dicolyledonous wood resembling willow or cottonwood, about four inches in diameter, and extending six or eight feet across the excavation, how much further was not ascertained.

§As already stated, places were shown me near the Kankakee where three years before a man would have mired down; now they are wagoning their hay from adjoining prairies safely across these formerly quaking bogs.

and watering of the sub-soil for the germination of trees is often effected near river courses by long continued moisture, being probably taken up through capillary attraction, is rendered probable from observing the rise and fall of water in wells near rivers,* and from our finding generally a belt of large timber on the water courses of treeless plains, as well as scrubby timber on the porous sand-ridges of prairies. The same is made further evident from the fact that artificial stirring of the soil soon enables the settler to have his orchards and walnut groves, &c., and the analytical chemist also shows that no essential ingredient is lacking in the prairie soil. In short the difficulty arises evidently from a mechanical not a chemical defect, consisting in the compact, impervious nature of these hydrographic-basin soils; plants deriving most of their nourishment through gases and the solvent power of water holding saline and other ingredients in solution. In the prairie soil, air and moisture penetrate this refractory stratum far enough to nourish grasses and some herbaceous plants, but not to a sufficient depth for the spongioles of a ramifying tree-root. In some wet prairies, where the moisture extends far enough, the growth is only favorable to certain aquatic or marshy plants, on account of the warm and stagnant water decomposing the seeds of many plants before they have time to germinate. Those who have drained thousands of acres of lake† or swamp land, give their testimony that, after a few years, they have exactly the appearance and character of low prairies.

Other vast regions are a drifting sand, which in certain latitudes and with sufficient atmospheric moisture will give rise to specific vegetation, but in climatological circumstances less favorable to evaporation and deposition, (such as a vast continent, absence of trees, the arrest of clouds by high mountains skirting the rainless district, or too much heat for the reduction of the surcharged cloud to the dew-point temperature, necessary for rain,) may often remain arid and uninhabitable deserts, only furnishing a few oases, at spots where water reaches the surface from some favoring cause.

The vast regions of sand seem sometimes to have been the shore or sea beach of the great ante-historic water-expanse, that formerly cov-

---

*A high freshet in the Wabash not only rises the water high in the New Harmony wells, but even filters into some of the cellars in the lower part of town 500 yards from the left bank of the river.

†See the remarks on the draining of Beaver Lake by Major Dunn, given among the details of Jasper and Newton counties.

ered the large valleys, and may be more usually an accumation, *rather South* of the extensive plains *than North*, from causes dependent either on prevalent winds raising sand banks, on southern elevations of land, or on a northern depression, or possibly may be influenced remotely by the earth's rotation and the oblate spheroidal form.

In our own State it is scarcely possible for any one, after closely examining the regions south of Lake Michigan, to deny that the lake is having its waters gradually dammed up on the southern shore, and that for ages past sand ridges parallel to the lake shore have been redeeming land from the lake partly as prairie swales, partly as more elevated arenaceous banks. Possibly, indeed probably, the chief outlet for these waters was formerly through the Mississippi valley south, and successive elevation through these ridges alone, from the prevalence of northern winds and waves piling up the sand on the south shore, or the above combined with a gradual rise of land, may have compelled the lakes to empty themselves by their eastern basin margin, overcoming the previous subordinate side slope elevations, and even damming up such streams as St. Mary's river and St. Joseph in north-east Indiana, forcing the waters through the Maumee river almost directly back upon their original course of descent. This may have occurred as early as the Drift-period, when vast accumulations of gravel and sand were deposited in the north and south lines of depression, subordinate valleys intervening in Indiana, between the strike of different geological formations, deepened by subsequent deundation.

The great deposits of sand south, usually somewhat west, of the plains or valleys and adjoining a height of land or mountain range, besides being partly caused by changes of level, may be also partly due in some parts of the globe to detritus from the mountain slopes, but generally appear more like the accumulations, which can be seen washed and drifted, as bowlders, gravel and sand, on the shore of Lake Superior, such as have been described in this Report,* as occurring on the south shore of Lake Michigan to the elvation of one hundred and seventy-six feet, and in ridges parallel to the lake shore in concentric curves, sometimes a few miles apart, with intervening prairie, &c., consequently in all probability the effect of wind, and water and ice combined.

Whether this drifted sand has been the result chiefly of former south-

---

*See the details of St. Joseph and LaPorte counties.

erly currents, or whether it is due to a still prevailing wind, or partly perhaps to laws connected with the earth's motion from west to east, and the equatorial configuration of our globe, or to all combined, remains for future examination and determination.*

The effect of such accumulations of sand barriers on the drainage of a country is well illustrated by reference to Mr. Ellet's valuable remarks on the Mississippi valley and its rivers, in which he shows that among the steep banks on the Allegheny river, a dam or barrier fifty-eight feet high would create a pond twenty-five miles long; of course the area in a more level district would be much greater.

If the head waters of the St. Croix, flowing into the Mississippi, were connected across the present short and low portage with the sources of the Bois brule or other Lake Superior tributary, the waters of that lake could be made to flow into the Gulf of Mexico; and if a canal some 1,300 miles long were constructed, there is fall enough, allowing an inch to the mile, to drain Lake Superior into the Gulf until less than 300 feet of water remained; inasmuch as the surface of the water is 618 feet above high tide in the ocean, and the greatest depth of the lake is 791 feet.

Attention is called to this fact not with the slightest intention of recommending such work as practical or useful, but simply as the most ready mode of giving a general idea of the relative levels, the former and present draining of our portion of the Mississippi valley, the gradual diminution of the waters on the south shore of Lake Michigan, and the probable increase of drainage through the river and Gulf of St. Lawrence.

It is thought there is strong proof offered in the details of the Report on the Indiana counties situated in the Drift or Early Quaternary Formation to show that if our valley were again depressed for a sufficient period we should have all the requisites to present at a subsequent elevation a genuine coal-field. But before proceeding to collate and compare some of those details, it is deemed advisable to recapitulate briefly the evidence that our prairies and plains are the result of a gradually dried estuary bed:

1. Drained lakes have very much the appearance and character of prairies, as proved by the experience of Major Dunn and others on Beaver Lake.

---

*If similar phenomena were observable in the southern hemisphere, with, however, the points of the compass reversed, the latter supposition would assume considerable probability.

2. The great North American valley, like those of Europe and Asia, exhibits the lowest ground centrally, and gradually rises on both sides of this central drainage to higher ground by successive terraces, which are evidently ancient water marks.

3. The lower prairies are often composed of gravel, sand and clay, with some bowlders, such as might be found at the bed of a lake or estuary. The higher prairies are more gravelly and have also bowlders like those of the lacustrine shores. The accumulation or piling up of materials on the lake-shores can be most satisfactorily studied around Lake Superior, between the headlands.

4. The great prairies have smaller arms of prairies running into the timber, exactly as we see the bays and coves of estuaries and lakes cutting into the low shores.

5. The timber groves are almost invariably more elevated and sandy than some near adjoining prairie, just as the islands and sandbars left, when a lake or river is drained, would be higher and more arenaceous than the dried, muddy ooze, compacted by dessiccation into a soil too impervious for the germination of forest trees.

6. The growth approaching to good sized timber, which we first see on drained prairies, is usually a willow, or one of the true poplar family, as cotton-wood or aspen, (Populus monilifera and P. tremuloides); this is also often the first growth on a river sandbar or dried up bottom.

7. Extensive borings exhibit on the prairies vast deposits of clay, with occasional beds of sand, the whole sometimes covered by a humus many feet thick. This is very much the character of the materials exhibited on boring in some drained lakes and water courses.

8. In nearly all the numerous observations made during this survey the prairies were found to some extent surrounded by sand rides with scrubby trees, and if another prairie succeeded at a greater elevation, it was also partially surrounded by sand-banks, and a third or even a fourth occurred similar in character, except being a little more elevated; just as the chain of lakes would appear if drained, which extends from Lake of the Woods, through Rainy Lake, &c., to the mouth of Pigeon river in Lake Superior; or from Lake Athabasca, through Great Slave Lake, into Great Bear Lake and McKenzie river of the Arctic ocean, or any similar terrace chain of fresh-water expanses.

9. Nearly all the low, flat, or gently undulating prairies of North America are in a belt extending south of our great American lakes, to the Gulf of Mexico, through Indiana, Illinois, Missouri, Arkansas and Texas, while east and west of those we have higher and more rolling

arenaceous prairies in Michigan, Wisconsin and Iowa, and cypress swamps in part of Louisiana, rising to Buffalo and Artemesia plains in Nebraska, Kansas and Indian Territory, and finally to the great table lands of Utah, with Pinus monophyllus and some Yucca growth, or the Mexican plateau of cactus, palmetto and date trees.

Let us now examine how far these vast natural meadows would resemble our coal fields, if the former were partially submerged for a long period, and subsequently elevated:

1. They would have the same basin-like form observed in all our extensive coal fields, with an increased rate of dip near the margins.

2. These herbaceous prairies and adjoining cypress swamps of Indiana and elsewhere, with occasional drifted timber swamps, would afford about the same character and amount of material, &c., viz: decomposing vegetable matter, which chiefly by the loss of its oxygen has evidently furnished the hydro-carbon coal beds, usually three or four feet to eight, ten and even occasionally thirty or forty feet in thickness, during the true Carboniferous era, as well as the concomitant fossil resins, bitumens, &c.

3. Underneath this ligneous fibre, humus, &c., would be commonly found a clay or aluminous deposit, just as we usually find a bed of fireclay beneath the coal-seam.

4. At a higher elevation than the swamp-muck exists often a sand ridge such as those immediately south of Lake Michigan, which during submergence would probably be somewhat leveled and more equally distributed over the ligneous detrital beds of the future coal, alternating with the mud deposited by the overflowing water, thus furnishing future roof shales, locally of sandstone, more generally of aluminous schists.

5. If prairie after prairie, with its groves, were submerged, and stratum after stratum received its dying vegetation, peat-mosses, ferns, grasses, shrubs, trees, as well as its overwhelmed inhabitants, mollusks, crustaceans, fishes, reptiles, birds and mammals, we would have a succession of beds and of palæontological variety not very dissimilar in their general character from those of the carboniferous flora and fauna, only that the vegetables and animals would usually be more highly organized.

6. Some of the beds (which then would resemble our brasher and more western coal,) might result chiefly from prairie humus with decaying ferns, cat-flags, rushes, rosin weeds, wild indigo and other shrubs, willows, aspens, &c., or further west, the woody fibre from countless gen-

eration of sage bushes, (Artemesia) and the like, the detrital accumulation of ages might furnish the predominant material. The animal remains would be mostly the land and fresh water shells, (such as those covering the prairie near the Kankakee river, described in the report on Lake and Porter counties,) crawfishes and, in places, a swamped elk or reindeer,* as in the Irish peat-bogs the Megaceros is found, possibly even a representative of the human family, misled by an ignis fatuus and irretrievably mired. In the great plains might occur the fossil bones of a buffalo, prairie dog, or occasionally, the glutton wolverene, (Gulo luscus.) More bituminous and anthracitic beds† would be likely to result from the chemical changes which the cellulose of the transported trees buried hundreds of feet in the delta of the Mississippi river, would furnish, or from the vast natural rafts of drift-wood in Red river, similar accumulations of less extent in the "bottoms" or lowest terraces of other tributaries; as well as in the northern cedar and tamarack or southern cypress swamps. Associated with the former are abundant Balsam Firs, (Abies balsamea,) the exudations of whose blistered bark, would readily furnish a transparent viscous resin to enclose a struggling mosquito, in lieu of any præ-historic insect, and render up, on examination, a fossil resin as pure as any amber that ever stranded on the shores of the Baltic. The cypress "knees," the piny needle-leaves, leaving in some species a double bark-scar, and the abundant fire-cones, would have some points of resemblance, at least externally, with the trigonocarpon, supposed to belong to a conifer, and the lycopodeaceous lepidodendron, carrying its pendant lepidostrobus, also with the somewhat allied sigillariæ and not yet fully assigned stigmarian roots.

It seems unnecessary to enumerate the animals which under the supposed circumstances might be found fossil, such as the alligator and chameleon, &c., in the south, perhaps the porcupine or lynx of Canada, an oppossum or raccoon, the pinnated grouse of the prairie, or the common meadow-lark.

7. Associated with some of these beds would be quantities of the bog iron-ore (hydrated peroxide) representing the iron ores usually found with the low coals, sometimes with the sub-conglomerate series.

8. The prairie deposits, such as those of the Grand Prairie near

---

*The horns of elk are found abundantly about Lake Superior, and traces of reindeer are by no means uncommon in the same region.

†See the numerous comparative analyses of coals in Bischof's Chemical and Physical Geology, vol. 1, page 258 *et seq.*

Mounts Gilbo and Nebo, described in the details of Jasper county, might often be found resting on the coarse gravel of older Quaternary age, just as the low coals rest on the true Carboniferous Conglomerate or Millstone Grit.

9. That such a deposit of fossil fuel should be formed at some future age from the submergence of our prairies, plains and swamps, or from the vast heaths and bogs, steppes and savannas of other regions, seems not improbable, when we consider that nearly every geological period has been ushered in or separated from the preceding by a deposit of drifted, rounded and sometimes re-cemented materials, to which we give the designation of conglomerate. Further, we have in each geological epoch successions of sandstone, limestone and shales, affording, after the earliest palæozoic periods, when some dry land made its appearance, evidences of basins of organic matter converted into coal, even, it is said, as early as the Devonian period. Somewhat thicker layers are found, as circumstances became more favorable, during the deposition of the sub-carboniferous sandstone and limestone series, culminating to its maximum during the epoch of the true Coal Measures. These sub-carboniferous coal deposits can be interestingly seen and extensively examined in Indiana. The Oolitic period also furnished its vegetable matter for conversion into coal, and some of that material mined in Eastern Virginia is assigned to the Oolite, as well as some coal and shales containing fossil ferns, found, on the western side of the Rocky Mountains, by Col. Fremont and described by Prof. Hall as probably of that age. One locality is near Fort Hall, the other near Utah lakes; both at an elevation of between four and five thousand feet above the sea. In the Upper Oolite of Europe, the Lower Purbeck exhibits, according to Sir. Chas. Lyell, "beneath a thin marine band, purely fresh-water marls." * * "Below the marls are thirty feet of blackish water beds." * * "The great dirt bed or vegetable soil, * * rests upon the lowest fresh-water limestone," which in its turn rests on the Portland stone of purely marine origin. The black dirt, according to the same author, "was evidently an ancient vegetable soil. It is from twelve to eighteen inches thick, is of a dark brown or black color, and contains a large portion of earthy lignites. Through it are dispersed rounded fragments of stone from three to nine inches in diameter, in such numbers that it almost deserves the name of gravel. Many silicified trunks of coniferous trees and the remains of plants allied to Zamin or Cycas are buried in the dirt beds." The above description fills in many respects the outline of the appearance which

some of our prairies and adjoining groves might be supposed to present after long submergence and partial fossilization, and subsequent elevation or at least dessication.

In the Tertiary Period there are abundant beds of good lignite or Brown coal, usually lighter, but sometimes scarcely distinguishable from cannel coal seams of the true Coal Measures.

If then we have beds of sub-carboniferous coal, coal fields of the true carboniferous fossil fuel, Oolitic coal and Tertiary coal, does it seem improbable when we observe similar agencies at work, that there should hereafter result a Quaternary coal. The marine accompaniments would be especially liable to occur in regions similar to that of the Great Dismal Swamp, supposing a slight subsidence of the land, or an unusual breaking down of sea barriers between the Chesapeake Bay and Albemarle Sound, or in the everglades of Florida, and the inundated prairies and marshes of Louisiana and Texas; but they might also exist through the great valley of the Mississippi, supposing a depression of a few hundred feet which would suffice to connect the waters of Hudson's Bay with those of the Gulf of Mexico. A similar result might occur from a rise of our Appalachian chain and Atlantic sea-board, to about an elevation equivalent to the above supposed inter-montanic depression, in which case we might expect to see extensive strata made up of consolidated oyster beds, and long reefs of coraliferous limestone with a due imbedding of New Jersey King-crabs, Georgia saw-fish, sting-rays, echini and star-fishes, Florida green-tortoises, &c.

The greatest depth of the Atlantic between New Foundland and Ireland is 12,740 feet, but a continuous internal force acting moderately for a lengthened period, or a sudden convulsion, such as has occurred more than once in the Historic Period, to elevate a Graham's Island or a Monte Nuovo, and such as all geognostic observations tell us must have happened frequently in the earth's history, might readily lay dry portions of the "telegraphic plateau," parts of which are at present one to two thousand feet below the surface, or upheave an island teaming with the fucus-vegetation of the Sargasso Sea, which obstructed the ships of Columbus, and which, although it does not grow at a greater depth than 200 feet, may for ages have been depositing its detritus at the bed of the ocean.

It may be proper to observe that some eminent writers (see remarks of Mr. Lesquereux in vol. 3, Kentucky Report, page 523,) deny any true marine formation of *coal*, adding that "there does not exist a bed of true marine peat, viz: peat formed entirely of fucoides and marine plants;" and that he has "never seen a piece of coal with evident marks of marine origin." But the same writer, at page 552, qualifies the above and admits *marine accompaniments*, by stating that "where a quiet water is high and the marine element predominating, a limestone may be formed, when at the same time, in more shallow marshes, the plants will grow and their remains make a deposit of coal or shales." Bischof, vol. 1, page 205, of his work already quoted, after contending that "all the hypotheses referring to the formation of coal from partial vegetable remains, agree in ascribing the origin of coal to materials, comparatively less abundant than is consistent with the immense quantities of this substance," adds on the next page: "Fuci, exposed to the influence of a high temperature and water, are decomposed after a few days; therefore it cannot be doubted that after their decomposition they would have sunk into the sea, and furnished material for the formation of coal; moreover, varieties of fuci actually occur in a fossil state in coal." As proof of this Bischof refers to Brown, Handbuch einer Geschichte der Natur, T. III, p. 61.

Without for a moment designing to decide between such high authority, I quote these observations mainly to show that, under either supposition, without any violent straining of probabilities, all the circumstances might readily exist, for the conversion of the vegetable detritus in our great prairies and vast plains into fuel for future epochs.

But in the same deposits we have meanwhile rich agricultural resources requiring only a judicious system of drainage to make these humus beds sources of immense wealth. We are therefore naturally led next to an examination of this eminently practical question.

NOTE.—It was the intention to subjoin a chapter on Drainage, recommending, where prices of agricultural products justified it, a system of underdraining; also, to give a chapter on Palæontology, systematically arranging the fossils of Indiana obtained from the different Formations; then to follow with an exhibit of the main facts collected regarding the localities, causes and other concomitants connected with milk-sickness, and finally to close with a miscellaneous chapter containing suggestions with regard to the best mode of prosecuting the Survey, the most usefully manner of arranging the State collection for reference, lithologically, palæontologically and zoologically, as well as recommendations regarding the formation of minor illustrative collections for public schools; but a call to serve my country in maintaining the Union and the Constitution precludes the possibility of completing that design, and compels me to close **the report.**

CAMP TIPPECANOE, 20th June, 1861.

A REPORT

OF THE

CHEMICAL ANALYSIS

OF

THIRTY-THREE SOILS OF INDIANA,

MADE FOR THE

STATE GEOLOGICAL AND AGRICULTURAL SURVEY.

BY ROBERT PETER, M. D.,
PROFESSOR OF CHEMISTRY, LEXINGTON, KY.

INTRODUCTORY LETTER.

CHEMICAL LABORATORY, LEXINGTON, KY., April 30, 1860.

DEAR DOCTOR:—I have the pleasure to transmit to you my report of the chemical analyses of thirty-three of the soils of Indiana, of those collected by your brother, Prof. Richard Owen, in his reconnoissance of a part of the State in 1859. They have been carefully analyzed according to the method described by me in my chemical report in the third volume of *Reports of the Geological Survey of Kentucky*, (Frankfort, 1857,) and, it is hoped, may throw some light on the chemical nature of the soils of Indiana, in the regions where they were collected, and aid enlightened agriculturalists in the culture of their lands.

The whole number of soils collected by your brother has not been examined, (only about half having been analyzed,) because of the limited means which could be devoted to this investigation; but I have selected some from each of the geological formations represented. The number now reported is, however, too small to enable us to make, at this time, a satisfactory comparison in this relation; if indeed the wide prevalence of the Quaternary deposits may not interfere with it.

Every person of reflection is no doubt convinced that agriculture lies at the very foundation of our national prosperity. Should our lands fail to support the population, no perfection in the scheme of our government, no honesty or skill in its administration, would preserve us from decay as a people. It is therefore our duty, as well as that of the government, to foster this indispensable branch of industry, and to aid, as much as possible in the establishment of a correct and philosophical system of agriculture.

Experience of the diminution of agricultural products, not only in the old and long worn lands of Europe and of the older States, but also in our new States, has shown that in the ordinary modes of cultivation of the soil, the land undergoes a gradual and sometimes a fatal deterioration; so that, even on this continent, lands which were abundantly fertile to the early settlers, and capable of supporting and enriching a dense population, are now wastes which no longer reward the labor of cultivation, or only yield useful crops as a result of the application of foreign manures, in the forms of guano, ashes, bone dust, &c., from abroad or from the neighboring cities. The cause of this impoverishment is now clearly made known by the aid of chemical analysis; which is the only mode by which it could be certainly ascertained. Chemical study of the soil and of plants and animals has demonstrated that certain elements, essential to vegetable and animal development, are gradually consumed from the soil in the crops—that the soil is not a *unit in composition*—that while the great bulk of it acts only *mechanically*, or *physically*, in the support of vegetables, the mineral elements which are essential for the nourishment and growth of organic beings, vegetable or animal, are found in it only in relatively small proportion, and must be carefully husbanded and restored to it in order to maintain constant fertility. Such a process as this, by which the land would be constantly *kept up to the height of fertility and would annually yield abundant crops without any diminution of its richness*, would be the perfection of agriculture. *Such a system is perfectly practicable* in an agricultural community, where the chemical nature of soils, of manures, and of vegetable and animal products have been studied and understood. The path of improvement in modern agriculture, therefore, lies in this direction; and it is the duty of our enterprising farmers to prepare themselves to pursue it, by the scientific study of their profession; and that of States and communities liberally to aid progress in this pathway.

The fundamental study in this relation is that of the chemical nature of soil; a study which is yet in its infancy, but which may be matured by judicious public patronage into a branch of science of extensive utility.

Yours respectfully,
ROBERT PETER.

D. D. Owen, M. D.

EXPLANATORY REMARKS.

The principal elements found in soils which are essential to vegetable nourishment, as well as to animal growth and development, are the following:

Carbon, \
Hydrogen, } *First*, those which are sometimes called the *organic elements*, because from the greatest weight of animal and vegetable bodies. These in the soil, are contained in the *organic* and *volatile matters*, so called.
Oxygen,
Nitrogen.

Second, What are called the *mineral* or *inorganic elements*; as follows:

Potassium, } which form the alkalies *Potash* and *Soda* when they are
Sodium, } united with oxygen.

Calcium, } which, combined with oxygen, form *Lime* and *Magnesia*.
Magnesium, }

Iron, } existing as *Oxides* in the soil and in the plants, &c.
Manganese, }

Silicon, existing in the soil and as *Silex* and *Silicic acid* or *Silica*, which is also an oxide, or a compound of silicon and oxygen.

Phosphorus, } which, combined with oxygen, form *Phosphoric* and
and Sulphur. } *Sulphuric Acids*.

Chlorine, Especially combined with *Sodium* as common salt.

The *Carbon, Hydrogen, Oxygen* and *Nitrogen*, are sometimes called the *Atmospheric elements*, because they exist abundantly in the air. The great bulk of the atmosphere being Oxygen and Nitrogen gases, whilst water, the vapor of which is always present in the air is composed only of Oxygen and Hydrogen; Carbonic acid also, always present there, consists of Oxygen and Carbon; *Ammonia*, which is never absent from the atmosphere, consists of Hydrogen and Nitrogen, and the traces of Nitric acid of the air and of the soil are composed of Oxygen and Nitrogen. These *Atmospheric elements*, forming the principal weight of animal and vegetable bodies, exist also abundantly in the dark-colored material called sometimes *Humus*, which blackens the garden mould, and is derived from animal and vegetable decomposition. Abundantly present wherever air and water exist, which is every where on the globe, we need never fear their exhaustion in any locality. The sole care of the agriculturalist, in regard to these, being to favor their application to his growing crops in an available condition; on which it is not our present design to dilate.

The other elements called the *Mineral elements* by distinction, never enter the atmosphere, except in dust, but are confined to the soil, where they all exist as oxides. They were orginal constituent elements of the rocks of which the globe was made, and from which the soil was derived by gradual disintegration. They may be called the *fixed elements* of organized beings, because when these decay, or are burned up,—whilst the atmospheric elements make their escape into the air, as gases and vapors—these remain, *as dust or ashes*.

The greatest bulk of all soils is what is stated as *Sand and Insoluble Silicates*, in the following analysis: Sand, more or less fine, the greater part being so very fine as to have been very generally mistaken by writers on the soil for clay, (Alumina). This sand is mainly mechanical in its action on vegetables; not only serving as a matrix through which the nutritive materials are diffused, through which watery solutions, gases and vapors penetrate by capillary action, but also a loose bed through which roots and rootlets of plants may ramify; giving them a fixed support. The *alumina* which is mixed with this sand, and which forms the basins of *clay*, exists in the soil in variable quantities, making it stiff, heavy and cold, of the nature of clay, when in superabundance, from the tenacity with which it turns water, but when in proper proportion, as in the rich loam, causing the soil to retain not only the moisture but fertilizing gases and the humus which is produced from vegetable and animal decomposition; for which, as well as for ammonia, it has a strong attraction; retaining these valuable materials on the surface, within the reach of growing vegetables, and preventing their too rapid waste by solution in the water which penetrates the soil. Alumina, derived no doubt, originally, from the *felspar* of the original granitic rock, in which it has always combined with much alkali, always holds in store a considerable proportion of *potash*. It is not known that this earth *Alumina* enters into the composition of either vegetables or animals, but its valuable properties make it an almost indispensable ingredient of the fertile soil. The oxides of iron and manganese which are almost always associated with sand and alumina in soils, act both mechanically and as actual nutritive ingredients. Some remarks on the chemical action of oxide of iron in the soil may be found below, under the head of analysis of soil No. 7.

The other constituents of soil, (except sometimes the carbonate of lime,) are in much smaller proportion and are all essential to vegetable nourishment; and those which are the most readily removed by cropping are the *organic matters*, the *phosphates*, the *alkalies*, (*potash* and *soda*,) and the *lime* and *magnesia*.

There are several modes in which these may be gradually alienated from the soil, leaving it incapable of supporting crops.

They may be rapidly carried off in heavy green crops, as garden products, tobacco, heavy crops of clover or grasses, or straw from large crops of grain, taken off from the soil; these products generally carry away a large proportion of the alkalies, and of lime and magnesia; whilst large grain crops, or animals fed on the ground, cause the alienation of more of the phosphates than of the alkalies.

The soluble ingredients of the soil may be actually washed out, to a considerable extent, by abundant rains washing through a soil which is kept bare of vegetation by stirring the surface, which also facilitates the removal of the finer earthy particles, which are the richest portion. Hence hoed crops, in the cultivation of which a larger space of surface is kept clear of any kind of vegetable growth, whilst it is freely exposed to the summer's sun and the washing of grains, cause a greater deterioration of the soil than can be accounted for by the growing crop alone. The heat and moisture, aided by the capillary attraction of the porous soil, not only cause a rapid decomposition of the *organic matters* of the soil, by favoring oxidation, but the water which penetrates through it, charged as it always is with carbonic acid, and aided by the organic acids present, always dissolves more or less of the soluble essential elements. Not only has it been ascertained by experiment that a field kept constantly stirred during the growing season, and on which nothing is permitted to grow, will be nearly as much reduced in fertility as a similar one which has supported a crop; but, notwithstanding the positive assertion of Liebig to the contrary, in his late *Letters on Modern Agriculture*, the writer has ascertained by numerous experiments, that water containing carbonic acid can not come in contact with fertile soil, without dissolving a small quantity, out, more or less, of its essential ingredients.

In the ordinary course of nature this loss is in a great measure prevented. The surface being covered with growing vegetable which gradually absorb the solution in the soil, exhaling the water rapidly from their green surfaces and retaining the nutritive mineral elements, and when they die, their decay leaves them upon the surface to nourish another generation of plants. So that the constant result of this natural fallow is to make the surface more and more fertile,—to bring the nutritive elements of the soil more and more to the surface.

The space allotted to this article precludes any detailed remarks on the best modes of renovating exhausted soils, &c. For full instruction in the chemistry of agriculture the enquiring student can be at no loss for able modern works. We commend him to those of Johnston, of Liebig, and to the modern elementary works on scientific agriculture generally.

CHEMICAL ANALYSIS OF INDIANA SOILS.

ARRANGED ACCORDING TO GEOLOGICAL FORMATIONS.

I.—SOILS OF THE LOWER SILURIAN FORMATION.

No. 1. Virgin soil from Mr. William H. Bennett's farm, northwest fourth of section 17, township 12 north, range 1 west, Union county, Indiana. Lower Silurian formation. Growth—Grey and Blue Ash, Poplar, and a few Black Walnuts. The dried soil is of a light umber color.

No. 2. Surface soil, forty to fifty years in cultivation, Wm. H. Bennett's farm, (same locality as above.) Yields forty bushels of corn, and fifteen bushels of wheat to the acre. Dried soil of a dusty grey-buff color.

No. 3. Sub-soil of the same old field, Wm. H. Bennett's farm, &c. Dried soil of a dirty-buff color.

No. 4. Virgin soil, from Mr. J. Hurty's farm, close to Liberty, Union county, Indiana. Lower Silurian formation. Growth—Beech, Poplar, Oak, and some Maple. The dried soil is of a grey-umber color.

No. 5. Surface soil, over thirty years in cultivation, Mr. J. Hurty's farm, &c. His land produces from thirty-five to forty bushels of corn, and from ten to twelve bushels of wheat to the acre; bears good grass, A chalybeate spring not far off. Dried soil of a dirty grey-buff color.

No. 6. Sub-soil of the same old field, Mr. J. Hurty's farm, &c. Dried soil of a dirty-buff color.

To ascertain the relative quantity of *soluble matter, immediately available for vegetable nourishment,* contained in these soils, one thousand grains of each was digested for a month, at the ordinary temperature, in a closely stopped bottle, in distilled water which had been charged with carbonic acid under pressure. After filtering off the clear solution it was carefully evaporated to dryness, weighed and analyzed. As the water which irrigated the soil always contains dissolved carbonic acid, which solution is a principal solvent of the nutritive ingredients which the earth yields to growing vegetables, it is believed that this process shows, to some extent at least, the comparative *immediate fertility* of the soils submitted to it.

The results of the process, as applied to the six soils above described, were as follows:

Extracted from 1,000 grains by the water charged with Carbonic Acid.

	No. 1. Virgin Soil.	No. 2. Old field Soil.	No. 3. Sub-Soil.	No. 4. Virgin Soil.	No. 5. Old field Soil.	No. 6. Sub-soil.
Organic and Volatile matters	0.850	0.366	0.266	0.900	0.333	0.166
Alumina and oxides of Iron, and Manganese, and Phosphates	.086	.063	.063	.163	.060	.130
Carbonate of Lime	.797	.630	.053	1.143	.510	.483
Magnesia	.228	.200	.111	.228	.159	.089
Sulphuric acid	.029	.029	.027	.052	.028	.029
Potash	.037	.035	.032	.077	.035	.019
Soda	.022	.007	.026	.033	.029	.020
Silica	.180	.286	.163	.247	.347	.146
Soluble extract dried at 212° F., grains	2.229	1.616	0.741	2.843	1.501	1.082

Although, other things being equal, the greater the quantity of this *soluble extract*, within reasonable limits, the higher should we estimate the immediate fertility of the soil; yet the *composition* of this extract must always be considered in this important estimate. Thus an excess of *carbonate of lime and magnesia*, or of *Alumina and oxides of Iron and Manganese* or *Silica*, or even of *organic matters*, might prove not only inert but injurious in their action on growing plants; while there might be a deficiency in those indispensable ingredients the alkalies, even where this excess caused the whole weight of the extract to be large.

These six soils, after having been carefully air-dried, were dried at 400° F., and the loss of weight noted as moisture. They were then submitted for analysis, with the following results:

Composition dried at 400° F.

	No. 1. Virgin Soil.	No. 2. Old field Soil.	No. 3. Sub-Soil.	No. 4. Virgin Soil.	No. 5. Old field Soil.	No. 6. Sub-soil.
Organic and Volatile matters	4.885	3.350	3.417	6.792	3.437	3.156
Alumina	4.730	4.158	6.030	} 5.495	7.060	8.140
Oxide of Iron	2.965	3.190	5.165			
Carbonate of Lime	.715	.540	.690	.345	.295	.470
Magnesia	.661	.536	.741	.470	.491	.393
Brown oxide of Manganese	.220	.145	.270	.095	.130	.170
Phosphoric Acid	.282	.189	.226	.173	.260	.178
Sulphuric Acid	.067	.058	.041	.067	·055	.075
Potash	.270	.237	.318	.147	.212	.246
Soda	.004	.027	.095	.049	.020	.061
Sand and Insoluble Silicates	85.190	86.815	82.679	86.790	88.165	87.715
Loss	.011	.755	.328			
Total	100.000	100.000	100.000	100.423	100.125	100.604
Moisture expelled at 400° F	4.675	4.200	7.100	4.300	2.975	3.375

Persons unaccustomed to the study of the results of soil analysis may be disappointed on seeing the relative small proportions of several of the essential ingredients in these soils, which may be considered quite fertile. It may be seen that in the 100. parts of the soil No. 1, dried at the temperature of 400° F., there exists only 0.27 per cent., (i. e. a little more than the fourth of one per cent.) of *Potash*, and 0.282 per cent. (which is but little more) of *Phosphoric Acid;*—ingredients which are amongst those so essentially necessary to vegetable growth that in their absence no plant, however simple and small, could be developed. These quantities appear fearfully minute when we reflect that all the crops we remove from the land carry off more or less of these valuable substances; and that, as they cannot come back again by means of the atmosphere, in the form of gases and vapors, and must be restored, if restored at all, to the soil in the form of solid manures, the probability of their final exhaustion from the land, by our common thoughtless methods of farming, becomes unpleasantly apparent.

But when we resort to figures we find the danger, although a real one, not so *imminent* as at first sight it seemed to be; and these quantities, small as they are compared with the 100. parts of the soil, swell to some magnitude when calculated in the large amount of earth which is found on an acre of ground, taken only to the depth of one foot. Some expriments made on Lower Silurian soils of Kentucky, which will apply sufficiently well to the present instance to serve the purpose of illustration, gave as the weight of the dry earth on an acre of land, to the depth of one foot, more than three million pounds avoirdupois, (3,116,413 80-100 pounds). Taking three million pounds then as the basis of our calculations, we find that the 0.27 per cent. of Potash is equal to eight thousand one hundred (8,100 lbs) pounds per acre, in each foot in depth of the soil; and the Phosphoric acid equal to eight thousand four hundred and sixty pounds (8,460 lbs); quantities which, although not inexhaustible by thriftless husbandry, may at least *give us some little time* to study the philosophy of agriculture, and to consider the ways and means of preventing their loss from their soil, or of their restoration if too much reduced by exhaustive farming.

The evidences of the deterioration of the soil by cultivation may be seen in these analyses; first in the dimished quantity of *soluble matters* extracted by the carbonated water from the soil of the old field as compared with that extracted from the Virgin soil; this quantity being 2.229 grains from soil No. 1, and 2.843 grains from soil No. 4, whilst it is only 1.616 grains from soil No. 2 and 1.501 from soil No. 5. It is shown, secondly, in the general analyses of these soils, especially in comparing those of soils Nos. 1 and 2; in the diminished quantities of *Organic and Volatile matters, Carbonate of Lime, Magnesia, Oxide of Manganese, Phosphoric and Sulphuric Acids*, and *Potash*, and in the increased proportion of *Sand and Insoluble Silicates* in the soil of the old field as compared with the Virgin soil. The latter will also be seen to have a greater power of absorbing and holding *moisture*, and being of darker color will absorb the heat of the sun with greater rapidity than the soil of the old field; a difference owing in part to the greater proportion of organic matters (*Humus, remains of decayed vegetable matters,*) which it contains, which also enable it to absorb vapors and gases with greater facility.

The differences are not so well exhibited in the analyses of soils Nos. 4 and 5.

The set of soils Nos. 1, 2 and 3, although containing a little less *Organic and Volatile matters* than Nos. 4, 5 and 6, yet are slightly more rich than these, in the essential mineral element of vegetable food.

II.—SOILS OF THE UPPER SILURIAN FORMATION.

No. 7. Surface soil, from an old field, on James Clayton's farm, three miles west of Winchester, Randolph county, Indiana. Upper Silurian formation. Noted for its productions and for its red color. Chalybeate springs and Bog Iron-ore abundant around. The dried soil is of a light reddish-brown color.

No. 8. Surface soil, from an old field, on William T. R. Ross' farm, S. W. half of section 14, township 27 north, range 7 east, Wabash county, Indiana. Upper Silurian formation. The dried soil is of a yellowish-grey color.

No. 9. Sub-Soil of the next preceding, &c. The dried sub-soil is of a yellowish-grey color; lighter than No. 8. One thousand grains of each of these soils thoroughly air-dried, gave, on digestion for a month, in water charged with carbonic acid, dissolved materials as represented in the following table:

	No. 7. Old field Soil.	No. 8. Old field Soil.	No. 9. Sub-Soil.
Organic and Volatile matters	0.883	0.566	0.283
Alumina and oxides of Iron and Manganese, and Phosphates	.080	.098	.090
Carbonate of Lime	1.430	1.677	.377
Magnesia	.233	.122	.055
Sulphuric acid	.073	.044	.028
Potash	.048	.054	.023
Soda	.030	.008	.033
Silica	.216	.180	.147
Extract, dried at 200° F., (grains)	2.993	2.749	1.036

In the following table is represented the *Chemical Composition* of these three soils, dried at the temperature of 400° F.:

	No. 7. Old field Soil.	No. 8. Old field Soil.	No. 9. Sub-Soil.
Organic and Volatile matters	6.331	3.740	2.430
Alumina	1.660	3.335	4.085
Oxide of Iron	7.210	2.740	3.790
Carbonate of Lime	.495	.495	.220
Magnesia	.537	.490	.644
Brown oxide of Manganese	.415	.395	.270
Phosphoric acid	.225	.161	.177
Sulphuric acid	.059	.059	.084
Potash	.300	.111	.212
Soda		.003	.074
Sand and Insoluble Silicates	82.890	89.690	88.765
Total	100.122	101.219	100.751
Moisture expelled at 400° F	3.450	2.425	2.250

The cause of the red color of soil No. 7 is found in the considerable quantity of *Peroxide of Iron* which it contains, (7.21 per cent.); and its productiveness might have been inferred from its proportions of *Carbo-*

nate of Lime, Magnesia, Phosphoric and Sulphuric Acids, and *Potash*. The large amount of *organic matters* aid to make it a fertile soil. Soil No. 8, containing less of these, as well as of *Potash* and *Phosphoric Acid*, will not probable yield as abundantly as this does.

Although but a very small proportion of oxide of Iron enters into the composition of vegetable and animal bodies its presence in the soil, in notable proportions, seems to be highly conducive to its fertility. The oxidation of metallic iron, in the presence of air and moisture, even its slow rusting in a moist atmosphere, causes the formation of a small quantity of ammonia, by the union of the Nitrogen of the air with the Hydrogen of the decomposed moisture; and the oxide of iron seems to have an affinity for ammonia, so that, as has been long known to chemists, almost all samples of iron rust, or oxide of iron are found to contain more or less of that compound. The oxide of iron of the red soils may not only thus absorb this nutritive substance from the atmosphere,—in which traces of it are almost always to be found—and in this manner may more strongly aid vegetable growth, by furnishing a greater supply of available Nitrogen than a light-colored soil, but it is found that it also powerfully aids in the decomposition of dead animal and vegetable matters; resolving them speedily, under favorable circumstances, into materials suitable for vegetable nourishment. A decomposing animal or vegetable substance mixed, in moist state, with the peroxide of iron, (or the red soil which contains it abundantly,) and kept at a favorable temperature for decomposition, has this process greatly accelerated by the oxygen, which is furnished to it by the peroxide; and when the decomposition is over, the protoxide of iron, resulting from this partial deoxidation, soon recovers its oxygen again by free exposure to the atmosphere. In this manner, not only may the substances be restored into those compounds which gave the dark color to vegetable mould, (*Humus*,) but a more complete oxidation may result, and the process of *nitrification*, and the production of carbonic acid and water be favored by the peroxide of iron in the soil.

The union of organic matters with the red oxide of iron changes it to a reddish-brown, or chocolate-brown color, of greater or less depth of tint. When a soil is of a pure red or brownish-red, brick or iron-color, it is evident that it is deficient in organic matter, and vegetable and animal manures may be advantageously applied to it.

Yellow and buff colored soils also frequently contain much peroxide of iron, in a condition resembling yellow ochre, but it is usually associated with more alumina than in the old red soils, and hence they may be stiffer and of a more clayey nature. In soils Nos. 8 and 9 this may be to a certain extent observed.

III.—SOILS FROM THE DEVONIAN FORMATION.

No. 10. Virgin upland soil, from Jacob Ruddell's farm, on Clarke's Grant, near Utica, Clarke county, Indiana. (Devonian formation.) Growth—Beech, Sugar-Tree, Black and White Walnut, Elm, Ash, Cherry and Buckeye. The dried soil is of a light brown color.

No. 11. Surface soil of an old field, thirty years at least in cultivation, Jacob Ruddell's farm, &c. Raises excellent wheat, (25 bushels); good corn, (50 to 55 bushels,) good clover and potatoes. Dried soil of a light reddish-brown color.

No. 12. Sub-soil of the next preceding. Jacob Ruddell's farm, &c. Dried sub-soil lighter colored and more reddish than the preceding.

No. 13. Virgin soil, from S. D. Irish's farm, near Pendleton, Madison county, Indiana. (Devonian formation.) Growth—Beech, Sugar-Tree, Ash and Black Walnut. The dried soil is of a light umber color.

No. 14. Surface soil from an old field, twenty-eight years in cultivation. S. D. Irish's farm, &c.* Dried soil somewhat lighter colored and more yellowish than the preceding.

No. 15. Sub-soil of the old field, S. D. Irish's farm, &c. The dried sub-soil resembles the next preceding in color.

One thousand grains of each of these soils, digested for a month in water charged with carbonic acid, gave up to that solvent the materials stated in the following table:

	No. 10. Virgin Soil.	No. 11. Old field Soil.	No. 12. Sub-Soil.	No. 13. Virgin Soil.	No. 14. Old field Soil.	No. 15. Sub-Soil.
Organic and Volatile matters	1.090	0.383	0.783	1.660	0.733	0.850
Alumina and Oxide of Iron and Manganese, and Phosphates	.127	.050	.046	.176	.173	.046
Carbonate of Lime	1.517	.160	.596	1.843	1.810	2.906
Magnesia	.107	.055	.183	.207	.277	.120
Sulphuric acid	.083	.029	.061	.050	.050	.120
Potash	.081	.054	.037	.048	.018	.035
Soda	.025	.073	....	.021	.021	.068
Silica	.257	.280	.163	.160	.180	.220
Loss	.096	.066	.114	....	....	.078
Total Watery Extract, dried at 200° F., grains	3.333	1.150	1.983	4.165	3.292	4.433

*The Devonian limestone comes to the surface, and bowlders are common in this field.

Almost always it is found that the sub-soil, when digested in the carbonated water, gives up much less *soluble extract* than the surface soil, although the former may be really richer in most of the essential mineral elements. But when the sub-soil contains more *Carbonate of Lime*, or as much or more *organic matters*, it may, as in the instance above of sub-soil No. 15, give up even more soluble matter than the surface soil. Because carbonate of lime is quite soluble in water containing carbonic acid, and because the *organic matters* act as solvents, and aid in the solution of other substances present in the soil.

The chemical analysis of these soils, dried at 400° F., gave the following results, viz:

	No. 10. Virgin Soil.	No. 11. Old field Soil.	No. 12. Sub-Soil.	No. 13. Virgin Soil.	No. 14. Old field Soil.	No. 15. Sub-Soil.
Organic and Volatile matters	6.677	4.095	2.883	6.827	5.849	5.357
Alumina	3.035	3.785	3.920	2.235	3.435	5.310
Oxide of Iron	3.015	3.065	3.815	2.850	3.150	3.700
Carbonate of Lime	.470	.220	.320	.745	.745	1.120
Magnesia	.451	.205	.356	.605	.594	.500
Brown oxide of Manganese	.390	.365	.365	.240	.240	.240
Phosphoric acid	.313	.260	.276	.197	.235	.215
Sulphuric acid	.178	.075	.072	.109	.067	.092
Potash	.308	.161	.142	.156	.169	.331
Soda	.073	.003	.092	.043	.017	.067
Sand and Insoluble Silicates	85.140	88.380	88.140	85.990	85.365	82.715
Loss				.003	.134	.353
Total	100.050	100.614	100.381	100.000	100.000	100.000
Moisture expelled at 400° F	4.150	2.950	2.900	4.070	3.875	4.800

The marked diminution in the quantity of the *Carbonate of Lime, Magnesia, Phosphoric and Sulphuric Acid, Potash* and *Soda*, as well of the *Organic matters;* and the increased proportion of the *Sand and Insoluble Silicates*, in the soil of the old field No. 11, as compared with the virgin soil No. 10, show clearly the deriorating influence of the thirty years cultivation. The much smaller quantity of soluble matters extracted by digestion in the carbonated waters from the former soil, as compared with the latter, exhibits the same fact.

Calculating on the basis given a few pages back, the *Potash* in the vigin soil (0.308 per cent.) amount to nine thousand two hundred and forty pounds (9,240 lbs) to the acre, in the depth of one foot; that in the old fild (0.161 per cent.) is only equal to four thousand eight hundred and thirty pounds in the same space; consequently, if these two fields were originally similar in composition, and no errors are made in computation, as much as four thousand four hundred and ten pounds

(4,410 lbs) of *Potash* have been removed from the superficial foot of soil, per acre, by the thirty years cropping. By carrying out the calculation to the *Phosphoric Acid* it will be seen also that a difference of fifteen hundred and ninety pounds (1,590 lbs) of this substance appears in the favor of the virgin soil.

It is not pretended that these figures are exactly correct; as the known difficulties which attend minute soil analyses, the slight differences in composition of the soil which might exist even in different parts of the same field, as well the fact that any small error which might occur in the analyses would be very greatly multiplied in our computations, are considerations which should prevent any dogmatism; yet these figures sufficiently well represent the fact, which has been verified in numerous instances by the author, in comparative analyses side by side, of old soils and virgin soils, especially for the Geological Survey of Kentucky; (See introductory remarks to the Chemical Report in the forthcoming fourth volume of Reports of Kentucky Geological Survey,) that by the cultivation of the soil, according to the ordinary system, however skillfully it may have been conducted, the *essential elements* of the soil are gradually diminished, if the crops or products are removed from it.

Most of this valuable matter which is thus removed from the soil is carried off in the vegetable and animal products, in which they are *essential constituents*. It is well known that the ashes of plants and grains contain these essential elements, and that the bodies and bones of animals could not be developed without them. In short we get all our *Potash* by the lixiviation of burnt vegetables, and all our *Phosphorus* from the bones of animals, and in which it exists as Phosphoric acid, combined with *Lime* and *Magnesia*; and all these fixed principals, as well as the *Sulphur*, the *Silica*, the *Oxide of Iron*, the *Soda*, even the *Oxide of Manganese*, which is said to aid in giving the dark color to our hair, were originally derived from the soil; having been constituents of the primeval rocks from which soil has been slowly formed by disintegration. It is fortunate for us that these valuable mineral substances are almost universally diffused—otherwise plants would refuse to grow and animals could not exist, except on imported food, on many parts of the globe; even common sand and white sand-rock, iron-ore, as well as numerous varieties of limestone, &c., analyzed by the author, have been always found to yield, at least traces of, *Potash*, of *Soda*, of *Phosphorus*, and of *Sulphur*; yet experience, as well as chemical analysis, have fully proved that even a fertile soil may be so far reduced, by thriftless cropping, as no longer to yield profitable crops to the husbandman.

The space allotted to this article will not allow the full quotation of authorities on this important subject, but we refer the reader to numerous writings and statistical reports* showing the diminution in the products of arable land, even in our new States; we refer also to the exhausted land of Virginia and northern Atlantic border of our continent, and the sterile wastes in Europe and Africa, which in former ages yielded rich harvests of grain and provender.

No question of greater importance to humanity can engage the attention and study of scientific agriculturalists, than *how to cultivate the soil and enjoy its products without robbing it of its essential elements.*

Returning to our table of analysis, we observe that the soil of the old field, No. 14, does not appear any poorer than the virgin soil of the same locality, No. 13; but noting the composition of the subsoil, which may have been turned up somewhat by the plow, it will appear probable that the original richness of this may have helped to sustain the surface soil during its twenty-eight years of cultivation, and to compensate somewhat for its losses in that period. Sub-soil plowing in this locality would be beneficial.

IV.—SOILS FROM THE SUB-CARBONIFEROUS FORMATION.

No. 16. Virgin soil, from six miles east of Corydon, Harrison county, Indiana. (Sub-Carboniferous formation.) Timber, chiefly Beech and Sugar-Maple. The dried soil is of a yellowish light-umber color.

No. 17. Surface soil of an old field, twenty-five to thirty years in cultivation; same locality as the preceding, &c. The dried soil is of a dirty brownish buff-color.

No. 18. Sub-soil of the old field, next preceding, &c. Dried sub-soil slightly lighter colored than the preceding.

Treated by digestion for a month with water charged with carbonic acid, one thousand grains of each of these soils, thoroughly air-dried, gave up of *soluble matters* as described in the following table.

---

*See Liebig's recent work, "Letters on Modern Agriculture," Klippart on the Wheat Plant, and Patent Office Reports, &c.

Extracted from 1,000 *grains by water charged with Carbonic Acid.*

	No. 16. Virgin Soil.	No. 17. Old field Soil.	No. 18. Sub-Soil.
Organic and Volatile matters..	0.833	0.733	0.333
Alumina and oxides of Iron and Manganese, and Phosphates.....	.157	.073	.090
Carbonate of Lime...	.766	.727	.627
Magnesia...	.077	.094	.144
Sulphuric acid..	.039	.033	.027
Potash..	.069	.064	.031
Soda..	.013	.016	.021
Silica..	.163	.163	.296
Extract dried at 212° F., grains..................................	2.117	1.903	1.569

The *Chemical Composition* of these three soils, dried at 400° F., is represented in the following table:

	No. 15. Virgin Soil.	No. 17. Old field Soil.	No. 18. Sub-Soil.
Organic and Volatile matters..	4.757	4.731	3.352
Alumina...	2.210	3.185	3.760
Oxide of Iron..	2.565	3.065	3.315
Carbonate of Lime...	.370	.385	.385
Magnesia..	.461	.452	.451
Brown oxide of Manganese.......................................	.165	.290	.290
Phosphoric acid..	.212	.261	.211
Sulphuric acid...	.084	.084	.067
Potash...	.168	.145	.174
Soda...	.054	.038	.003
Sand and Insoluble Silicates.....................................	87.240	86.265	87.615
Loss...	1.714	1.099	.377
Total..	100.000	100.000	100.000
Moisture expelled at 400° F......................................	3.325	3.300	3.050

V.—SOILS FROM THE COAL MEASURES GROUP.

No. 19. Virgin soil, in a grove adjoining a prairie; Wagner's Grove, Warren county, Indiana. (Coal Measures.) Growth—Bur Oak, Hickory, Grey Ash, Walnut, Buckeye, Red Elm, Cherry, Sassafras, Red Bud, Hazel and Elder bushes. Dried soil of a very dark mouse color, or yellowish-black.

No. 20. Prairie surface soil from rising ground, about twenty-five years in cultivation, near Wagner's Grove, Warren county, Indiana. (Coal Measures.) Dried soil mouse-colored, a little lighter than the preceding.

No. 21. Prairie surface soil, from a bottom near Wagners's Grove, &c. Dried soil darker-colored than the two preceding; almost black.

No. 22. Prairie sub-soil, at one foot depth, near Wagner's Grove, &c. Dried soil of a dark ash-grey color.

No. 23. Prairie sub-soil, at two feet depth, near Wagner's Grove, &c. Dried soil of an ashey-grey color, lighter and more yellowish than the preceding.

No. 24. Prairie sub-soil, at three feet below the surface, near Wagner's Grove, &c. Dried soil of a dark ash-grey color; a little darker than the two preceding.

The soluble matters, extracted from a thousand grains each of these soils, thoroughly air-dried, by digestion for a month in water charged with carbonic acid, are stated in the following table, viz:

	No. 19. Virgin soil in grove.	No. 20. Prairie soil in old field.	No. 21. Bottom Prairie soil.	No. 22. Sub-Soil, at 1 foot.	No. 23. Sub-Soil, at 2 feet.	No. 24. Sub-Soil, at 3 feet.
Organic and Volatile matters...........	1.266	0.983	1.127	0.800	0.500	0.550
Alumina and oxides of Iron and Manganese, and Phosphates...............	.603	.323	.090	.247	.073	.130
Carbonate of Lime.......................	1.793	1.477	1.610	.641	.410	1.127
Magnesia.................................	.380	.383	.321	.144	.100	.227
Sulphuric acid...........................	.079	.113	.107	.104	.129	.068
Potash...................................	.064	.147	.109	.145	.177	.060
Soda.....................................	.122	.132	.209	.187	.037	.161
Silica....................................	.380	.420	.337	.330	.247	.347
Loss.....................................			.307		.227	
Extract, dried at 212° F., grains......	4.687	3.978	4.217	2.598	1.900	2.670

The *Chemical Composition* of these soils, dried at 440° F., is represented as follows:

	No. 19. Virgin soil, grove near Prairie.	No. 20. Prairie soil, rising gr'd, old field.	No. 21. Prairie soil, bottom.	No. 22. Prairie Sub-Soil, at 1 foot.	No. 23. Prairie Sub-Soil, at 2 feet.	No. 24. Prairie Sub-Soil, at 3 feet.
Organic and Volatile matters...	8.286	5.473	8.851	2.805	2.654	2.931
Alumina..........................	2.010	2.610	4.335	1.810	2.460	2.985
Oxide of Iron...................	3.365	2.740	3.315	2.150	3.765	4.540
Carbonate of Lime..............	.945	.645	1.545	.270	.395	.895
Magnesia.........................	.753	.795	.878	.519	.599	.901
Brown oxide of Manganese......	.215	.115	.190	.090	.215	.190
Phosphoric acid..................	.255	.198	.237	.194	.161	.214
Sulphuric acid...................	.153	.100	.127	.062	.084	.050
Potash...........................	.256	.125	.309	.235	.272	.360
Soda..............................	.038		.086	.041	.036	.056
Sand and Insoluble Silicates....	82.615	86.565	80.515	91.490	88.065	86.066
Loss..............................	1.109	.634		.334	1.294	.812
Total.............................	100.000	100.000	100.388	100.000	100.000	100.000
Moisture expelled at 400° F.....	7.375	5.000	7.075	2.850	2.975	4.475

The rich prairie soils, Nos. 19 and 22, have a remarkably large proportion of *organic and volatile matters* in their composition, to which they perhaps owe their high *hygroscopic* power; more than 7 per cent· of moisture being retained by these soils after being thoroughly dried in a room daily heated with a stove. The bottom land contains more *Alumina, Carbonate of Lime, Magnesia and Potash* than the soil from the grove, and both are superior in richness to that from the rising ground; which contains more *Sand and Insoluble Silicates*, less *Organic Matters*, and less *Lime, Magnesia, Oxide of Manganese, Phosphoric* and *Sulphuric Acid, Potash and Soda* than these. These essential mineral elements of vegetable nourishment being abundant in this, with plenty of organic matters to aid in their solution; these soils ought to be quite productive.

The sub-soil seems to be poorer at the depth of one foot than at a greater depth; the valuable mineral ingredients increasing in proportion as we descended from that depth to three feet below the surface. Whether this increase continued still further as we descend is of course not ascertained.

No. 25. Virgin soil, from Mr. Delamater's farm, close to Dover, Martin county, Indiana. (Coal Measures.) Upland; near the locality of natural paints, a coal seam, fire-clay and iron ore. Timber—Chestnut, Oak, Poplar, Hickory, some Beech and Sugar-Tree, White and Black Walnut, Sycamore, Red Bud, Pawpaw and Persimmon. The dried soil is of an umber-grey color.

No. 26. Surface soil, thirteen years in cultivation; Mr. Delamater's farm, &c. Dried soil of a dirty grey-buff color.

No. 27. Sub-soil of the next preceding. Dried soil of a buff color.

No. 28. Virgin soil, from Mr. J. D. Williams' land, south-east quarter of section 14, township 2 north, range 8 west. White River bottom, Knox county, Indiana. (Sandstone near.) Timber—Black Walnut, Burr Oak, Spanish Oak, Elm and Sassafras. (Coal Measures.) The dried soil is of an umber color.

No. 29. Surface soil, twenty years or more in cultivation, Mr. J. D. Williams' farm, &c. Produces from sixty to one hundred bushels of corn, and twelve to thirty-nine bushels of wheat to the acre; was three years in clover and produces it well. Dried soil umber-colored; a shade lighter than the preceding.

No. 30. Sub-soil of the next preceding, &c. Dried soil lighter colored and more yellowish than the preceding.

The soluble matters entrusted by digestion in water charged with carbonic acid, one thousand grains of each of these soils, after thorough drying in the air of a room warmed with a stove, are stated in the following table:

	No. 25. Virgin Soil.	No. 26. Old field.	No. 27. Sub-Soil.	No. 28. Virgin Soil.	No. 29. Old field.	No. 30. Sub-Soil.
Organic and Volatile matters..........	0.516	0.776	0.233	1.163	0.483	0.376
Alumina and oxide of Iron and Manganese, and Phosphates............	.263	.263	.057	.230	.096	.033
Carbonate of Lime	1.643	1.593	.110	3.177	1.177	.827
Magnesia................................	.223	.233	.220	.311	.155	.167
Sulphuric acid.........................	.039	.033	.027	.033	.027	.033
Potash..................................	.106	.087	.103	.056	.022	.023
Soda....................................	.028	.049	.037	.048	.016	.025
Silica...................................	.263	.320	.180	.313	.230	.230
Loss....................................	.036	.012	....	....	.194	....
Extract dried at 212° F., grains......	3.117	3.366	0.967	5.331	2.400	1.714

The *Chemical Composition* of these six soils, dried at 400° F., is as follows:

	No. 25. Virgin Soil.	No. 26. Old field.	No. 27. Sub-Soil.	No. 28. Virgin Soil.	No. 29. Old field.	No. 30. Sub-Soil.
Organic and Volatile matters..........	5.814	3.851	3.032	13.443	7.917	5.348
Alumina................................	2.515	3.765	5.805	6.565	5.265	6.665
Oxide of Iron..........................	2.515	2.965	5.240	5.405	5.190	5.590
Carbonate of Lime.....................	.520	.370	.120	1.670	1.145	.820
Magnesia...............................	.596	.567	.668	1.021	.936	.852
Brown oxide of Manganese..............	.230	.295	.145	.345	.370	.235
Phosphoric acid........................	.162	.194	.212	.461	.327	.320
Sulphuric acid.........................	.076	.056	.056	.115	.093	.067
Potash.................................	.110	.135	.236	.381	.285	.328
Soda...................................	.016	.038	.058	.065	.065	.056
Sand and Insoluble Silicates...........	87.590	87.315	84.490	71.690	78.840	79.790
Loss...................................	....	.449	....	....	....	....
Total..................................	100.144	100.000	100.122	101.191	100.433	100.131
Moisture expelled at 400° F...........	3.250	3.015	4.525	9.250	6.225	6.050

In the soil of the field which has been in cultivation for thirteen years, (No. 26) we notice the influence of the sub-soil in sustaining the surface soil under cultivation; for whilst the proportions of *Carbonate of Lime*, *Magnesia*, *Sulphuric Acid* and *Organic matters* are less in this than in the Virgin soil, the *Potash* and *Phosphoric acid* seem to have been increased by the admixture of it with the sub-soil by the operations of the plow.

The three soils from White River bottom, Nos. 28, 29 and 30, are of extraordinary richness; not only is the proportion of *Organic and Vol-*

atile matters enormous, especially in the virgin soil, (13.443 per cent.,) but they contain more than the usual quantities of *Carbonate of Lime, Magnesia, Phosphoric* and *Sulphuric Acids*, and *Potash*, and exhibit a very high *hygroscopic power*. If these lands are well drained they must certainly be very productive.

It is instructive to observe, even in this rich land, the influence of ordinary cultivation in producing the gradual deterioration of the soil. On comparing the two neighboring columns of figures it will be seen that all these essential ingredients of the soil, above stated, are in diminished proportion in the soils of the old field, whilst the *Sand and Insoluble Silicates* are in larger amounts.

The sub-soil, although quite rich, is not more so than the virgin surface soil.

VI.—SOILS FROM THE QUATERNARY FORMATION.

No. 31. Virgin soil, from Mr. J. D. G. Nelson's farm, near Fort Wayne, Allen county, Indiana. Second Maumee bottom. Timber—Beech, Sugar-Maple, some Poplar and Black Walnut. (Drift Period.) The dried soil is of an umber color; containing much sand.

No. 32. Surface soil, thirty years in cultivation. J. D. G. Nelson's farm, &c. He found Plaster of Paris wonderfully to improve his clover crop. Raises fair wheat and corn crops. Dried soil of a brownish-grey color, containing more sand than the preceding.

No. 33. Sub-soil of the preceding, &c. Dried soil brownish-buff color, principally impure sand.

The digestion, in water charged with carbonic acid, of these soils, after being thoroughly air-dried, gave the following results—to one thousand grains of each, viz:

	No. 31. Virgin Soil.	No. 32. Old field Soil.	No. 82. Sub-Soil.
Organic and Volatile matters	0.650	0.666	0.410
Alumina and oxides of Iron and Manganese, and Phosphates	.196	.113	.097
Carbonate of Lime	.860	.795	.693
Magnesia	.230	.177	.076
Sulphuric acid	.025	.022	.022
Potash	.042	.032	.027
Soda	.011	.009	.021
Silica	.143	.130	.130
Loss	.126		
Extract, dried at 212° F., grains	2.283	1.882	1.476

These porous, calcareous, sandy soils are more productive than their chemical composition would seem to indicate, (see following table of their composition,) not only because the essential ingredients are returned with but small force by the sand, and hence are easily dissolved and appropriated by growing vegetables; but also because of the freedom with which atmospheric air penetrates them, bringing carbonic acid, gas, vapor of water and ammonia to the vegetable roots, and favoring oxidation and *nitrification*.

In some recent remarks of M. Boussaugault, made to the French Academy, in commendation of an elaborate work, by M. Barral, in four volumes, on the subject of *Drainage, Irrigation and Liquid Manures*, he gives to that author the credit of having been the first to discover that the water which flows out of the drains, of land, contain a quantity of *nitric acid*, greater in proportion as the drainage is more perfect, the soil more aerated, and the manure more abundant; from which it is necessary to conclude, says that distinguished chemist, "that the principal effect of drainage is to determine oxidation, the transformation into nitrate, of the nitrogenous principles of the air and of the manures." The free penetration of air in the moist, sandy soil may produce a similar result.

It has been ascertained by the application of the flame of a candle to the mouth of drains in the summer time that a continual current of air sets in through them to rise through the heated soil. Doubtless in winter the heavier cold air above penetrates downwards through the soil and flows outwards from the mouths of the drains.

The crops on these soils may also be benefitted by the ease with which water penetrates through them, carrying, perhaps, under favorable circumstances, dissolved nutritive materials, by capillary attraction, from other neighboring localities or from a richer sub-soil. But yet they cannot be classed as very rich or durable soils, *per se;* and would require the careful husbanding of manures.

Their *Chemical Composition*, dried at 400° F., is as follows:

	No. 31. Virgin Soil.	No. 32. Old field Soil.	No. 33. Sub-Soil.
Organic and Volatile matters	3.829	1.667	0.856
Alumina	1.410	1.535	1.187
Oxide of Iron	1.160	1.360	1.360
Carbonate of Lime	.515	.490	.415
Magnesia	.312	.269	.312
Brown oxide of Manganese	.140	.165	.165
Phosphoric acid	.217	.166	.158
Sulphuric acid	.066	.032	.032
Potash	.067	.058	.042
Soda	.006	.032	.095
Sand and Insoluble Silicates	92.365	94.960	96.140
Total	100.087	100.734	100.672
Moisture expelled at 400° F	2.725	1.050	0.725

The *Organic matters, Carbonate of Lime, Magnesia, Phosphoric and Sulphuric Acids,* and *Potash* are all in smaller quantities in the soil of the old field than in the virgin soil; but the sub-soil contains still less of these essential ingredients. The *hygroscopic* properties of these soils are but low.

TABLE I.—SOILS FROM THE LOWER SILURIAN FORMATION.

No. in report.	County.	Moisture.	Extracted from 1,000 grains by carbonic acid water.	Organic and Volatile matters	Alumina.	Oxide of Iron.	Carbonate of Lime.	Magnesia.	Brown oxide of Manganese.	Phosphoric acid	Sulphuric acid.	Potash.	Soda.	Sand and Insoluble Silicate.	Remarks.
1	Union	4.675	2.229	4.885	4.730	2.965	0.716	0.661	0.220	0.282	0.067	0.270	0.004	85.190	Virgin soil.
2	Union	4.200	1.616	3.350	4.158	3.190	.540	.536	.145	.180	.058	.237	.027	86.815	Old field soil.
3	Union	7.100	0.741	3.417	6.030	5.165	.690	.741	.270	.226	.041	.318	.095	82.679	Sub-soil of old field.
4	Union	4.800	2.843	6.792	5.495		.345	.470	.095	.173	.067	.147	.049	86.790	Virgin soil.
5	Union	2.975	1.501	3.437	7.050		.295	.491	.130	.260	.055	.212	.029	88.165	Old field soil.
6	Union	3.375	1.082	3.166	8.140		.470	.393	.176	.178	.075	.246	.061	87.715	Sub-soil of old field.

TABLE II.—SOILS FROM THE UPPER SILURIAN FORMATION.

No. in report.	County.	Moisture.	Extracted from 1,000 grains by carbonic acid water.	Organic and Volatile matters	Alumina.	Oxide of Iron.	Carbonate of Lime.	Magnesia.	Brown oxide of Manganese.	Phosphoric acid	Sulphuric acid.	Potash.	Soda.	Sand and Insoluble Silicate.	Remarks.
7	Randolph	3.450	2.993	6.331	1.660	7.210	0.495	0.537	0.415	0.225	0.059	0.300		82.890	Old field soil.
8	Wabash	2.425	2.749	3.740	3.335	2.740	.495	.490	.395	.161	.059	.111	.003	89.690	Old field soil.
9	Wabash	2.250	1.036	3.430	4.085	3.790	.220	.644	.270	.177	.084	.212	.074	88.760	Sub-soil of old field.

TABLE III.—SOILS FROM DEVONIAN FORMATION.

No. in report.	County.	Moisture.	Extracted from 1,000 parts by carbonic acid water.	Organic and Volatile matter.	Alumina.	Oxide of Iron.	Carbonate of Lime.	Magnesia.	Brown oxide of Manganese.	Phosphoric acid	Sulphuric acid.	Potash.	Soda.	Sand and Insoluble Silicates.	Remarks.
10	Clarke	4.150	3.333	6.677	3.035	3.015	0.470	0.451	0.390	0.313	0.178	0.308	0.073	85.140	Virgin soil.
11	Clarke	2.950	1.150	4.995	3.785	3.065	.220	.205	.365	.260	.075	.161	.003	88.380	Old field soil.
12	Clarke	2.900	1.983	2.883	3.929	3.815	.329	.356	.365	.276	.072	.142	.092	88.140	Sub-soil of old field.
13	Madison	4.070	4.165	6.827	2.235	2.850	.745	.606	.240	.197	.109	.156	.043	85.990	Virgin soil.
14	Madison	3.875	3.292	5.849	3.465	3.150	.745	.594	.240	.235	.067	.169	.017	85.365	Old field soil.
15	Madison	4.800	4.433	5.357	5.310	3.700	1.126	.500	.240	.216	.092	.331	.067	82.715	Sub-soil of old field.

TABLE. IV.—SOILS FROM SUB-CARBONIFEROUS FORMATION.

No. in report.	County.	Moisture.	Extracted from 1,000 parts by carbonic acid water.	Organic and Volatile matter.	Alumina.	Oxide of Iron.	Carbonate of Lime.	Magnesia.	Brown oxide of Manganese.	Phosphoric acid	Sulphuric acid.	Potash.	Soda.	Sand and Insoluble Silicates.	Remarks.
16	Harrison	3.325	2.117	4.757	2.210	2.565	0.370	0.461	0.165	0.212	0.084	0.168	0.054	87.240	Virgin soil.
17	Harrison	3.200	1.903	4.731	3.185	3.065	.385	.452	.290	.261	.084	.145	.038	86.265	Old field soil.
18	Harrison	3.050	1.569	3.352	3.769	3.315	.385	.451	.290	.211	.067	.174	.003	87.615	Sub-soil of old field.

268 GEOLOGICAL RECONNOISSANCE.

TABLE V.—SOILS FROM THE COAL MEASURES FORMATION.

No. in report.	County.	Moisture.	Extracted from 1,000 grains by carbonic acid water.	Organic and Volatile matters	Alumina.	Oxide of Iron.	Carbonate of Lime.	Magnesia.	Brown oxide of Manganese.	Phosphoric acid	Sulphuric acid.	Potash.	Soda.	Sand and Insoluble Silicates.	Remarks.
19	Warren	7.375	4.687	8.286	2.010	3.365	0.945	0.753	0.215	0.255	0.153	0.256	0.038	82.615	Virgin soil.
20	Warren	5.000	3.978	5.473	2.610	2.740	.645	.795	.115	.198	.100	.125	.086	86.565	Old field Prairie soil.
21	Warren	7.075	4.217	8.851	4.335	3.315	1.545	.878	.130	.237	.127	.309	.011	80.515	Virgin Prairie soil.
22	Warren	2.850	2.598	2.806	1.810	2.150	.270	.519	.090	.194	.062	.235	.036	91.490	Prairie sub-soil, 1 foot
23	Warren	2.654	1.900	2.654	2.460	3.765	.395	.599	.215	.161	.084	.272	.036	88.065	Prairie sub-soil, 2 feet
24	Warren	4.475	2.676	2.931	2.985	4.540	.895	.901	.130	.214	.050	.360	.055	85.066	Prairie sub-soil, 3 feet
25	Martin	3.250	3.117	5.814	2.515	2.515	.520	.696	.230	.162	.076	.110	.016	87.590	Virgin soil.
26	Martin	3.015	3.366	3.851	3.765	2.965	.370	.667	.295	.194	.056	.135	.038	87.315	Soil 13 y'rs in culti'n.
27	Martin	4.525	0.967	3.032	5.864	5.240	.120	.658	.145	.212	.056	.236	.058	84.490	Sub-soil.
28	Knox	9.250	5.331	13.443	5.565	5.405	1.670	1.021	.345	.461	.145	.381	.065	71.690	Virgin soil.
29	Knox	6.225	2.406	7.917	5.265	5.190	1.145	.986	.370	.327	.092	.285	.065	78.840	Old field soil.
30	Knox	6.050	1.714	5.348	6.665	5.590	.820	.852	.295	.320	.067	.328	.056	79.790	Sub-soil of old field.

TABLE VI.—SOILS FROM THE QUATERNARY FORMATION.

No. in report.	County.	Moisture.	Extracted from 1,000 grains by carbonic acid water.	Organic and Volatile matters	Alumina.	Oxide of Iron.	Carbonate of Lime.	Magnesia.	Brown oxide of Manganese.	Phosphoric acid	Sulphuric acid.	Potash.	Soda.	Sand and Insoluble Silicates.	Remarks.
31	Allen	2.725	2.283	3.829	1.410	1.160	0.516	0.312	0.140	0.217	0.056	0.067	0.006	92.365	Virgin soil.
32	Allen	1.050	1.882	1.667	1.535	1.360	.490	.269	.165	.166	.082	.058	.032	94.960	Old field soil.
33	Allen	0.725	1.476	0.856	1.187	1.360	.415	.312	.165	.158	.082	.042	.005	96.140	Sub-soil of old field.

REPORT

ON

THE DISTRIBUTION

OF THE GEOLOGICAL STRATA IN THE

COAL MEASURES

OF INDIANA.

BY PROF. LEO LESQUEREUX.

INTRODUCTORY LETTER.

COLUMBUS, OHIO, February 5, 1861.

Prof. Richard Owen:

DEAR SIR:—According to the directions of Dr. D. Dale Owen, I was charged last spring to examine, with your kind assistance, a part of the Coal Measures exposed in each of the counties comprised within the limits of the coal field of Indiana. The purpose of this examination and the manner in which it was performed, is briefly stated in the beginning of the accompanying report.

By the death, so universally and so deeply lamented, of your brother, I have now to present you this report, which contains the result of five weeks of explorations in the coal fields of Indiana. In doing it permit me to publicly acknowledge my obligations to you for friendly and most valuable services received during my Geological tour, and also the high regard with which I am, sir, most sincerely yours,

LEO LESQUEREUX.

REPORT.

I.—INTRODUCTORY REMARKS.

Before entering into the examination of the Coal Measures of Indiana, I have to mention, in a few words, the purpose of the explorations which I had under my charge, the manner in which they have been performed, and the benefit which may result for the people at large from the Geological data which have been collected and are exposed in this report.

The director of the Geological State Survey of Indiana, my lamented and much respected friend, Dr. D. Dale Owen, thought advisable to direct, as a preliminary step to a future detailed survey, a general reconnoissance of the distribution of the coal strata of the State, in order to ascertain, if possible, the number and position of beds of coal probably attainable and workable in every one of the counties included within the limits of the coal fields of Indiana.*

Such a geological reconnoissance can be pursued in two ways: 1st. By a stratigraphical survey, in following the dip or inclination of the strata, as well as the exposure of the rocks will permit, and drawing from their general inclination conclusions about the horizontal position which the strata ought to occupy at a given place. A survey of this kind demands a great deal of time, and is only practicable in detailed explorations. In a country, like Indiana and Illinois, where the rocks

---

*It is perhaps unnecessary to remark that under the name of Indiana coal fields I truly mean the area, belonging to Indiana, of the coal basin covering the greatest part of the State of Illinois, the south-western corner of Indiana, and a north-western part of Kentucky. Isolated or connected spurs of this basin extend to the West and South, through Iowa, Missouri, Kansas, Arkansas, Texas and even New Mexico. I consider this western coal basin as a detached part of the great Apalachian coal fields, from which it is separated by a Devonian and Silurian ridge, passing through western Ohio, eastern Indiana and middle Kentucky to Tennessee, in a direction nearly parallel to the Allegheny Mountains, and of a contemporaneous upheaval. (Silliman's Journal, July, 1859, page 28.)

of the old formations are overlaid by the Drift, it is often impossible.
Moreover, even in the most favorable circumstances, owing to the nature, thickness and inclination of the strata, it is subject to many errors.
2d. By Palæontological evidence, or from the examination of the fossil remains of rocks, each bed of coal, with its accompanying strata of fire-clay, shales, sandstone, limestone, &c., may be considered as a peculiar or separate formation, which has some fossil remains, fishes, shells or plants peculiar to it, and different, specifically or numerally, from those of higher or lower strata. The determination of these peculiar fossil remains, indicating the geological horizon of a bed of coal or of any other strata, is the palæontological evidence. According to this, the Palæontological Geologist, coming to an outcrop of coal, is expected to fix at once its geological horizon from the fossils of its shales, though he may know nothing about the dip or direction of the strata. Such a proceeding does not take much time and when practicable gives more reliable results than the other. I say *when practicable*, because coal banks and their connected strata are not always exposed and worked in such a way that their fossil remains can be found and examined. Even when the strata are sufficiently exposed it sometimes happens that no fossil can be found on account of the local barrenness of the shales. Moreover, the fossils preserved in the upper part of the Coal Measures are mostly crushed and undeterminable shells. Their number is very great and the species have not, up to the present time, been studied well enough to ascertain, with any degree of certainty, those which are peculiar to various geological horizons. Thus only the fossil plants give, as yet, reliable palæontological evidence. But when the strata are of marine formation, as is often the case with the roof shales, no fossil plant is found with them. The geologist is thus in many cases obliged to try to determine his horizon from the appearance and nature of the rocks, (*lithological evidence*,) which, especially for the strata of the Coal Measures, is extremely variable and affords only an unreliable guidance.

I do not say this to lessen the value of the conclusions which I was enabled to take from my short Palæontological survey of the coal fields of Indiana; but to show why the geological horizon of a few of the examined coal strata has not been definitely fixed. In spite of these few exceptions I consider the general distribution of the coal strata as certainly established, in this report, for every one of the counties where coal can be found in Indiana. Until a detailed survey can be made, the remarks concerning each county are sufficient to direct researches for coal at any place.

To understand the horizontal position of a coal bank, following the indications of this report, it is only necessary to keep in view the general section of the Coal Measures of Indiana, as it has been established by the State Geologist. The number of a vein indicates its horizon, and shows at once if there is any chance to find a workable bed of coal lower than the one exposed or mentioned, and at what depth it may be found. Researches for coal do not appear to be now of great interest to the proprietors of coal lands in Indiana, because the combustible mineral when found out of the principal lines of communication, (the Ohio river and the railroads,) is not of great value at the present time. But as the combustible is every year in greater demand, researches for coal will soon prove a remunerative investment of money, and will become more active. It is thus proper to give some directions which may facilitate these researches.

II.—DIRECTIONS FOR SEARCHING FOR COAL.

The coal beds are the remnants of ancient marshes or peat-bogs, where successive generations of plants have heaped their remains. These, mostly woody materials, have been preserved and diversely modified by a slow process of decomposition in water, and have been changed into coal, anthracite and other mineral combustible, bitumen, &c. A bed of coal is thus a more or less expansive and thick sheet of combustible matter, covering a certain area as does a marsh or a lake; and not a vein of mineral meandering in the rocks like a river in its bed, and which can be struck and followed, at some places, without regard to a peculiar level or a peculiar horizon. Of course the extent of the primitive marshes was very variable. They were separated by patches of dry land or surrounded by water, or cut by hills of sands, just as our peat-bogs are now. The area occupied by each of them has been also modified after its formation, or after the deposit of combustible matter; hence we can not be surprised to see a coal bed losing itself, or more or less abruptly thinning out, in passing to shales, sandstone or other kind of rocks, to reappear again at the same horizon in some other part of the country. According to this remark, then, in a broken country, a bed of coal is found cropping out on the slope of a hill, it can sometimes be followed all around the same hill, and if the strata overlying it were taken out it would be found to cover entirely the surface of the truncated cone of the hill. It is also generally found in the same hills at the same level or at a level corresponding with the dip.*

The dip of a geological stratum, or of a bed of any kind of rocks, is about the same thing as the slope of an open surface. It is the declination or descent towards some point of the compass. Supposing the strata to have been bent in successive undulations, the dip is contrary on both sides, to the highest point of the flexure, which is called an *anticlinal axis*. The base of the undulation, at its point of flexure, forms a *syclinal axis*, with the dip directed on both sides to it. It is evident that the dip of a stratum is but in one direction, till it is changed by a flexure or an axis. It is evident also that the same bed of coal exposed on different and somewhat distant hills will occupy different levels, except when there is no dip in the country, or when the hills are placed on a line perpendicular to the dip, which is called the strike. Thus, also, if the two borings are made at some distance, the dip, according to its direction, will bring in both places the coal at a different level. Supposing, for example, the dip to be towards the west, and to measure twenty feet per mile, a boring made one mile westward of another must be twenty feet deeper to reach the geological horizon of the same coal.

It is clear, from what is said above, that searching for coal, either by leveling or by boring, can not be pursued with security before the general dip of the country is ascertained; neither can a coal bed be worked with every advantage till its local dip is known. The drainage, the air shafts, the breasts and coal chutes of a mine follow the dip; the gangway and the galley follow the strike.†

The dip of a bed of coal is not always easily ascertained from the examination of its outcrop. On the slope of a hill the stratum of soft fire-clay, which underlies every bed of coal, has been often disintegrated and washed away either by erosion or by percolating springs, which find their way between the coal and the fire-clay. As the coal itself is overlaid by easily broken or bent shales, which are themselves covered with strata of heavy materials, sandstone, limestone, &c., the pressure

---

*When coal is searched above the general level of the country in hills, mountains, or on the slopes of valleys, a pocket level becomes extremely useful. It can supply topographical measurements in many cases.

†Lesley's "Manual of Coal and its Topography," page 39. This manual should be carefully studied by every proprietor of coal lands; it is a faithful guide in the explorations for coal.

of the overlying strata generally causes the exposed borders of a coal bank to bend downwards, or even to slip slowly upon the moistened clay and thus to accidentally change its original level. Accordingly, the coal banks at their outcrops, even for a distance of twenty feet, appear sometimes as dipping in the direction of the slopes of the hills. This partial deflection is entirely independent of the dip and may be contrary to it.

Sometimes the strata exposed or cut along a river, or a creek, indicate at once the direction of the dip. But generally, especially in Indiana and Illinois, where recently deposited materials cover the rocks, the ascertaining of the dip is extremely difficult.

Outcrops of coal are discoverable by coal dirt, or by small parcels of coal, which are found in the beds of the creeks, or on the slopes of the hills, where the rocks are exposed or where the turf is overturned by some accidental cause. These outcrops are caused in two ways. Either by the dip in a flat country, where it may alternately bring to the surface the different strata of the measures, according to its direction, or, in most cases, by erosion or deundation. In Indiana, as in Illinois, water has been the only force in activity to change the uniform level of the country. It has plowed valleys and hollows, and transporting away the loosened materials, embossed the country with hills of various forms and of various elevations, according to the nature of the rocks. Hard limestone and sandstone, have been sometimes vertically cut like high walls. If this work of deundation had not acted upon the surface, as the old rocks are generally overlaid by recent deposits, the coal could not have been discovered in Indiana but by borings.

The coal strata covering very variable areas in extent and in outline, it becomes evident: 1st. That borings for coal can not be begun with the certainty of being productive, or remunerated by the discovery of a coal bank.

2d. That, if by a boring, coal is not found at its indicated horizon, it is not a reason to suppose that it can not be found in the neighborhood. The bore can pass near the tail of a coal bed, through shales or other rocks, and the coal can be found of a good workable thickness at a few feet from the place. From five or six borings made at a distance from each other around and at Uniontown, Kentucky, the coal was found four to six feet thick at three of them, while at both the others, shales and fire-clay only were reached at the place of the coal. At Uniontown the coal is worked six feet thick, just on the left side of the Ohio river, while on the other side, in Illinois, the boring opposite Uniontown is unproductive.

3d. That it is impossible, from a single boring, to know the true thickness of a bed of coal, when it has been found. It is always desirable to have as many borings as possible, not only to ascertain the place of the greatest thickness of the bed but to find out the general dip and to become acquainted with the nature of the strata intervening between the surface and coal.

It is very difficult to give general directions about the place where borings should be made. External circumstances, the vicinity of a line of communication, easy access to the mouth of the shaft, facility of drainage, are essential matters to be considered. The first boring for an exploration is generally made near the deepest part of a ravine, especially in order to lessen the cost of the labor. It is only where a bed of coal of workable thickness has been found, at a depth where it can be attained by a shaft, without too great expense, that borings are continued at different places to ascertain the best way of coming to the coal and of building a shaft.

It is always neccessary to make a careful record of the strata passed in each boring, of their thickness, their nature, &c. For it is only by comparison of the records, that the exact position which the coal ought to occupy can be fixed. Moreover, the acquaintance with the nature of the strata is necessary to enable the proprietors to prepare the best materials for the building of a shaft and to make a valuation of the cost.

When a boring has reached the depth where a coal bed ought to be found, and when, at this geological horizon, it comes to black shales, overlying fire-clay, this can be considered as the place of the coal, and the best is to stop the boring at once and to begin it at another place, unless one should desire to go as low as the lowest coal of the measures. Generally the workmen do not not like to transport their tools far, and are constantly asserting that coal can be found a few feet lower. In Indiana and in Illinois, the general distribution of the strata of the Coal Measures is such, that the space between two beds of coal does not vary much in the same county.

In any case the boring for coal must be stopped when they reach the Millstone Grit. This sandstone formation is generally very hard, mixed with small pebbles of quartz, easily recognized by miners. I mention this precaution because, as will be seen hereafter, the average thickness of the only bed of coal underlaying the conglomerate in Indiana is from two to three feet. A coal bed of this thickness can not pay the working by a shaft.*

It has been remarked before that the original strata of the Coal Measures are generally covered with recent deposits, Drift, Quaternary or Alluvial. In northern Indiana, as in Illinois, the Drift composed of sand, gravel, bowlders, &c., covers the Coal Measures, on a thickness varying from fifty to more than one hundred feet. Along the Ohio and the Wabash rivers a peculiar Quaternary formation, evidently anterior to the Drift, a compound of clay, sand and soft materials, is exposed in the base of the hills, along the bottoms, and sometimes also overlies the coal strata. It is, of course, necessary to ascertain, either in the ravines cut through these formations by creeks, or by borings, the exact thickness of the loose materials, through which shafting is always difficult and expensive.

From the outcrops of a coal, or from what is called its dirt, it is rarely possible to determine at first sight what may be the average thickness of the bank, and what is the quality of the coal. According to a former remark, a coal bank is generally compressed and somewhat displaced where it outcrops by the weight of the overlying strata, and it is often thinned and broken. On another side the percolating water, removing small loose parcels from a thin bed of coal, may strew them on the slopes and makes a dirt deposit indicating apparently a thick bed of coal. To try a bank the miners ordinarily content themselves with making a short entry, and as soon as they come to solid coal they form their opinions, (always too favorably,) about the average thickness and quality of the combustible matter. For proprietors, who wish to begin the working of a coal, and for companies, that are interested in bargaining for coal lands, such a superficial examination often causes ruinous bargains and useless expenses. The average thickness of a coal bed is rarely fully exposed except by an entry thirty to forty feet deep into the bank. The same remark can be applied to the quality of the coal.

III.—QUALITY OF THE COAL AND ITS VALUE.

The western coal fields have only bituminous and cannel coal. Bituminous coal is fat or dry, according to the greater or less quantity of

---

*The cost of boring varies according to the difficulty of transporting the tools, and also to the nature of the strata, being softer or harder. It is generally from 75 cts. to one dollar and fifty cents per foot.

bitumen which it contains. This quantity varies from ten to forty per cent. A fat coal burns with a dark yellow flame, emitting a thick, black, strongly scented smoke, and generally runs and coalesces like melted metal. It is then a caking coal. To burn freely it wants to be stirred, or needs the action of the bellows. It is thus good for blacksmithing, but is not as desirable for the grate; and still less for steamboats or engine furnaces. For this last use a dry coal is preferable. According to its compactness and crystallization a dry coal becomes *cherry coal* or *splint coal*. It readily burns without agglutination of its fragments, emits less smoke, gives a light yellow flame, but has often in its compound a large amount of earthy and mineral (sulphur and iron) matter, which causes heavy cinders.

Cannel coal is generally of a dark brown or black color, very compact, of a fine homogenous texture, with smooth conchoidal fractures, sometimes susceptible of a fine polish and resembling jet. It contains a great proportion of bitumen, and burns with a bright flame, like a candle. It is used mostly now for obtaining by distillation its bitumen, which produces the numerous varieties of coal oil. When used for the grate or for the furnaces it is generally mixed with dry coal on account of its flame, which may become too intense and dangerous.

An exact classification of the different species of coal is impossible. The chemical compound of this combustible matter, as well as its external or apparent characters, is extremely variable, not only in the strata at different places and different geological horizons, but even in pieces taken from the same bank. According to the species of plants which have formed it and to their nature, it contains more or less bitumen, and also more or less carbon. According to circumstances which have accompanied and followed its formation, it is mixed with more or less of earthy matter, and impregnated with a variable quantity of mineral compounds, especially sulphur and iron. According to the position of the bed to its superposed strata and its liability to percolation by water, the coal is more or less oxidated with iron, it is of various degrees of capacity and its chemical elements combine in different manners, producing compounds of various kinds. All these influences have acted or are still acting in a different way, on different parts of the same coal strata, and for this reason, it is seldom indeed that we find the coal, worked out of a bank, homogenous material in its whole extent.

Fully to understand the causes of this diversity in the composition of a coal bed, it suffices to examine the formation of a peat-bog, or the appearances and vegetation of its surface, together with the perpendicular section of a bank where it is cut for obtaining peat. On the surface of the bog, at any time of its growth, we find some parts overgrown with a kind of pine* from whose branches and leaves pitch constantly exudes and drops in such abundance that it crusts the ground, sometimes a few inches in thickness. Near by, the surface is occupied by small, very shallow ponds, of various forms, without any vegetation whatever. By the evaporation and percolation of their water (only a few inches deep) they become entirely dry in the summer months, and the surface of the peat, thus exposed to atmospheric influence, is decomposed and changed to a thin layer of half muddy, half coaly, matter. Near by again, the mosses have overrun the ground, covering it entirely with such a soft carpet that passing through it you sink knee-deep into the spongy and humid mass of this peculiar vegetation. Here and there some tufts of rushes or of sedges pierce the mosses, forming a hard woody knot, where the foot can rest securely. At a few paces distance the ground suddenly becomes compact by the vegetation of some small species of the heath family, the cranberry, the cowberry, the bog-bilberry, &c. At some other places the ground is covered by lichens; at others still, with grasses, sedges and rushes. Indeed there is not, on the whole surface of a peat bog, a space of twenty square feet where the same vegetation and the same general appearance can be seen. In examining the perpendicular section of a bank of peat, cut for the extraction of the combustible matter, the same extraordinary variety is seen in its compounds. To soft, spongy layers of scarcely decomposed mosses, from one to twelve inches thick or more, succeed, in descending order, thin layers of muddy, black, hardened matter; then a compact thin stratum of interwoven rootlets and stemlets of small woody plants; then bunches of half decomposed grass, overlying trunks and roots of prostrated (rarely standing) pines, generally imbedded in black, bituminous, compact peat, and thus, by continuous and alternating changes, to the base of the bank. In following the same section in a horizontal direction, or from one point to another of the bank, on the same level, the same changes appear in perfect accordance with what we have remarked on the surface. The layers extend in a kind of homogeneousness for a few feet, and are then suc-

---

Pinus pumilio. It is a European species. In the bogs of the North of the United States it is replaced by the Tamarack (*Larix Americana*) and the Cypress (*Cupressus thyoides,*) and in the bogs of the South by the Bald-Cypress, (*Taxodium distichum.*)

ceeded by others of another compound. Thus it happens that two pieces of peat, taken from the same bank and subjected to chemical analysis, rarely show the same proportion in their chemical compounds.

In the coal we find exactly the same varieties of appearances as those remarked in the peat. The top-coal, the middle-coal, the bottom-coal, are terms generally employed by the miners to indicate different quality of matter in the same bank. And in following an entry or a tunnel every miner knows that the coal sometimes, either at once or by slight transitions, becomes a better or a poorer quality. The best places in the bank are carefully looked for, and the worst portions obtained are thrown away with the rubbish. It is then easy to understand how repeated analyses of the same coal bank generally show differences of some kind. When the Breckenridge coal was first examined it was pronounced free from sulphur, and leaving scarcely one per cent. of ashes. The average of four analyses, reported on page 177 of the first volume of the Kentucky Survey, gives 7.96 per cent. of ashes and 62.40 per cent. of volatile combustible matter. In the analyses of Dr. Peter's report, vol. 2, pages 211 and 212 of the same survey, the averages of repeated analyses show 54.40 per cent. of volatile combustible matter and 12.30 per cent. of ashes. On examining different portions of a large piece of this coal, about five inches thick, which had been sent for analysis, there was found a considerable difference in the proportion of the compounds. For example, the proportion of total volatile matter was found to vary from 55.70 per cent. to 71.70; of coke from 28.30 to 44.30, and of ashes from 7 to 13.30 per cent.

This can not in any way discredit the value of chemical analysis of the coal, but only show how careful one must be in selecting specimens for the laboratory. It is evident that the success of an enteprise for the working of coal depends as much on the quality of the matter as on the thickness of the bank. There is certainly no country where chemists are so often called in to give an opinion about the value of a coal bed, where chemical examinations have been pursued with more conscientious care, and none also where so many fallacious and deceptive valuations on the quality of coal banks have been published. Many proprietors and companies have sustained great losses, and some failed, only from this cause. Their coal was, in the average, (as it is when delivered to the market,) far inferior to what chemical analyses had led them to expect. Of course, proprietors and miners are all interested in giving a good name to their coal, and are all apt to boast of having the best coal in the country. When they send specimens of coal for examination to a laboratory they pick up, always, the best pieces. If in the vein there is occasionally a thin layer of pure coal, free of sulphuret, of shales, and of charcoal, of course it is this part which is usually selected as showing the probable average quality of the newly opened coal bank. It is a voluntary cheat, which helps nothing and deceives badly the proprietors themselves. For, if the true value of a coal was fairly ascertained, it would be easy to know for what purpose it might be used to the best advantage. It would thus be possible to find a market even for an inferior quality, and measures would be taken accordingly. But as it is now, every coal being proclaimed from chemical evidence, of the very best quality for every purpose, proprietors and companies send their combustible indiscriminately to every market, make bargains with gas works, iron furnaces, steamboat landings, coal merchants, &c., investing large outlays for their workings. When after awhile the coal is pronounced unfit for a single one of the purposes for which it is used, it loses at once its name and is declared good for nothing whatever. Thus the works are stopped, the money lost and a coal valuable perhaps for a purpose different from the one to which it has been applied, is abandoned as worthless.

How then can the average value of a coal bank be fairly estimated? By all the ordinary methods, chemical analyses, trial in the forge, in a furnace, in the grate, &c.; every kind of examination may be satisfactory if only samples for examination are fairly selected, and according to the following very simple rules:

1. Specimens should not be taken from the outcrop of a coal bank or from its proximity. When the roof of a coal bank is not of solid stone, an entry of at least twenty feet is necessary to find the matter in its normal state and its average quality.

2. Samples of coal, for any kind of experiment, ought to be selected at various places in a tunnel, and taken from different parts of the whole thickness of the bank. If even there should be, in the bank, streaks of sulphuret or of shales, which are too thin to be easily separated by the miner, pieces of these matters ought to go with the specimens for examination.

3. Generally the miners know the coal from its looks; but their opinion is influenced by personal interest, and is somewhat unreliable. If an experienced person is called to examine a coal, he can make his conclusions in a far better manner by carefully looking at a few car loads, or at a heap of coal taken from the different parts of the mine, than by going himself into it and examining the entries.

18

When the value of a coal bed has been ascertained, it is to the advantage of the proprietors to direct the mining in the fairest possible way, and thus to order the careful cleaning of the coal from every impure matter, shales, and especially sulphuret of iron, when mixed in bands with it. Thin bands of charcoal, and repeated bands of opaque shaly matter, streaks of sulphuret, too thin to be separated from the coal in cleaning out, ordinarily indicate a coal of poor quality.

The compactness of the coal is of great advantage, but it can not be exactly ascertained before the coal has been exposed for sometime to atmospheric influence. When a coal, though compact, contains a certain proportion of sulphuric acid and salts, the efflorescence of these causes a rapid disintegration, transforming the hardest blocks into powder. Such coal when stored is exposed to spontaneous combustion. Although the quality of the coal is very variable, each bed of coal, according to its geological position shows an average amount of similar compounds or peculiar general properties which it may be advantageous briefly to examine, in order to direct the researches, to some lower bed, when the exposed one can not be worked with as much advantage as desirable.

SEC. 1.—SUB-CONGLOMERATE COAL.

This bed of coal underlying the Millstone Grit formation is rarely thicker than two feet. In some places in Indiana, however, it attains three and even four feet, including a clay parting. The coal is generally very hard and compact, dry, burning with a bright yellow flame, without caking, being thus one of the best coals for the forge. It is generally free from sulphuret of iron, but a little shaly, and covered on the top with bands of brashy or slaty coal. It is sometimes impregnated by percolation with oxide of iron. This coal with the two next in ascending order, is generally accompanied with iron, in one form or other, the shales being sometimes entirely oxidated, sometimes intermingled with pebbles of carbonate of iron, generally overlaid with a bed of conglomerate iron ore, which immediately covers the coal in some places where the shales are absent. The compactness of this coal is due to the weight of the great conglomerate formation overlying it. When this bed is found exposed near the surface of plains, as in Arkansas, and is only covered with shales, it becomes brittle by atmospheric influence and oxidated by infiltration.

SEC. 2.—COAL 1 A.

The first coal above the conglomerate is rarely worked on account of its vicinity to No. 1 B., which is much thicker. Owing to the materials forming its roof, its coal has different appearances. When it is overlaid by soft, black, bituminous shales, it is brittle, easily decomposed by atmospheric influence, and marked with bands of sulphuret. When it is overlaid by a bank of compact, coarse, hard sandstone, it appears on the contrary hard and compact, resembling the sub-conglomrate coal; in this case it is a dry coal, in the other case it looks like a fat coal. Although I have seen this bed worked at some of its outcrops in Kentucky, I never had an opportunity to see the coal burning, or to make a fair trial of its quality.

SEC. 3.—COAL 1 B.

From its average thickness and quality, this coal is one of the best of the measures. It has a great tendency to pass to Splint coal and to Cannel coal. Sometimes, as at Breckenridge and Greenup counties, Kentucky, its whole thickness is Cannel coal; at other places, one half only of the bank, generally the upper part is bituminous. In many of the localities, where it outcrops or is worked, it has only a few inches of Cannel coal on the top of a bank of four to five feet of dry Bituminous coal. Its Cannel coal is rich in oil. Most of the oil factories of Kentucky, at Breckenridge, Maysville, Greenupsburg, Ashland, &c., of Ohio, at Newark, &c., use for distillation the Cannel coal from this bed.

Contrary to the assertions of some Geologists, it is certain that from this coal bed is mostly derived the mineral oil which is now pumped out in large quantities from different places on the borders of the coal fields. An active and peculiar decomposition of the woody matter, or of other substances of the plants, caused by atmospheric action, has separated the bitumen, which, after percolating through the coarse underlying sandstone, has been arrested and gathered in subterranean reservoirs, at the surface of lakes and pools of water. This process can be followed at different stages around the Breckenridge coal mines. The sandstone underlying coal No. 1 B. is still so much impregnated with coal oil that oil drops out of the pieces exposed to the sun. All the springs percolating through this sandstone, and gushing out at the base of the hill around, are called oil springs, or bring with the water drops of oil, which may be gathered at the surface. No doubt borings in that country would cause the discovery of oil wells as rich as those of Ohio, Pennsylvania and Virginia. No doubt also that these deposits of oil, like worked beds of coal, must be exhausted in a given time. Though this is not directly related to our examination of the coal, the practical part of the question in regard to the origin of these oil springs, may be treated in a few words, to satisfy the inquiries often made by proprietors of coal lands. In my explorations through the whole extent of the Coal Measures of the United States, I have seen only two geological strata producing oil. The one is coal No. 1 B., the other the Marcellus or black shales of the Devonian. The sulphur springs of Bath county, Kentucky, emerge from the bottom of a small funnel-like valley at the base of hills one hundred and fifty-five feet high, all composed of these Devonian bituminous shales. Here, as at Breckenridge, the process of percolation of the oil can be followed from the base of the shales, through a bed of hard porous sandstone underlying them. Springs come out of this sandstone and bring drops of oil with the water. Such a locality promises also profitable results to explorations and the borings for oil. Of course nobody can assert that subterranean reservoirs ought to exist at a given place; and if there is none, the oil percolating for centuries through the sandstone may have been carried away by the water of the springs and of the rivers. But from geological evidence the indications for large subterranean deposits of oil are as favorable for this place and for Breckenridge as for any other locality where oil is now obtained from wells. I am satisfied that this oil coming from the Devonian shales, like that pecolating from the coal, is of a vegetable origin. Only the plants living in connection with the formation of the Marcellus shales were marine plants, and could not form any coal by their remains, because they have no woody fibre. Marine plants especially, decomposed under certain peculiar influences, have then produced the mineral oil.

The deposits of coal oil are mostly found along the true borders of the Coal Measures. The cause of this peculiar disposition can not be discussed here. The absence of oil springs is remarked all along and on both sides of the Silurian and Devonian axis, which separates the coal fields of Ohio from those of Indiana. I find in this phenomenon a new proof that both coal fields were originally formed in a single basin, and that the Silurian axis is due to an upheaval posterior to the formation of the coal.

Returing to coal No. 1 B, we find its Cannel coal of a generally coarse texture, and of a brown-black color. It leaves for residue a great amount of ashes, the quantity varying from six to fifteen per cent. This fact proves that Cannel coal is not a compound of purely woody fibre as some believe. Though the Cannel coal of No. 1 B., has generally much sulphur, the bituminous coal of this bed is free from this mineral matter, or at least does not contain a great quantity of it. It is mostly a dry, hard, splint coal, finely crystallized, leaving after combustion, white ashes and cinders in small proportional quantity. This coal is good for every purpose. Along the Ohio river it is reputed as equal to the best Pittsburg coal. It is the Bell, Carey, Hawesville, and Cannelton coal of Kentucky and Indiana; and the Hocking, Cuyahoga and Sharp county coal, of Ohio. At any portion of the Coal Measures where working for coal is desirable, and where this coal can be reached at a moderate depth, it is always safe to try to find it by borings. For besides the general good quality of its coal, the bed is one of the most extensively formed and thus one of the most reliable.

SEC. 4.—COAL 1 C.

This bed gives on the contrary one of the worst coals of the measures. It is mostly an agglomeration of stems, transformed into coal and mineral charcoal, intermixed with shales and impregnated with sulphuret of iron. Sometimes the bank is formed only of very bituminous shales, which burn, but do not consume. In this case they contain a large proportional amount of bitumen, and produce oil by distillation. They are ordinarily mixed in the retorts with Cannel coal of No. 1 B. When coal 1 C is well developed, its thickness varies from three to five feet, rarely more. The coal, from oxidation of iron has sometimes a rusty color; sometimes also it is very black and veined by numerous streaks of sulphuret. It is generally brittle, fat, caking, with a strong bad scented smoke. In Indiana and in Illinois, it is better than in Kentucky, and it is thus extensively worked in some places; but is never demanded when another coal can be obtained. It is worked a great deal in St. Clair county, Illinois, opposite St. Louis, on the eastern banks of the Mississippi river.

It is somewhat remarkable that coal 1 B and 1 C, which are sometimes either united in one or placed in close proximity to each other, are scarcely found equally well developed at the same place. Where coal 1 C is thick and of tolerable quality, coal 1 B is thin, and on the contrary coal 1 C is not found workable where coal 1 B is of nominal thickness.

SEC. 5.—COAL NO. 2.

It is generally four feet thick, separated by a thick parting of shales, clay or sulphuret of iron. The coal, when taken out of the mines, is hard and apparently compact, but it contains sulphuric acid and salts in abundance. When exposed to atmospheric influence, it is, in a short time, covered with a white efflorescence, and, by and by, crumbles to powder. It is a caking coal, at least generally so. It is worked in many places along the Ohio river, at Ironton, Hanging Rock, Amanda furnace, &c. Although the bed is generally well formed on extensive areas, it entirely disappears in some places. I have not seen it of a good workable thickness in the Western coal fields of Kentucky, Indiana and Illinois.

SEC. 4.—COAL NO. 3.

This bed is, like No. 1 B, occasionally a Cannel coal. Its bituminous coal is very black, so black, indeed, that at some places the miners call it black diamond. It is generally free from sulphur and from iron, but has occasionally its faces or lines of cleavage, covered with lamellæ of selenite, or sulphate of lime. It is finely crystallized, has not much shale, very little mineral charcoal, and is compact, though the cleavage breaks it in pieces of medium size. As it burns without caking and leaves no cinders it is in great demand for furnaces and for gas works. Its Cannel coal is not as bituminous or as good for oil as that of coal No. 1 B; but it is of a much finer texture, and of a darker color, looking much like polished ebony. Its thickness is not as great as that of No. 1 B, though it attains, rarely indeed, five feet; and it is also much less reliable in its horizontal development. Nevertheless it is worked over the whole extent of our coal fields, from the slopes of the Alleghany Mountains, in Pennsylvania, to the Mississippi river. Its external characters are remarkably uniform, and one somewhat acquainted with coal will easily recognize either its bituminous or its cannel coal at first sight.

SEC. 7.—COAL NO. 4

Is placed under a thick stratum of coarse sandstone, (the Mahoning,) resembling the Millstone Grit, though less gritty and more micaceous. The coal, according to its position, is a very hard and compact coal. It is dry, sometimes splint coal, never cannel coal, containing less of bitumen or more of pure carbon than any other coal bed; and thus not very good for gas works, but excellent, indeed the best coal for coke. It is generally known as the Pomeroy coal. The thickness of the bed averages four feet. In some places of the Coal Measures the coke of this, No. 4, has been used instead of charcoal in the iron furnaces. It is generally free from sulphuret and shales, and the bank is rarely divided by a parting. The roof shales are thin and sometimes entirely absent, the sandstone often forming the roof.

SEC. 8.—COAL NO'S 5, 6, 7 AND 8.

The coal of these beds is not known to me but from chemical analyses, and from exterior appearance. I never had an opportunity of judging by comparison, at different places, the value of the mineral of these places as a combustible. These coal strata occupy, in the western coal fields of Kentucky, Illinois and Indiana, a space generally barren of coal in the east, and not very reliable. The coal of No. 5, at the few places where I have seen it worked, appears nearly as black as that of No. 3; but it is shaly and mixed with streaks of sulphuret. Probably a caking or a fat coal. The bank is often divided by shales or partings. The coal of No. 6, like that of No. 1 C, is liable to be disintegrated and rusted with oxides of iron, and like No. 2, when exposed to atmospheric influences, it is soon covered with a white efflorescence, and crumbles to pieces. It is never in great demand. Coals Nos. 7 and 8 are generally thin. I have not seen them worked.

SEC. 9.—COAL NO. 9

Is covered by a thick bed of black bituminous shales. Its coal is of a good quality, more fat than dry, intermediate between the coal of No. 3 and No. 2. It is extensively worked in both banks of the lower Ohio, at Mulford, Curlew, Newburg, Shawneetown, Evansville, &c. Its thickness varies from three to six feet, and is scarcely, if ever, cut by a parting, a peculiarity which serves sometimes to distinguish it from coal No. 11. Its coal has locally some sulphuret. It is in good demand for steamboat furnaces.

SEC. 10.—COAL NO. 10.

This coal is nearly unknown to me. I consider it an irregular member of coal No. 11, which is itself very irregular in every point. I have seen this coal No. 10 only at Shawneetown, where it was exposed but not worked.

SEC. 11.—COAL NO. 11.

The combustible mineral of this bed is sometimes very black, finely crystallized, very bituminous or fat, of good quality, somewhat caking, excellent for gas, sometimes shaly, full of sulphuret and of mineral charcoal, very poor indeed; sometimes also, in part, a fine Cannel coal, rich in oil. It is mostly separated in two, three, or many more members, by partings of shales, and sometimes of limestone; and thus, by successive alternations of coaly matter and bituminous shales, becomes a very thick bank of little value and of difficult working. This coal has been formed over vast areas. It is the same as the great Pittsburg coal, which covers a surface of hundreds of square miles. Sometimes coal No. 9, No. 11 and No. 12 are united in one.

SEC. 12.—COAL NO. 12.

Like No. 10, it might be considered as a member of coal No. 11. It is generally a thin bed, placed at the base of the Anvil-Rock sandstone, separated from the former coal by a limestone. It thickens sometimes to four or five feet. The coal is extremely shaly, mostly what is called brash by the miners, and is unfit for use. I have never seen it worked.

The coal strata above the Anvil-Rock Sandstone are worked at some places in Indiana, and will come under examination in considering their geological horizons. They are thin beds, mostly two feet thick, nearly unknown, and worked only for the use of some blacksmiths. I am not yet well enough acquainted with their coal to be able to make even an approximate valuation of their quality.

IV.—GEOLOGICAL HORIZON OF THE COAL STRATA OF INDIANA.

A few of the following remarks about the horizontal position of some of the coal banks of Indiana, do not agree, perhaps, with the indications which I may have given to the proprietors, when I first visited and examined their coal beds. I have to explain first the cause of this discrepancy, for fear that it should cause the whole of my assertions to be considered hazardous and unreliable.

In exploring the western coal fields of Kentucky for the geological State Survey of that State, where the coal beds of the lower division of the measures, (except No. 1 B,) are scarcely exposed, I never had an opportunity to study and note the remarkable changes to which the shales of some of these coal strata are exposed, from the influence of a marine formation. I was thus wrongly led to admit, that a certain nature of black, very bituminous shales, whitish spotted and generally containing fins and teeth, of small species of fishes of the shark family, accompanied extensively the two coal beds of the upper measures, No. 9 and No. 11. But after the exploration of the coal fields of Indiana, having passed to Illinois, to follow the same researches, I had opportunity to remark, at some well exposed coal banks, of which the stratigraphical position was evident, that some coal strata of the lower measures, especially No. 1 C and No. 3, are sometimes overlaid by black bituminous and fossiliferous shales, and by limestone, in apparently just the same manner as coal No. 9 and No. 11, of the upper measures. This led me to doubt my former assertions about the position of a few coal banks that I had admitted, from lithological evidence only, and to re-examine and carefully compare all the collected specimens. I had also to take into account all the data of another nature recorded in my notes, the general dip, the relation of the strata, their relative distances, &c. Thus my first impression had to be changed in a few cases; and I was left in doubt for a few others. For this reason, I can not now decide the geological horizon of coal strata marked only by the lithological characters of the shales.

It is very much to be regretted that the animal palæontology of the coal is still nearly unknown, and that not only the numerous species of shells of the Coal Measures have not been carefully determined, but that nothing has been done to ascertain if there is not, with each different geological horizon, some species of shells which, like the plants, are peculiar to a certain geological level. It is only when we possess such a guide that the history of the Coal Measures will be fully understood, and will give us certain indications to determine the exact position of each coal bank.

POSEY COUNTY.

Six miles north of New Harmony, in the range of hills three-fourths of a mile north of Mr. Jos. Calvin's house, a coal is exposed, about nine inches thick at its outcrop. The coal is shaly, in soft layers, intermingled with streaks of mineral charcoal, especially remains of *Calamites*. It is overlaid by a thick bed of grayish soft shales, containing fossil shells, in a poor state of preservation, especially a species of *Lingula*. In the upper part of this bank of shales I have found also a specimen of the bark of a *Sigillaria*, of the same species as some fossil trees found at Blairsville, in an erect position. The section of the strata overlying this coal is:

	FEET.	INCHES.
Covered space, tops of the hills	70	
Limestone without fossils	2	
Shaly sandstone in bank	5	
Grayish, yellow, soft shales, with plants and shells	21	
Coal		9
Fire-clay	2	

On the Mackaddo creek, eight miles north-east of New Harmony, two thin beds of coal crop out near the bed of the creek. The lower one, six inches thick, is covered by black, coarse, micaceous, sandy shales, passing to sandstone, and full of broken remains of plants. In ascending the creek, the strata, above this coal and sandstone, give place to a soft, buff-colored, fossiliferous, shaly sandstones, intermixed with streaks of limestone, or passing to a bank of limestone. This limestone and the buff-colored shales contain a great abundance of fossil shells, among which are a species of *Gervillia* and some *Trilobites*. The coal connected with the limestone, apparently overlying it, is opened behind the hill at some distance off the creek, about fifty feet higher than the former. It is said to be twelve to eighteen inches thick, roofed by black, bituminous, soft, fossiliferous shales. This coal could not be closely examined.

Near Springfield, a bed of coal, ten to twelve inches thick, is somewhat worked for the forge, on the land of Mr. W. C. Pitts. It is overlaid by a coarse, soft sandstone, easily disintegrated. This coal is, in my opinion, the equivalent of the lower bed of coal of Mackaddo creek, but there is no other reason for this assertion than the identical and apparent nature of the sandstone overlying the coal in both places.

On Big creek, near the road from New Harmony to Mt. Vernon, a bed of coal, eight to ten inches thick, is exposed and worked for burning lime. This coal has a roof of black, very bituminous and fossiliferous shales, about one foot thick. Among the fossils there are very small comb-like shark's teeth, of a species identical with those of the shales of Rush creek and Grayville, and a small *Aricula*. There are also broken remains of plants, especially broken pieces of *Calamites* and *Licopodites*, a fine new species with a stem (or branch) half an inch thick, forking near the top and bearing a few small drooping branches. The leaves are sessile on the stem, about half an inch long, lanceolate, pointed and concave. This species has more the appearance of a true *Lycopodium* (*Club-Moss*) than any of the plants found in the Coal Measures until now. Above these black shales, there is, at Big creek, a bank of sandstone eight feet thick. It is soft and sometimes entirely absent or replaced, like the coal, by a fossiliferous hard limestone, worked for burning lime. Among the numerous fossil shells of this limestone there are some Trilobites, and apparently some of the species of the limestone of Mackaddo creek. Near the mill, below the bridge and at a short distance from the place where the coal and the limestone are worked, these strata are replaced by yellow, ferruginous shales, underlaid by a bed of shaly sandstone, probably marking the geological horizon of the coal. About ten feet lower, under the dam of the mill, a bed of soft grey shales, covered with remains of plants, especially leaves of *Neuropteris hirsuta*, is exposed at low water level. This section is approximately thus.

	FEET.
Grayish, yellow, ferruginous shales, with pebbles of carbonate of iron	50
Coarse, soft, shaly sandstone	6
Gray, soft shales with plants, at low water	0

At Rush creek, near its mouth, there is a thin bed of coal, twelve to eighteen inches thick, overlaid by a bank of sandstone four to six feet thick, and on the border of the Wabash, at low water, a bed of yellow soft shales is exposed, and contains the greatest abundance of fossil plants, especially of *Neuropteris hirsuta* and of *Pecopteris polymorpha*. Somewhat higher up the Wabash a bank of black shales, with some coal, is seen exposed, apparently overlying, by a few feet, the geological horizon of the former yellow shales. These black shales contain, with a great many shells, the same comb-like small teeth of sharks as those of Big creek, and also apparently the same species of shells. The bank is broken and disturbed; but it is a mere local distubance, probably caused by erosion of the soft underlying beds of fire-clay, by the action of the Wabash river. A slip or disturbance like this is exposed at a somewhat higher level, below Grayville, on the Illinois side.

This bluff at Grayville, though it does not belong to Indiana, ought to be examined and described in this report. Indeed, the remarkable modification of the strata exposed there, and the whole section of the bluff, is a representation of nearly all the main part of the Coal Measures which we have seen exposed in Posey county, above the great limestone of West Franklin.

Near the landing of Grayville a bed of coal, ten inches thick, has for its roof a bank of black, very bituminous shales, containing shells of the same species as the black shales of Rush creek and Big creek, with the same small comb-like teeth of shark. The coal is here underlaid by shaly fire-clay, black shales, shaly sandstone, passing at its base to a bed of yellow shales, covered with remains of the same species of fossil plants, as those mentioned in the shales at low water level of the Wabash river, below the mouth of Rush creek, and under the dam at Big creek. A little lower on the Wabash, below the landing of Grayville, a thin stratum of ferruginous limestone makes its apprearance, above the coal, and increases in size further down, replacing, by and by, part of the shales and the coal. The section varies accordingly at each part of the bluff. The average distribution of the strata is as follows:

	FEET.	INCHES.
Covered space..	8	
Black, bituminous, fossiliferous shales	4	
Fossiliferous, ferruginous limestone................................	1	
Black, soft, bituminous shales..		3
Brashy coal, full of calamites..		3
Coal slaty and intermixed with charcoal........................		3
Fire-clay..	1	
Shaly sandstone and gray shales with plants, to the level of the river..	12	

Near the lower part of the bluff, down the Wabash, the limestone becomes two feet thick, is overlaid by six feet of black bituminous shales and a coarse sandstone. This limestone is there extremely fossiliferous, being in part a compound only of this shell, (*Gervillia*,) seen in the limestone of Mackaddo creek. From this it appears that considering the extraordinary likeness in the distribution and the nature of the strata, and also the identity of their fossils, especially of the fossil plants, the upper coal of Mackaddo creek, the coal at Big creek, at Rush creek, and at Grayville, are equivalent, or belong to the same geological horizon.

The strata exposed at the bluff of Grayville are there overlaid by a thick bed of hard sandstone, (about 30 feet,) which I consider as the equivalent of the Cut-off sandstone of New Harmony, and as the highest part of the Coal Measures exposed in Posey county. This is, accordingly, the central part of the synclinal axis from which, on both sides, the measures are slowly uprising eastward and westward. Thus, as was surmised by my friend, Prof. E. T. Cox, New Harmony is favorably placed for the boring of an Artesian well.

The strata exposed on the bank of Big creek, at Blairsville, have the following section:

		FEET.	INCHES.
1.	Alluvial soil and clay	5	
2.	Shales and shaly sandstone	15	
3.	Coal brash		3
4.	Fire-clay and broken plants	6	
5.	Sandstone in bank	6	
6.	Fire-clay and trace of coal		3
7.	Shales and shaly sandstone to level of the creek.		

The shales and shaly sandstone, No. 2, contain many broken remains of fossil plants, especially *Calamites;* they resemble the shales underlying the coal at Grayville, at the level of the Wabash river. In the sandstone No. 5 of the section, and perhaps in the fire-clay underlying it, very remarkable fossil remains of standing trees were discovered by Dr. D. Dale Owen. One of the largest specimens, preserved in the cabinet of this celebrated geologist, is two feet three inches high, from the base of the root, perfectly cylindrical, and thirteen inches in diameter at its top where it is broken. The roots still attached to the trees and obliquely directed are about fourteen inches long, from six to nine inches in diameter at their point of divergence from the trunk, and

one inch only at their broken extremity. These trunks (many of the
same species were found together,) were first considered as the remains
of Palm trees; but specimens found with their bark, still preserved,
showed them to belong, or at least to be nearly related, to the genus
Sigillaria, which has furnished to the Coal Measures a great many species of large trees, thus greatly contributing to the formation of the
coal by their heaped remains. The Blairsville species, named *Sigillaria
Owenii* from its discoverer, has its bark marked by double oval scars,
placed, two by two, in a quincunxial order, at about one inch distance,
and joined together by a deep line or groove, which gives them the
form of a pair of spectacles. Each scar is one-fourth of an inch broad
in a horizontal direction, marked by a ring parallel to its border, and
by a central vascular point, like that of *Stigmaria*. The scars of the
root are numerous, more irregularly placed, triangular or round, marked also with a ring and a central point, and thus a true *Stigmaria*.

From the plants and nature of the shales overlying the coal of Blairsville, I consider it a lower level than the coal of Big creek and Rush
creek, and probably as the equivalent of the coal of Springfield and of
the lower Mackaddo creek. At Mr. Calvin's coal, in the hills north of
New Harmony, I found a piece of the bark of a *Sigillaria*, which belongs to the same species as the trees of Blairsville. But as I do not
find, in the composition of the strata, any likeness whatever, I do not
believe that the geological horizon is the same, and I would rather suppose that Mr. Calvin's coal is a higher coal than that of Big and Rush
creek. Its position is still uncertain.

In boring for a well at Speck and Hoffman's brewery, ten miles east
of Evansville, on the road to New Harmony, a coal has been found, at
a depth of fifty feet below the surface. The materials taken from the
shaft are only gray shales, containing remains of the same plants as
those of Blairsville; especially pieces of *Calamites*. The coal found at
this place is probably the equivalent of the Blairsville coal.

At West Franklin, the Ohio river is bordered by high banks of limestone, sandstone and shales, exposed and quarried a little above the
town. The limestone is in two banks, separated by black shales, containing sometimes a thin coal, six to eight inches thick. The upper
bank is of a fine, smooth fracture, hard, compact and fossiliferous.
The lower bed is yellowish, less compact, sometimes nodular and also
fossiliferous.

The general section of the bluff is as follows:

		FEET. INCHES.
1. Covered space, alluvial..	12	
2. Yellow, ochrous, coarse sandstone, mixed with clay iron ore.	10	
3. Coal..		10
4. Hard yellow sandstone...	4	
5. Black or gray, sometimes micaceous, mostly argillaceous shales, with broken plants..	15	
6. Black, hard, compact limestone.................................	4–5	
7. Coal 6 inches, with shales..	3	
8. Coarse limestone, passing to chert.............................	6–8	
9. Covered space to level of the river............................	20	

The thin coal, No. 3, of this section is apparently the equivalent of the lower coal on the Mackaddo creek. At least the sandstone which accompanies it has just the same composition.

The conclusions arrived at till now regarding the comparative geological horizon of the coal strata of Posey county, agree well enough with those taken by Dr. D. Dale Owen from stratigraphical observations, and published, pages 5 to 8, of the first Geological Report of Indiana. But there is, concerning these coal strata, a far more important question, viz: that of their exact place in the Coal Measures; in order to ascertain if coal of a greater thickness can be reached in Posey county, and at what depth from the surface. The question can be answered only after the examination of the strata of Vanderburgh county, and by a comparison of the palæontological data of sections, made at various places of the Coal Measures, along the Ohio river.

VANDERBURGH COUNTY.

On the road from West Franklin to Evansville, about two miles west of this last place, a bank of flinty limestone is exposed, just at the base of the hills bordering the bottom of the Ohio river. The limestone is evidently the equivalent or the continuation of the lower bank of the West Franklin limestone. Consequently, the mouth of the Bodiam shaft which appears at a short distance, near the Ohio river, is at or near the base of the same limestone. Now if we consider the relation of the strata, as we have seen it and established it in Posey county, according to lithological and palæontogical evidence, we can fix a general and reliable section of the measures of Posey and Vanderburgh counties, as follows:

		FEET.	INCHES.
1.	Sandstone of the Cut-off and of Grayville...............	40	
2.	Black argillaceous, bituminous shales, with fossil shells and comb-like teeth of shark, at Grayville, Rush creek, Big creek, &c., sometimes passing to or replaced by limestone	8	
3.	Coal, ranging from 10 to 18 inches........................	1	
4.	Fire-clay, passing to limestone..............................	1	
5.	Shales, with fossil plants at low water level of the Wabash, &c., at Grayville, below mouth of Rush creek, and at Big creek..	30	
6.	Soft sandstone, at Springfield, West Franklin, and shaly sandstone at Blairsville..	10	
7.	Coal		10
8.	Shales and sandstone exposed at West Franklin...........	20	
9.	Hard, black, fossiliferous limestone, at West Franklin.....	5	
10.	Coal, thin a few inches, with argillaceous shales, at West Franklin...	3	
11.	Coarse fossiliferous limestone	8	
12.	Top of Bodiam shaft, covered space, alluvial, probably shales and a thin coal*...................................	30	
13.	Slaty clay...	68	
14.	Sandstone..	12	
15.	Slaty rock...	43	
16.	Shales, with iron stone......................................	4	
17.	Sandstone (Anvil Rock?)..................................	16	
18.	Coal Nos. 11 and 12.......................................	3	
19.	Fire-clay..	1	
20.	Limestone...	8	
21.	Sandstone..	18	
22.	Slaty clay, with iron stone, &c............................	70	
23.	Main coal No. 9, at...	399	10

In the third volume of the reports on the Geological State Survey of Kentucky, Dr. D. Dale Owen has established, (page 18 to 24,) from data collected by borings and from stratigraphical evidence, a connected section of which a part may be compared with the former, and shows a striking analogy. It is subjoined.

---

*At places this space is occupied by two thin strata of coal, placed at a short distance from each other. The balance of this section is that of the Bodiam shaft. It was kindly furnished to me by Dr. Richard Owen.

OF INDIANA. 299

Connected Section of Coal Measures, copied from Dr. D. Dale Owen's Geological Report of Kentucky, 3d vol., page 18.

Space between coal.	Feet.	Inches.	Kind of Rocks.	Feet.	Inches.	
Space.	60			50		Soft sandstone and shale.
					8	Thin coal, No. 1 B.
				50		Sandstone and shale.
			L L	8		Carthage limestone.
			L L	2	8	Thin coal No. 17, of 8 inches.
Space.	35	8		8	4	Soft shale.
				2	6	Sandstone.
				24	10	Soft shaly rocks and bands of sanistone.
					8	Thin coal, No. 16.
				3	3	Fire clay.
			L L	2	0	Soft and hard sandy limestone.
				10	5	Hard shaly sandstone,
				19	4	Soft slaty sandstone.
Space.	102			34	2	Argillaceous shales.
				11	10	Brown shales.
					9	Hard limestone.

19

Connected Section of Coal Measures.—Continued.

Space between coal.	Feet.	Inches.	Kind of Rocks.	Feet.	Inches.	
				1		Soft shales.
				2	6	Coal No. 15.
				4	6	Fire clay.
				2		White limestone?
				5	10	Brown shales.
				3	11	Limestone.
				9	4	White sandy shales.
Space.	115	4		50	8	White sandstone.*
				38		Brown shales.
				1	1	Hard black shale.
				2		Coal 1 foot, No. 14. One foot of fire-clay.
				1		Hard limestone.
				1	1	Hard stone.
Space.	77	6		23	6	Brown shale.
				4		Dark brown shale.
				5		Black shale.
				1	6	Soft gray limestone.
				3		Hard limestone.
				4	6	Blue and light shales.

*The Indiana section, in Posey county, begins at the top of this sandstone.

OF INDIANA. 301

Connected Section of Coal Measures.—Continued.

Space between coal.	Feet.	Inches.	Kind of Rocks.	Feet.	Inches.	
				11	9	White limestone.
				16	2	Bluish shale.
					4	Thin coal, No. 13.
				7		Fire clay and red oxide of iron.
Space.	100	2		10		Shaly sandstone.
				18	6	Hard gray sandstone.
				14	7	Soft gray sandstone.
				19	15	Bluish shale.
				7		Micaceous shale.
				12		Hard gray sandstone, Anvil Rock.
				8	8	Coarse sandstone.
				3		Hard sandstone.
					3	Thin coal, No. 12.
				12	8	Hard limestone, bituminous shale.
Space.	21			8	1	Bluish limestone and clay.

Connected Section of Coal Measures.—Continued.

Space between coal.	Feet.	Inches.	Kind of Rocks.	Feet.	Inches.	
				5		Coal with clay parting, No. 11.
				5	6	Fire-clay and pyriteferous sandstone.
Space.	46			40	4	Thin bedded sandstones, with hard bands intercalated.
				3		Coal 2 to 3 feet, No. 10.
				2	6	Fire-clay
				1	2	and } In all 7 feet 8 inches.
				4		Shales,
				10		Sandstone.
				5		Shale and thin sandstone.
				5		Sandstone.
Space.	67			36	4	Indurated argillaceous shale, with clay iron stone basis.
				3		Avicula shale.
				5		Main Mulford coal, No. 9.
				2		Fire-clay.
				4		Shale.
				25		Sandstone.
Space.	86			8		Shales, with coal and argillaceous iron ore.
				10		White and pink sandstone.
				2	6	Well coal, No. 8.

Connected Section of Coal Measures.—Continued.

Space between coal	Feet.	Inches.	Kind of Rocks.	Feet.	Inches.	
Space.	43			16		Sandstone.
				27		Sandstone and shales.
				2		Coal No. 7, and ferruginous limestone ?
Space.	84			42		Impure limestone ferruginous shale.
				24		Shale.
				18		Thin bedded sandstone, with shale partings.
				3		Three foot or Little Coal, No. 6.
				3		Fire clay.
Space.	65			30		Soft sandstone.
				25		Micaceous sandstone.
				7		Shale, with carbonate of iron.
				4		Coal No. 5.
				3		Fire-clay.
				20		Shales.

Connected Section of Coal Measures.—Continued.

Space between coal.	Feet.	Inches.	Kind of Rocks.	Feet.	Inches.	
				42		Shale with segregations of iron stone.
Space.	116			50		Massive sandstone. Mahoning sandstone.*
				4		Gray shales with plants.
				4		Coal No. 4, four to five feet thick, with parting.
				15		Shales.
Space.	34		L L L	4		Curlew limestone.
				15		Black bituminous or gray soft shale.
				4		Coal No. 3, underlaid with fire-clay.
				62		Shale.
Space.	102			10		Sandstone.
				30		Shale or sandstone.
				3		Coal No. 2, with clay parting.
				2		Fire-clay.
Space.	62			40		Sandstone.

*This part of the section is modified by Mr. Lesquereux according to the measures of Indiana, from the Mahoning sandstone down to the upper Archimedes limestone.

Connected Section of Coal Measures.—Continued.

Space between coal.	Feet	Inches	Kind of Rocks.	Feet	Inches	
Space.	53		L L L	20		Black bituminous shale, with flint and limestone, or soft stone, Burstone.
				4		Coal No. 1 C.
				3		White fire-clay.
				50		Micaceous grey shale or grey metal with sandstone and black shale.
Space.	32			5		Coal No. 1 B.
				2		Fire-clay.
				30		Hard gritty sandstone and Shales.
				2		Coal No. 1 A.
Space.	60			50		Conglomeratic or gritty sandstone, Millstone Grit.
				10		Black, soft or grey shales, with clay iron ore.
				3		Coal, (sub-conglomerate.)
			L L L L			Archimedes limestone.

I consider it evident, that the great bank of sandstone, 50 feet 8 inches thick, marked on his section between coal Nos. 14 and 15, is the equivalent of the sandstone of the Cut-off on the Wabash, near New Harmony Beginning thus from this sandstone and descending to coal No. 9 the section of Dr. D. Dale Owen is:

		FEET.	INCHES.
1.	White sandstone	50	
2.	Brown shales	38	
3.	Hard black shales	1	1
4.	Coal No. 14	1	
5.	Fire-clay	1	
6.	Hard limestone	5	
7.	Hard stone	1	1
8.	Brown shale	23	6
9.	Dark brown shales	4	
10.	Black shales	5	
11.	Soft gray limestone	1	6
12.	Hard limestone	3	
13.	Blue and light shales	4	6
14.	White limestone	11	9
15.	Bluish shale	16	2
16.	Thin coal, No. 13		4
17.	Fire-clay and red oxide of iron	7	
18.	Shaly sandstone	10	
19.	Hard gray sandstone	18	6
20.	Soft gray sandstone	14	7
21.	Bluish shales	19	5
22.	Micaceous shales	7	
23.	Hard gray sandstone, (Anvil Rock)	12	
24.	Coarse sandstone	8	8
25.	Hard sandstone	3	
26.	Thin coal, 3 inches, No. 12		3
27.	Shales and sandstone	12	8
28.	Hard limestone, bituminous shales, &c	8	1
29.	Coal with parting, No. 11	5	
30.	Fire-clay and pyritiferous sandstone	5	6
31.	Thin bedded sandstone, with hard bands	40	4
32.	Coal 2 to 3 feet, No. 10	3	
33.	Fire-clay and shales	7	8
34.	Sandstone	10	
35.	Shales and thin sandstone	5	
36.	Sandstone	5	
37.	Indurated argillaceous shales, with clay, &c	36	4
38.	Avicula shale	3	
	To coal No. 9, 5 feet thick	408	11

In this section, hard sandstone, five feet thick, (No. 6 of the section,) represents the limestone under the coal at Big creek, a limestone sometimes replacing the coal, and thicker. At Grayville it is underlaid by 30 feet shales with fossil plants, and at other places by a sandstone containing a thin coal, which in this section is replaced by five feet of black shales. The two great banks of limestone separated by argillaceous shales, agree perfectly in both sections, and at the base of this limestone, begins the Bodiam shaft. If we consider that both these sections have been made at far different places and on different principles, their coincidence can not but appear striking.

At Lasalle, Illinois, a place which, from my observations, occupies just the same geological horizon as Evansville, the two strata of limestone are overlaid by thirty feet of shales. Above these shales there is a coal one foot thick, overlaid by a very fossiliferous limestone three feet thick, containing the shells of the same species as some of those of Big creek, Rush creek and Grayville, especially the *Gervillia*. The measures at the border of the basin are, at Lasalle, somewhat reduced; nevertheless the proportion in thickness of the strata is well enough preserved. The two banks of limestone, the upper one twelve feet thick, the lower one fourteen feet, are separated by five feet of black, argillaceous, bituminous shales, with traces of coal. The space from the base of the lower limestone to the middle coal of the shaft of Lasalle, which, from the abundance of *Avicula* in its shale is coal No. 9, is 228 feet. At the Bodiam shaft it is 230 feet. The difference results from the reduction of the measures between coal No. 11 and No. 9, which has nothing extraordinary whatever, and is sometimes much greater. At the Bodiam shaft, coal No. 11 is 172 feet from the base of the limestone, and at Lasalle it is 175 feet.

According to the data which have been examined and discussed above, it follows: that borings for coal, in the central and western part of Posey county, would not offer any chances of a remunerative investment of money. Along the Wabash river, from Grayville to its mouth, coal No. 11 would be reached at 300 feet, and coal No. 9, a far more reliable coal and of a better quality than No. 11,) at 400 feet deep. On the south-eastern side of the county, at the base of the bluff of West Franklin, the same coal No. 9 would be found at about the same depth as at the Bodiam shaft, or at from 280 to 300 feet. In the western part of Vanderburg county this space would be reduced to from 100 to 150 feet.

WARRICK COUNTY.

The coal bank worked at Newburgh, with the section of the shaft, has been already reported in the first Geological Report of Indiana, (pages 10 and 11). The main coal of Newburg is No. 9, the equivalent of the main coal of the Bodiam shaft. It has, in its underlying shales, the Avicula in great number. The same bed was formerly worked near the mouth of Pigeon run, at high water level of the Ohio river. The bank is covered and its character could only be examined from a few shales, left outside of the tunnel. The distance from the base of the Anvil Rock sandstone to the coal is only seventy feet. At Newburg this space is one hundred feet, which is about the maximum. It is sometimes reduced to fifty feet.

From indications received at Newburg from Mr. G. I. Hutchinson, the director of the mines, there are exposed in the hills, near Taylorsville, two workable beds of coal. We had no time to examine this part of the county. From the direction of the general dip to the west, a little south, it is probable that both these veins are the equivalents of coal No. 4 and No. 3, or the same horizon as the coal banks of the hills around Rockport.

SPENCER COUNTY.

Just on the limits of the county, on Pigeon creek, section 6, township 16 south, range 7 west, the top of a bank of hard, somewhat conglomeratic and ferruginous sandstone, is exposed at high water level of the creek. It is referable to the Mahoning sandstone, and both coal beds, No. 4 and No. 3, should be met with at and near its base.

Four miles north-east of Rockport, on Mr. B. Shrode's property, section 3, township 7 south, range 6 west, a bed of coal five feet thick is worked under a bank of hard sandstone and shales, thirty-five feet thick. The coal is very compact, finely crystalized, free from sulphuret and shales, a dry splint coal, one of the best I have seen in Indiana. Thirty-five feet lower there is another coal bed two feet thick, overlaid by very black bituminous shales, containing fossil remains of fishes, of shells, and of plants. Between the upper coal and the sandstone a thin layer of gray soft shales, (brash) covers the coal. It contains in abundance species of plants characteristic of coal No. 4. From the nature and lithological composition of the black shales overlying the lower coal, which resemble those of coal No. 9, I was inclined to refer these coal strata to a higher level, and to consider the sandstone as the Anvil Rock. But neither the stratigraphy and the direction of the dip of the measures, nor the palæontology of the shales, could support such a conclusion. Even the black shales of the lower coal contain, at Mr. Shrode's bank, fruits and some broken species of fossil plants belonging to coal No. 3. As subsequent explorations in the coal fields of Illinois have proved with entire evidence that coal No. 3 is occasionally overlaid by black bituminous shale, of the same appearance and nature as those of coal No. 9, there is no doubt whatever that, following palæontological evidence, the coal strata, at Mr. Shrode's knob, are the equivalents of No. 4 and No. 3, and that the hard sandstone quarried above the coal (and very good indeed for building purposes) is the Mahoning sandstone. The fine quality of the upper coal, which is especially good for coke, and its short distance from No. 3, would not agree with the quality and the place of coal No. 11.

At the place where we examined it, coal No. 3 is only two feet thick, but it may be found all around in the hills of a greater thickness, and overlaid, as it is generally, with a limestone or a calcareous iron ore.

Seventy-five feet lower than coal No. 3 there is, on the same property, a bed of fire-clay containing crystals of gypsum. This fire-clay apparently marks the place of coal No. 2, which has not been found here. It is a fine white clay, used at some places for pottery, and valuable as a fertilizer of poor lands.

Though the top of the bluff at Rockport is, from barometrical measures, about 120 feet lower than the top of the knob above Mr. Shrode's coal, I can not but consider the sandstone exposed at both places as equivalent, or as of the same geological horizon. From the direction of the dip these strata should occupy about the same level, but along the Ohio river there have been many local disturbances, which can not be accounted for by general laws.

The sandstone, at Rockport, is softer than that of the knob, but it is cut in horizontal strata, of two to three feet thick, by a thin bed of soft shales with plants, in the same manner as the former. The section at both places is also somewhat different; the distance between the two coal beds, under the sandstone, being greater at Rockport. According to data, which were kindly furnished to me by Dr. Richard Owen, the section at Rockport, from the top of the bluff to the botton of a shaft dug at its base, is:

		FEET.	INCHES.
Shales and sandstone		37	
Slate and coal		2	
Fire-clay		10	8
Fine sand rock		6	6
Blue rock		30	
Dark shales		22	10
Coal		2	6
Dark shales		1	6
White rock		2	

There is probably some mistake in the figures of the above section. At Rockport the sandstone, from its base to its top, measures at least thirty-seven feet. As I have not seen any of the materials taken from the shaft, I am, of course, unable to discuss the question concerning the place of these coal strata with any reliable data. Coal No. 4 and No. 3 are, (as remarked before,) generally separated by a limestone, the Curlew limestone, varying from one to eight feet in thickness. Could the blue rock, marked *thirty feet* in the section, be a limestone? Probably not; for no limestone of such a thickness has ever been seen at this geological horizon. Where the Curlew limestone is absent, as at Mr. Shrode's knob, its place is taken by argillaceous, ferruginous shales, or clay iron ore, and the distance between both coal strata, which averages thirty-five to forty feet, is still reduced. The sandstone of the bluff of Rockport might be, by its position, the equivalent of the sandstone generally overlying coal No. 2, and the lower coal marked in the section under twenty-two feet ten inches of dark shales, could belong to No. 1 C. This question can be decided only by the examination of the shales overlying the coal strata, and of their fossils, if they contain any, or by a deeper boring. If the Rockport sandstone is the Mahoning, the Cannelton main coal can be reached at 300 feet from its base; and at about 150 feet, if it belongs to coal No. 2. In any case Rockport is placed in a very favorable position for reaching coal by shafts, all around the place.

About four and a half miles south-west of Rockport, on 'Squire James Stuteville's property, section 9, township 8 south, range 6 west, a coal 28 to 30 inches in thickness is exposed at different places, and overlaid by shales, presenting various lithological appearances. At one place, the black shales which cover it are of the same nature and have the same fossil remains as those of the lower coal at Mr. Shrode's. At another place, where the coal has been reached by a shaft, the shales are grayish, soft, a kind of soap stone, with remains of plants. That both these shales belong to coal No. 3, is proved from what is seen at some localities in Illinois, where this bed of coal is overlaid by two strata of shales, the upper one of black, laminated bituminous shales, like those of the Knob; the lower one of soft soapstone shales, full of plants like those of the shaft at 'Squire Stuteville's. The distance of the coal here from the base of the Mahoning sandstone, exposed at the top of the hills, is about 50 feet. From the direction of the dip, the sandstone occupies here its exact place, while at Rockport, if it is the same, it is about one hundred feet too low.

PERRY COUNTY.

A detailed private examination of this county being purposed, for a future time, by the State Geologist, Dr. D. Dale Owen, my explorations were not directed to it. Nevertheless, and from an examination of the coal strata of Cannelton, during my connection with the Geological State Survey of Kentucky, I am able to report the position of the coal strata of the county in a general way.

The borders of the coal measures follow the western borders of Perry county, and, accordingly, the sub-conglomerate coal banks of about two feet thickness ought to be found along the line of Deer Creek, and eastward of it, on Poison creek, &c. At Cannelton, and eastward, coal 1 A. 1 B. and 1 C. are found at different levels above and below the general line of surface of the country. The main coal of Cannelton has all the characters, and the quality also, of coal No. 113. It is in part Cannel coal, and even at some places entirely Cannel. It is mentioned in the first report of Dr. D, D. Owen, page 49. Westward of Cannelton, coal 1 C. and coal 2 may be exposed at some localities; but where a bed of coal is desirable, in that part of the country, it will be more profitable to search for coal No. 1 B. by a boring.

It is much to be regretted that the proposed survey of Perry county could not be performed by the Director of the State Geological Survey of Indiana, who was so well acquainted with the geological formations of the county. By its position, by the quality and the abundance of its combustible mineral, this county is one of the most interesting in the State. According to its geological horizon, some rich deposits of iron ore may be found in its eastern borders.

If now we consider the dip of the coal measures along the Ohio river, according to the data obtained by our examination up to Perry county, we come to find a remarkable agreement with the conclusions taken from stratigraphical and palæontological evidence.

The distance from Cannelton, where coal 1 B. is worked, to the base of the Mahoning sandstone, at Rockport, is about 16 miles.* The average vertical space from coal No. 1 B. to the base of the Mahoning sandstone is 320 feet, thus showing a dip of 20 feet per mile. From Rockport to above Newburg, where coal No. 9 crops out at the level of the river, the distance is nineteen miles. As there is an average space of 370 feet from the base of the Mahoning to coal No. 9, we have thus a dip of nineteen feet and a half per mile. The place of coal No. 9 is in the Bodiam shaft, at 280 feet from the surface, while at the Henderson boring it is only at 160 feet. I do not know the cause of this difference. At Henderson the space between coal No. 4 and No. 9 is somewhat too short; at the Bodiam shaft, the bank of shales overlying the anvil-rock sandstone is too thick. There may be some errors in the measures, or some local disturbances. For, counting the average depth of coal No. 9, between both places, we found it to be 420 feet for a distance of eleven miles, indicating a dip of twenty-two feet per mile. As I consider Mt. Vernon and New Harmony as placed just along the line of the great synclinal axis to which the dip converges on both sides of the basin, if we admit the same ratio of 20 feet per mile for the dip from Evansville to New Harmony, a distance of about 20 miles, we find that, at this last place, the depth at which coal No. 9 could be reached would be 400 feet—just the same conclusion to which we arrived by considering the succession of the strata and the local sections.

The line of the principal axis apparently passing, as previously remarked, from Mt. Vernon to Grayville and Lasalle, (Illinois,) is in a direction parallel to the border of the eastern coal fields of Indiana, and to the great anticlinal axis of the Silurian ridge, which divides the coal basin of the East from that of the West. As Dr. D. Dale Owen has remarked in his report, the dip appears continuous and remarkably uniform, and unbroken in Indiana, varied only by slight undulations. My explorations for the Illinois and Kentucky coal fields, forced me to admit the same conclusion for the whole area of the western coal basin. This is of importance for directing the researches for workable coal strata. As the eastern borders of the coal measures of Indiana are well

---

* These measures are approximative. The declivity and current of the Ohio river is not taken into account. An exact appreciation of the dip at divers places can not be taken but by a topographical survey.

marked, the distance of a certain place westward of their line may approximately indicate its geological horizon.

DUBOIS COUNTY.

On our way from Rockport to Jasper, we passed, in the northern part of Spencer county, a band of broken hills, formed by the hard sandstone of the Mahoning. Coal No. 4 may be found at the base of these hills, of a good workable thickness. On section 19, township 4 south, range 5 west, a bed of coal about three feet thick is worked under a stratum of one foot of sandstone, which makes its roof. The coal has, at its top, a few inches of brashy coal, with undeterminable remains of plants, and at its base a band of sulphuret, with large pieces of fossilized wood, especially *Sigillaria Menardi.* This species, the only one which could be determined, is especially abundant with coal No. 3. From the assertion of the miners, the roof of sandstone is often replaced by calcareous concretions, clay, iron ore or bitumen. This fact confirms the palæontological evidence.

Near Huntingburgh, I examined at a forge, specimens of an excellent coal, worked in the vicinity of Ferdinand P. O. This coal bears on its horizontal surface a quantity of leaves of *Lepidodendron* and blades of *Lepidostrobus*, preserved in charcoal. These are characteristic species of coal No. 1 B. This coal is a compact, very hard, dry splint coal, free from sulphuret. It is much valued in the country for the forge. At Huntingburgh I saw, at another blacksmith's, specimens of another coal opened six miles west of that place. This coal has much sulphur, and is shaly. It looks like a poor quality of coal No. 4, or perhaps No. 3.

From Huntingburgh to Jasper, the country is broken by hills of a hard sandstone, containing petrified trunks of fern trees, (*Psaronius*,) which, until now, have been found only in connection with the Mahoning. As this sandstone continues to Jasper, the coal opened or exposed at the base of the hills, at and around this place, should be referable to No. 4. This conclusion does not agree with the position of Jasper in regard to its distance from the border of the coal basin. It would lead us to expect here rather a lower coal. The two banks opened near Jasper, only 18 inches thick, did not show any characters in the accompanying strata by which their geological horizon could be ascertained. They were covered with water at the time when we passed them. The sandstone overlying the coal is blackened by broken fragments of plants, transformed into charcoal, as is sometimes the sandstone overlying coal

No. 2. At about one mile distance from the coal, and nearly at the same level, I was shown a thin bed of fossiliferous hard limestone. As the coal at Jasper is thin, and interposed between two banks of sandstone, its working can not be remunerative. A better coal might be found by boring at a lower level; but, from the impossibility of ascertaining the exact horizon of Jasper, I can not tell with reliability at what depth the coal No. 1 B. would be found.

From Jasper to Portersville, the sandstone disappears, and the country is more even or less broken. The dip between these two places appears to be a little more directed to the south. Hence the coal, at Jasper, ought to be at a higher geological horizon than at Portersville. The coal, at this last place, is exposed in the creeks, a little above high water level of White river. Just near the town, in the bottom of the creek, the coal two feet thick is overlaid by five feet of black, micaceous, coarse shales, generally very bituminous, sometimes passing to blackened shaly sandstone. From the few pieces which could be loosened and examined, they do not appear to contain any fossils. These shales are overlaid by a thick bank of limestone, very variable in its composition and appearance but always fossiliferous. It is either black, compact, of a smooth even fracture, or mixed with large pieces of flint, or even entirely transformed to flint. At other places it is gray, coarse, and somewhat crystalline. It generally passes to, or is even entirely replaced by a soft, argillaceous, buff colored, compact clay or shales, containing the same fossils as the limestone. These numerous transformations, with some others, still characterize the limestone and burstone of Ohio and Kentucky, which overlies coal No. 1 C., to which the coal around Portersville is referred, without doubt, from palæontological and lithological evidence. This coal, near the creek, is at some places overlaid with a bed 40 feet thick of buff-colored clay shales, mixed with fossils and pieces of silex. On section 27, township 1 north, range 5 west, one and one half miles east of Portersville, the same coal bank is three and a half to four feet thick, and immediately covered by a few feet of flaggy, gray, coarse, fossiliferous limestone, good for lime, and apparently also for constructions. The coal is hard, black, and of fine appearance, but it is mixed with sulphuret, bands of mineral charcoal and sulphur. By exposure it is covered with a white efflorescence and decomposes. Above the limestone there is a bed of shaly sandstone, (partly covered,) thirty to forty feet high. From the assertion of the miners, there is still a thin coal higher than the sandstone. But it was covered, and could not be examined. It is referable to coal No. 2, perhaps the same as the coal of Jasper. At another exposure, near the ferry, the coal two feet eight inches thick is worked for a saw mill. The coal is still poorer than the former, full of sulphuret and of charcoal, oxidated by infiltrations of iron and shales. Its roof is formed by four to five feet of black, coarse shales, of the same nature and appearance as those of the first exposure in the creek, and the black shales are overlaid by forty feet of sandstone, separated in numerous layers by their bands of clay, and without any trace of limestone. The distance from this outcrop to the former, in the creek, is no more than one fourth of a mile.

We have thus, for a bed of coal evidently on the same geological horizon, such differences in the overlying strata, that, judging from their appearance, it would be impossible to consider the different coal banks as equivalent. At one place, black shales, limestone, chert, and buff colored clay, with fossils and no sandstone. At another, coarse, gray limestone, without black slabs, and with a high bank of sandstone; at a third exposure, black slabs and sandstone, without a trace of limestone. Coal No. 1 C. is, from its accompanying strata, a true Proteus, and, as the shales do not contain any fossil plants, it is one of the most difficult to identify from lithological appearances. Its only constant character is the inferior quality of its coal. It was formed at a time of repeated local disturbances, mostly in deep marshes, often inundated by marine water. Nevertheless, it is very extensively formed, and sometimes preserves a workable thickness of four to five feet, over vast areas, and without any change, except in the strata overlying it.

From what we have seen of the measures of Dubois county, it appears that, in its south-western part, coal No. 4 can be found at the base of the hills of sandstone, but that only the sub-conglomerate coal, with coal No. 1 A, 1 B and 1 C, crop out in the north, according to different meridians. Of course the sub-conglomerate coal is seen above the sub-carboniferous limestone, along the eastern limits; coal No. 1 A, 1 B and 1 C, occupying the middle part, and the north-western borders of the county. It is even possible that the distribution of the coal measures might be the same in the southern parts of Dubois county.

PIKE COUNTY.

The coal banks exposed or opened in the northern part of this county, near Kinderhook, are referable to the same horizon as those of Portersville, or to coal No. 1 C. The first bank examined is two miles south-west of Kinderhook, on the property of Mr. James R. Thomas. The coal, three to four feet thick, has a roof of black, bituminous, fossiliferous shales. It is not of an inferior quality, has a great deal of charcoal and sulphur, and is easily disintegrated when it is exposed to atmospheric influence.

One and a half miles south-west of Kinderhook, the same coal is exposed on the property of Dr. Posey. At this place, and along the creek, the coal is apparently six to seven feet thick; but it is mixed with bands of shales or shale-partings, of various thicknesses, and also with bands of sulphuret. It is of the same quality as the former. It is first overlaid by a few inches of soft, crumbling shales, entirely formed of small shells and sulphuret of iron; and above this by three or four feet of very black, bituminous, coarse grained and micaceous laminated shales. These insensibly pass, above, to an argillaceous, buff-colored clay, just like that of the bank at the creek of Portersville. The clay has also the same shells, and passes to a blue, coarse grained limestone, or is intermixed with bowlders of Septaria, mostly resting upon the black shales.

The same coal is still opened at Mr. G. Fecklin's, on section 18, township 1 north, range 7 west, one mile from Kinderhook, and near by, on the canal, and where it is worked on Mr. Rhode's property. At this last place, the coal has a better appearance and is of a better quality. Its sulphuret, mixed with shales, forms layers thick enough to enable the miners to separate it, and thus to clean the coal. But it is also covered with a white efflorescence of sulphur, when it is exposed to atmospheric influence for a long time. The bank is five feet thick, and overlaid by soft, grayish shales, a compound of small shells, and by a few feet of black shales, of the same nature as those of Dr. Posey's coal.

Near the town of Petersburg, two beds of coal crop out at the base of the hills. They were covered by water at the time of our visit, and they could not be examined. From the nature of the overlying shales, I refer them to the same geological horizon as that of the former banks. The limestone connected with coal No. 1 C is found in the hills around, with its various appearances, sometimes black, hard and compact, sometimes blue and coarse, sometimes flinty.

Between Petersburg and Highland, two beds of coal are said to be exposed, on the banks of White River, the one a foot and a half, the other four feet thick. High water prevented an examination of them.

From the direction of the dip of the measures, coal No. 2, No. 3, and perhaps No. 4, can be found in the southern and western parts of Pike county. The eastern half of the county is mostly occupied by coal 1 C. It is better here than anywhere else in Indiana, especially at Mr. Rhode's bank. A far better coal, No. 1 B, can be found by borings, at a lower level, from 25 to 55 feet deep below No. 1 C.

It is probable the coal on the Patoka, on section 4, township 2 south, range 8 west, mentioned by Dr. D. Dale Owen, (in his first report, page 40,) as being ten feet thick, is the equivalent of coal 1 C. This coal, especially where it becomes united to coal 1 B, or is parted by shales or clay, becomes very thick; but it is only a good workable coal for a part of its thickness. I have seen it in Kentucky about 80 feet thick, composed of alternate layers of very black bituminous shales and coal, where the thickest workable band of coal was only two feet; and thus a nearly useless mass of matter. I regretted much that the coal on the Patoka could not be examined. But, at the time of our explorations, White River was out of its banks, and some of the most interesting coal banks of the country, especially along White River, were covered by water and unapproachable.

DAVIESS COUNTY.

Four miles south-east of Washington, on Mr. Nelson Jackson's property, a bed of coal, eighteen inches thick, has been somewhat worked by stripping. When we visited the place, the trenches were under water, and I could not even see any piece of the shales. As no rocks are exposed in the vicinity of this coal, its position is of course undetermined. It is probably coal No. 1 A; at least as much as can be seen from the direction of the dip, and from the topographical place of this coal.

At Washington, coal No. 1 B is worked by shafts about twenty feet deep, at some places; or at the base of the hills, by tunnels entering the exposed bank. It is overlaid by a bank, (22 feet thick,) of soft, gray, laminated shales, a kind of soapstone, containing remains of fossil plants. The species found are few, only blades of *Lepidostrobus* and leaves of *Lepidodendron*, which mostly characterize the coal of this geological horizon. The too soft nature of these shales probably prevented the preservation of the fossil plants which at some other places are very abundant. The coal is four feet thick, a fine, hard, compact, dry, nearly splint coal, free of sulphuret, and in great demand, especially for the grate and the furnaces. It is only to be regretted that the shales overlying the coal are here so soft. For, at the places where the roof is not thick, the percolation of water has caused the disintegration and the oxidation of the coal, which then loses a little of its value. This bank is pretty extensively worked on Mr. S. B. Legg's property, and also on Messrs. Church & Co.'s. It is underlaid by a thick bed of hard, black fire clay, good for pottery. It is said that limestone is found in the hills around Washington. In that case, coal No. 1 C can be found in connection with it.

A boring made a short distance from Mr. Legg's coal, passed through ten feet of yellow soft clay, eight feet of sandstone, and 132 feet of soapstone, to an eighteen inch coal. Perhaps this low coal might be referable to No. 1 A. But the distance to 1 B, which, at Mr. Legg's, is twenty-two feet below the surface, would be about 130 feet, or far too great, according to the average distance, which is no more than thirty feet. As no written and exact records have been taken of this boring, and as there may be some mistake in the figures, it is useless to speculate about the position of this thin coal.

From information received at Washington, it appears that eight miles north-east of this place, on the road to Dover Hill, there is a coal eighteen inches to two feet thick, considered excellent for the smith. From the direction of the dip and the meridian of this coal, it is referable to the sub-conglomerate bed; but it was not examined. Another bank, referable to the same geological horizon, is exposed on the Ohio and Mississippi railroad, eight miles east of Washington. Fourteen miles north of this same place a coal bank, said to be six feet thick, is exposed and overlaid by eight feet of limestone. It is referable from its limestone and its position to coal No. 1 C.

From these data, and from our examination in Daviess county, it appears that the sub-conglomerate coal bed is found along the eastern margins of the county; that the central line, from north to south, marks the general out-crops of coal 1 A and 1 B, and that coal No. 1 C, with No. 2 and No. 3, are found exposed along the western boundary line.

MARTIN COUNTY.

Mount Pleasant is built at the top of a thick, hard, gritty sandstone, about 140 feet high, which, at first sight, I was disposed to refer to the Mahoning sandstone, on account of the total absence of conglomerate or of pebbles in the sandstone. But in comparing the position and the lithological appearance of the various coal strata, examined before and after, and in closely examining the gritty compound of the coarse hard sandstone at the base of the hills of Mount Pleasant, it became evident that either the whole thickness of this formation, or at least its lower part forming a bank or escarpment of seventy-five feet high belongs to the Millstone Grit formation.

Just at the base of this bank of sandstone, on Mr. T. B. Bryant's property, a bed of coal, eighteen inches to two feet thick, has been opened for examination. It has the character of the sub-conglomerate coal, being shaly; but very compact, bituminous, and excellent for the forge. Another bed of coal is said to have been opened at the top of this sandstone; but the place is now covered and plowed up. If a coal is found there it is probably No. 1 A, thin coal.

On another side of the hills, near White river, on section 6, township 2 north, range 4 west, and on Mr. Reilley's property, two coal beds are exposed at the base of the hills. The upper one is a streak only a few inches thick, with some black shales; the lower one is one foot thick, and separated from the former by four feet of sandstone. These are evidently two members of the same coal locally divided. At some places, where this coal has been marked by stripping, it has been found two feet thick in its greatest development. The coal is exactly the same in appearance as the sub-conglomerate coal, very compact, somewhat shaly, free from sulphuret and especially valuable for the forge. It is generally overlaid by soft black ferruginous shales, passing occasionally to a yellowish soapstone, containing pebbles of carbonate of lime, and marked generally with the fossil remains of leaves of Lepidodendron. The sub-carboniferous limestone is exposed four miles east of this place.

From Mount Pleasant to Dover Hill the sandstone becomes more and more conglomeratic, or mixed with pebbles of quartz. On section 16, township 3 north, range 4 west, we find the base of the Millstone Grit, a coarse conglomerate, and under it a coal bed one foot thick, separated from the conglomerate by four feet of yellow soapstone, or soft shales, with the fossil plants of this horizon. The base of the Millstone Grit is here formed of a bed of carbonate of iron, in irregular agglutinated pieces, which is about four feet thick, and appears a valuable ore. Near the out-crop of the coal, on the same section, and on Mr. O'Brian's property, a fine chalybeate spring comes out from under a band of fifty feet of conglomerate sandstone.

Dover Hill is like Mount Pleasant, at the top of hills of the Millstone

Grit formation, about 130 above the sub-conglomerate coal, opened at their base. Here the coal is three feet thick; but its top is brashy, and its whole mass still more shaly than it is generally at the other places. It is nevertheless very good for the forge. The section from Dover Hill to the coal is as follows:

	FEET.
1. Covered space..	10
2. Millstone Grit, conglomerate...	70
3. Sandstone and ferruginous shales..	20
4. Carbonate of iron, (conglomeratic iron ore,).......................	3
5. Chocolate colored soft shales, with plants...........................	7
6. Coal somewhat shaly, (some layers very fine,)....................	3
7. Ferruginous fire-clay...	10
8. Sub-carboniferous, oolitic limestone, to creek.....................	5

On both sides of the place where this coal is worked there is a bank of very soft, ochrous clay, a true powder, as fine as flour, without any trace of coal, though occupying exactly the same horizon. It is overlaid by a clay iron ore, which looks as if it had been roasted. I consider this local formation as the result of the burning of the bank of coal at places where it was exposed along the creek. The destruction by fire of banks of coal, sometimes on a large area, has left more frequent traces than it is generally supposed. I have seen, in Indiana, Illinois and Kentucky, many localities where the peculiar nature of the strata, a kind of metamorphism, local displacement, land slips, hollows, &c., could be explained only by the agency of the fire, in destroying the coal banks. Such conflagrations are easily accounted for in a country where coal banks are often exposed on the slopes of the hills, or covered only by a thin bed of shales, where the surface is repeatedly set on fire.

The soft ferruginous shales, overlying the coal at Dover Hill, are marked with numerous remains of some species of fossil plants characteristic of this low coal, especially the bank and the leaves of *Lepidodendron*. In ascending the creek the bed of conglomeratic iron ore, which overlies these shales, takes its normal appearance and becomes two to three feet thick. As it is mixed with sandstone, the ore (a hematite) is probably of no great value. But it may be found better at other places in the neighborhood.

About 130 feet above the sub-conglomerate coal, at the level of the village of Dover Hill, another bed of coal, about two feet thick, is exposed under a bank of hard gritty sandstone, (quarried for buildings). It overlies a bed of hard fire-clay, which looks like a bastard limestone, and appears to be separated from the sandstone by one or two feet of soft shales, having the color and the appearance of those of the lower coal. As I could obtain only for examination a few bits of shales, I was unable to ascertain whether the coal is the equivalent of No. 1 A, the first coal above the conglomerate, or is a coal locally formed in the Millstone Grit formation. I rather think that the first supposition is right, because no coal like this has been observed in the western coal fields, in the strata of the true Millstone Grit formation. Coal No. 1 A is often overlaid by a hard gritty sandstone, and generally it has in its shales, remains of plants, leaves and cones of *Lepidodendron*, some of which are of the same species as those of the sub-conglomerate coal.

I was very anxious to ascertain the formation of a bed of coal under the sub-carboniferous limestone. By the kindness and in the company of Dr. Delamater and his brother, we had an opportunity of visiting, on White river, high banks of sub-carboniferous limestone, formed in part of a beautiful marble. Just under the upper banks of this Archimedes limestone there is a thin bed of coal, said to be about one foot thick, which, unfortunately, was covered, and of which we could see only a few pieces. That coal has been formed at this geological horizon is out of doubt. In Illinois numerous and beautiful specimens of fossil plants of the Coal Measures have been found in the sub-carboniferous sandstone, just under the upper Archimedes limestone. And in Kentucky, traces of coal have been seen in pockets at the same geological horizon. But those rare remains of the primitive vegetation of the Coal Measures can not, in any age, contradict the assertion of Dr. D. Dale Owen, that no workable bed of coal can be found below or within the sub-carboniferous limestone. On the contrary, the recent and careful explorations of most of the coal fields of the different States have confirmed an assertion which he regarded as a geological axiom.

From this it is evident that though the sub-conglomerate coal is well developed at some places in Martin county, and can be found from two to three feet thick over its whole western range, (range 4 west,) there is no chance whatever to reach any lower coal by borings From the higher coal strata, No. 1 A, and perhaps No. 1 B, may be found near the top of the hills, along the western borders of the county. The limit of the Coal Measures follows mostly along the western line of range 3 west.

GREENE COUNTY.

In entering this county the same limit of the Coal Measures makes a bend to the west, reaching, near Bloomfield, the western line of range 4 west, and bends again to the eastward, entering Owen county at the same parallel as it occupies in Martin county.

Six to seven miles south-west of Bloomfield the sub-conglomerate coal is opened at many places, near the top, and in some crevices of high hills of the sub-carboniferous and Millstone Grit formation.

On the property of Mr. W. M. Combs, where I first examined the coal, it is eighteen inches thick, overlying a bed of sub-carboniferous limestone, from which it is separated only by a stratum of fire-clay. It is covered by a bank of sandstone, two to three feet thick, mixed with bowlders of sandy iron ore. It has broken remains of fossil plants. The coal is not worked, and its appearance is the same as at other places. Its distance from the top of the hill is 130 feet.

Three and a half miles north-west of Scotland, on section 16, township 6 north, range 4 west, the same coal is worked by a tunnel. It is here in two banks, separated by a clay parting. The section is:

	FEET.	INCHES.
Hard, gritty sandstone, with remains of plants	20	
Coal	2	3
Shaly fire-clay	1	3
Clay	1	3

Fire-clay to the bed of the creek.

The coal is very fine, but has all the characters of the sub-conglomerate coal, being somewhat shaly. The sandstone overlies the coal immediately, and, though hard, contains streaks of sandy iron ore, with broken remains of large plants.

About one mile south, on section 21, township 6 north, range 4 west, the coal on Mr. Phelps' property is opened at four different places, at a very short distance from each other. The coal is generally three to four feet thick in a single bank; but its top, five to six inches of thickness, is only brash coal, and it is intermixed with some streaks of sulphuret. At the first opening the coal is overlaid with soft shales, marked with fossil plants and a band of clay iron ore, just as at Dover Hill. At another opening the coal is covered by a ferruginous conglomerate, or rather a pudding stone, mixed with remains of broken stems of *Calamites* and other plants. All the coal banks examined on these hills are exactly, by barometrical measures, at the same level, and there is no doubt that all belong to the same geological horizon, though every one of them is overlaid by a different kind of rock, by shales, or by sandstone, or by conglomeratic iron ore. All the modifications of the roof of the sub-conglomerate coal may be recalled to two distinct forms. Either the shales are present, between the sandstone base of the Millstone Grit and the coal, or these shales are not found or only very thin, a kind of brash coal, and the sandstone immediately overlies the coal. In the first case the shales are soft, more or less marked with remains of fossil plants, more or less bituminous and laminated, and accordingly of various color and hardness. Iron ore is always found in connection with these shales, either overlying them in a separate bed, or mixed with them in pebbles of carbonate of iron. Sometimes the bed of iron ore covers the coal, in the absence of the shales in the form of a conglomeratic and sandy iron ore. In the second case, when the sandstone immediately overlies the coal, it is ferruginous at its base, contains broken remains of plants, and when it becomes shaly forms a whitish agglomeration of carbonized plants, sand and iron. Of course these appearances are locally modified in many ways.

The thickness and the good quality of the sub-conglomerate coal, in this county, is the best proof that the base of the Coal Measures ought to be marked in the west by the sub-carboniferous limestone and not by the Millstone Grit formation. Coal strata above the conglomerate are generally thicker, but scarcely more reliable and more extensively formed than this sub-conglomerate coal.

In descending the hills, on the road to Bloomfield, the entire section of 300 feet, shows only alternate strata of shales and sub-carboniferous sandstone, without a trace of limestone, while at some other places, as in Martin county, the sub-carboniferous limestone is forty to fifty feet thick. The sub-carboniferous measures are still far more variable in their extent, their thickness, and the compound of their material than the true Coal Measures.

Near Bloomfield, one-fourth of a mile north-east of Richland furnace, a coal bed has been opened two to three feet thick. It has a roof of soft chocolate shales, with remains of fossil plants. At some places the coal is overlaid by a bed of iron ore, which, in the hills around, becomes four to five feet thick, and is worked for the furnace. The coal is, in places, overlaid by a kind of bastard limestone, or hardened fire-clay, which much resembles that of the upper coal at Dover Hill. As it was covered at the time of our visit, I was unable to ascertain its true geological horizon. It is still probably the sub-conglomerate coal. Pebbles of quartz, evidently derived from the Millstone Grit, are found in the creeks which run at the level of this coal bed.

The result of our researches for coal in Greene county may be summed up as follows: The sub-conglomerate coal crops out on a band on the outside of a line passing from Scotland to Bloomfield, and north to Freedom. Coals Nos. 1 A, 1 B and 1 C have their line of outcrop a little westward of the canal from Chesterfield to Worthington, and northward along the road to Bowling Green. On the western part of the county, within the limits of range 7 west, coal No. 2, No. 3, and perhaps No. 4 may be found in the hills, and, of course, borings there may be made to coal No. 1 B, at favorable places.

OWEN COUNTY.

Near the southern limits of the county, on the road from Bloomfield to Freedom, about three miles south of this last place, the sub-conglomerate coal, two feet thick, is exposed on the property of Mr. Bird Light. It is overlaid by six to seven feet of white sandstone shales, full of broken stems, just as those of the Phillips' coal. The sub-carboniferous limestone is in place just below this coal, and crops out all around in the hills.

On the south-west quarter of section 33, township 9 north, range 4 west, on the south side of White river, and on Mr. Henry Jackson's property, I discovered the same coal at its normal geological horizon, and found it here a true, hard, somewhat coarse cannel, apparently very bituminous and rich in oil. At the outcrop this coal is about eighteen inches; but the proprietor asserts that it had been worked or discovered by a miner, and found to be four feet thick; but that since that time it had been lost. The difficulty of finding this coal again by following the direction of the top of the sub-carboniferous limestone was not great indeed. But the search for the coal had been made by boring and shafting at the base of the hills, where occasionally some pieces of coal, brought down the steep declivity of the hill, had been picked up. The horizon of the coal is about fifty feet higher than the high water level of the river. This Cannel coal is separated from the sub-carboniferous limestone by four to five feet of hard sandstone. It is overlaid by one or two strata of hydrate of iron, or iron ore of the same appearance, and perhaps of as good a quality, as that used for the Richland furnace. If the mineral should prove, by chemical analysis, to be rich, the position for a furnace at this place, just on the border of White river, with an abundance of fine limestone and coal, would be far more advantageous than that of Richland.

On the property of Mr. James Haton, south-west quarter of section 2, township 9 north, range 4 west, the same coal has been found also. But it is covered and its thickness unknown. It is accompanied, like the former, with iron ore above it and sub-carboniferous limestone below.

After passing Freedom I examined, at a forge, samples of a very fine and excellent coal obtained from Arney's bank, six miles west of the place. This coal is in large, compact, hard blocks, finely crystallized, free from sulphuret and charcoal, with its faces obscured only by thin lamellæ of selenite. As it is one of the best coals that we have seen till now in Indiana, I much regretted that time did not permit us to examine the locality, and to ascertain its geological horizon. From its appearance the coal is the same as the Ferdinand coal, and referable to No. 1 B. It is said to be only two feet thick, with a soft shale for its roof.

The road from Spencer to Greencastle is nearly always on the sub-carboniferous limestone, which generally crops out at the base of the hills; while their top is composed of the Millstone Grit or of the Drift which there becomes pretty thick.

Near Cataract Mill, where the sub-carboniferous sandstone in all its beautiful varieties, oolite, marble, lithographic limestone, &c., attains a thickness of about seventy feet, the sub-conglomerate coal is formed just at the top of the limestone. It is two feet thick, overlaid by shales and iron ore. It has been opened on sections 35 and 36, township 12 north, range 4 west.

From information received, a bed of coal two feet thick is exposed or worked near Vandalia, and another, four feet thick, near Lancaster. The first is probably still the sub-conglomerate coal, the other coal No. 1 B. At least, judging from the direction of the boundary line of the Coal Measures in Owen county, which indicates the same distribution of the coal strata as in Green county. The borders of the measures pass a little outside of the western line of range 3 west, and thus the sub-conglomerate coal occupies range 4 west, and coal Nos. 1 A and 1 B range 5 west.

PUTNAM COUNTY.

Cloverdale is just on the outside limits of the Coal Measures; Greencastle, on the contrary, a little inside. At this last place the sub-conglomerate coal is found near the top of the hills, one and a half miles south-east of the town, where it crops out in a ravine. The section there is:

	FEET.	INCHES.
1. Top of the hill, gritty sandstone	25	
2. Coarse shaly sandstone	6	
3. Coal		1
4. Coarse white shaly sandstone and broken plants	3	
5. Black, bituminous, micaceous shales	8	
6. Yellow, soft, micaceous shales	10	
7. Shaly sandstone? covered space	9	
8. Soft grayish shales, with pebbles of carbonate of iron, and some plants	10	
9. Limestone (sub-carboniferous) to level of creek	10	

This section is very interesting. From the base of the Millstone Grit to the top of the sub-carboniferous limestone, we have here forty-six feet of measures, which show at different geological horizons some of the lithological characters which have been observed at different localities over the bed of the sub-conglomerate coal. Though this bed has always been referred to the same horizon, it may be that two or more coal strata can be formed below the conglomerate at a different level, according to the place occupied by the accompanying strata. This is not very probable, at least we have seen, close to each other, some openings of the sub-conglomerate coal, evidently made at the same topographical level or in the same bank, and with the overlying strata entirely different. Thus, at an opening of Mr. Phillips' bank, in Greene county, the coal is covered by a shaly sandstone, resembling the stratum No. 2 of the above section, while at another opening it is overlaid by shales resembling No. 5 and No. 6 of the section. It is certain that where the measures, especially the shales intervening between the base of the Millstone Grit and the top of the sub-carboniferous limestone, thicken much, two or even three beds of coal may be found in the space. But the palæontological remains of each of these coal strata have not been studied, their relation to the coal is mostly unknown, and where a single bed of this sub-conglomerate coal is found it is not possible to fix its geological horizon, according to strata which are not present. This question is merely a scientific one, and has not any importance for practical researches in Indiana.

We did not see any other coal in Putnam county but that of Greencastle. Coal No. 1 B is the only higher coal that may be found in the county, just along the boundary line of Parke and Clay counties. It may be found of a good workable thickness on Walnut and St. Croix creek, at the base of the hills.

CLAY COUNTY.

At Brazil depot coal is worked by a shaft sixty feet deep, with following section:

	FEET.	INCHES.
1. Soil and clay...............	20	
2. Limestone................	4	
3. Blue shale and soapstone............	28	
4. Hard sandstone...............	28	
5. Coal........................	3	6
6. White fire-clay...............	6	

This section already shows that the coal worked at Brazil belongs to one of the members of No. 1, probably No. 1 A. No materials taken out of the shaft can give any indication. But nearly the whole thickness of these strata can be studied in some ravines one and a half miles north of the depot. At this place the limestone (No. 2 of the section) is underlaid by two to three feet of black bituminous shales, overlying coal No. 1 C about two feet thick, and perfectly well characterized by fossils, and by the poor quality of its coal. It is here only a compound of broken carbonized stems, cemented with oxide of iron, and a few inches of very bituminous cannel shales, which burn easily but do not consume. From this bed to coal No. 1 B, of which the top only is seen, at the level of the creek, there a bank of soft shales or soapstone, containing a few fossils, just of the same nature and appearance as those overlying the coal at Washington, and the same also as the twenty-eight feet of blue shale and soapstone of the section of the shaft at Brazil. But it is evident that the coal at Brazil is still lower than this No. 1 B, and thus is referable to No. 1 A. The white fire-clay underlying coal 1 C is much used in the country for excellent pottery. This coal still crops out at a creek near the depot at Brazil, somewhat lower than the level of the railroad.

From Brazil, to Williamstown and Highland, the grade of the railroad is upward, contrary to the dip, which appears thus somewhat stronger than it is generally. Though the distance between both places is only six miles, coal No. 3 is exposed at Williamstown and coal No. 4 at Highland, where it is scarcely thirty feet above No. 3. At Williamstown coal No. 3 is overlaid by a bank of its characteristic black shales; it is three to four feet thick, but as it was not marked, I could not examine its quality. At Mr. Wallace's bank, on section 9, township 12 north, range 7 west, coal No. 4 is six to seven feet thick, immediately overlaid by a high bank of sandstone. At Highland mines the coal is about of the same thickness, and also overlaid by sandstone. The top of the roof shales of coal No. 3 crops out in the creek just below the main coal, and in the same section. Mr. Nathan Williams has passed both coal beds, No. 3 and No. 4, in a shaft only forty-two feet deep. The coal No. 4, especially at Highland, is not of as good a quality as it is generally found. It has more of sulphuret and of acid, and decomposes under atmospheric influence. This is probably due to the softness of the Mahoning sandstone, which here is thin, having been reduced by erosions, and has not all its ordinary consistence. Where coal No. 4 is too poor, as it is at Highland, it would be profitable to try the thickness and value of coal No. 3 by a shaft.

At the shaft of Mr. Carter, named the Staunton bank, the coal is much better and has nearly the true character of coal No. 4; but it is reached at twenty-three feet from the surface, and the sandstone which covers the coal is ten feet thick.

Between Staunton and Cloverport, near the railroad, a boring was made through forty feet of hard sandstone, and then abandoned. From the direction of the dip this sandstone is evidently the Mahoning, which being lower and covered there, has preserved its normal thickness of fifty to sixty feet. The boring has been stopped apparently at a short distance above the place of coal No. 4, which probably would have been found there of a good thickness and of an excellent quality.

At Cloverland, a coal much intermixed with shales and horizontal layers (partings) of laminated clay, is exposed under a bank of fifteen to twenty feet of soft, black, bituminous shales, containing some remains of fishes. I could find no trace of plants in the shales, and no indication whatever which could fix the geological horizon of this coal. From the dip of the strata or from stratigraphical evidence, and from the great thickness of the shales, it is referable to coal No. 5, or to the *five foot* coal of the Kentucky survey. This coal does not appear to have been formed over great areas; and, except near Greenville, Kentucky, I have never seen it worked with advantage. The shaly nature of the coal, and the numerous partings which divide the bank, have rendered the mining of the bank unprofitable at Cloverport.

The dip of the measures at Highland is said to average fifteen feet per mile to the west. By ascertaining the grade of the railroad, from Highland to Cloverport and to Terre Haute, the dip can be measured easily. As far as I could judge it is continuous and without variation. From this it is evident that the geological position of Clay county is extremely favorable for obtaining coal from the best and most reliable coal strata of the measures. From the eastern to the western limits, coal No. 1 A to coal No. 5, or at least six beds of coal, can be found cropping out at the surface at different meridians. The line passing from Middleburg to Brazil is nearly the middle line along which coal No. 1 C, and in the deepest hollows No. 1 B, crop out. Coal Nos. 3, 4 and 5 are exposed near the western boundary line.

VIGO COUNTY.

The shales overlying the coal of Cloverland follow about the same dip as the grade of descent of the railroad to Terre Haute. At least the upper part of the black shales still crops out in a ravine on the turnpike, about three miles east of Terre Haute. From approximative barometric measures, this place is twenty feet higher than the prairie at Terre Haute. If then we admit the dip to be fifteen to twenty feet per mile westward, as it is marked at Highland, the coal underlying the black shales, which have an average thickness of twelve to fifteen feet, should have its geological horizon thirty-five to forty feet below the Terre Haute prairie. As the drift here is more than fifty-three feet thick the place of the coal is marked within this formation, and consequently has been washed away and replaced by materials of recent formations. Hence, to get coal at Terre Haute, it would be necessary first to sink a shaft through the whole thickness of the drift, fifty feet at least, to the top of the Mahoning, and then to cut through at least fifty feet of this generally hard sandstone to reach coal No. 4, which underlies it. If it were even certain that coal No. 4 is of a good workable thickness, and formed its supposed geological horizon at Terre Haute, it would be scarcely remunerative to have the coal worked at a depth of more than one hundred feet, when the canal and the railroad bring to Terre Haute an abundance of coal worked under the most favorable circumstances. In any case it would be worth ascertaining, by a boring, at what depth and of what thickness the coal No. 4 is under the strata of Terre Haute.

On the western side of the Wabash river two banks of coal are exposed in the hills along the river. Their geological horizon could not be satisfactorily ascertained. The upper one is at the top of a quarry of sandstone and limestone below Harrison, near the Wabash. The section is:

	FEET.	INCHES.
Covered space, drift, top of the hills	10	
Coal	3–4	
Fire-clay	6	
Shaly soft blue sandstone	8	
Hard gray limestone	2	
Shaly blue sandstone	4	
Hard limestone, in bank	2	6
Sandstone shales or hard gray metal	3	6
Limestone	2	
Shaly blue sandstone	10	

Covered space to the river bottom, where a lower coal crops out.

This alternation of banks of sandstone and of limestone is truly remarkable, and I have seen no where in the Coal Measures such a formation as the one indicated by this section. The limestone has no fossil, or at least I could not find any. The shales of the upper coal are replaced by the Drift, and the lower coal, worked by stripping at McQuilkin's, in the Wabash bottom, is also deprived of its upper shales. Hence it was impossible to get any directions either from palæontology or from lithological evidence. In supposing that the dip of the measures follows the same direction and grade as indicated at Highland, it would bring here, at the high-water level of the river, the same coal No. 5 as at Cloverport. But the course of the Wabash has perhaps caused some change, or followed a depression formed by some disturbance, and thus the supposition must be confirmed by detailed explorations. The coal of McQuilkin is five to six feet thick, with some brash material at its top, and is overlaid by black shales or a gray bluish sandstone, passing to limestone and forming the base of the former section. The coal has some sulphuret and shales, and is not very compact, owing, perhaps, to its proximity to the surface; but it is nevertheless of a good quality.

As remarked in the first report of D. D. Owen, pages 38 and 39, coal is found in abundance on Honey creek, township 14 north, range 8 west. From the position of the coal, and the strata accompanying it, it is referable to coal No. 5, or perhaps No. 4.

PARKE COUNTY.

Near Clinton Lock, on the canal, (south-western corner of the county,) a coal bed is opened at the base of the hills and extensively worked by divers proprietors. It is four to five feet thick, of inferior quality, mixed, like the coal of Cloverland, with streaks of sulphuret, shales and mineral charcoal. It decomposes easily under atmospheric influence. It looks like a fat and caking coal, and is mostly used at Lafayette for the furnaces of steam engines. The coal is overlaid by a bank of five to ten feet thick of black, bituminous, laminated, hard shales, containing fossil shells and some remains of fishes. Except *Stigmaria*, I have not found any plants in the shales. They have the same appearance and peculiar thickness as those of coal No. 5, and also of coal No. 9. I refer the coal with some doubt, from the absence of any reliable character, to the same horizon as the Cloverland coal, or to No. 5. The section of the strata accompanying this coal bed is as follows:

	FEET.
Drift, top of the hills	50
Sandstone and shales	40
Shales and fire-clay	20
Black shales, with septaria	10
Coal	5–6
Fire-clay	5
Gray metal to creek	20

The septaria or bowlders of carbonate of lime and iron abound in the shales of Cloverland. But as they are found in connection with all the black shales of blackish or marine origin, they do not afford a reliable character.

In examining the outcrop of this coal behind the hills, on the property of Mr. G. M. Griffith, I found at one place the coal bank burnt to some extent, and the strata, above and below, charged by a kind of Metamorphism, as if they had been exposed to volcanic action.

One mile east of Rockport, the county seat, a high bank of hard sandstone is quarried. It is apparently the Mahoning sandstone.

Five miles south-east of the same town, on little Raccoon creek, the coal is exposed at different places. At Mr. John W. Campbell's bank, on section 34, township 15 north, range 7 west, the coal is worked just under a hard fossiliferous limestone, four to six feet thick, divided into two strata. The coal, three feet ten inches thick, is of a beautiful appearance, very black, compact, free from sulphuret and shales, a dry excellent mineral combustible. From the quality and composition of the coal, and from the fossils of its overlying limestone, it is evidently coal No. 3. It is seen at some other exposures in the vicinity.

About one mile north of Mr. Campbell's bank the coal is exposed along the creek on a high bank, where the section is:

	FEET.
Coarse, micaceous, soft sandstone	35
Shales? (covered space)	8
Coal No. 4	4
Soft ferruginous shales, with plant of coal No. 3	6
Coal	1
Gray micaceous shales and fire-clay	8

At this exposition coal No. 3 and No. 4 come in close proximity, separated by shales only, and without a limestone. The distribution, though abnormal, is seen sometimes, and it even happens that both these coal strata become united, and separated only by a parting of a few inches of shales and shaly clay.

On the same creek, somewhat further up, coal No. 4 is immediately overlaid by sandstone, and divided in two or three thin veins, which run and ascend in the sandstone where they become lost. Still further on, the sandstone descends to the level of the creek, and coal No. 4 is worked by trenches or by stripping.

On Sugar creek, north and north-west of Annapolis, the country is broken by hills of the Mahoning sandstone, which forms high banks along the creek and its affluents. At its base coal No. 4, and sometimes coal No. 3, are exposed at many places with their general characters, and mostly covered with shales containing fossil plants. Thus the coal six miles below the narrows, mentioned by Dr. D. Dale Owen (first report, page 34,) as *yielding the best coke seen in the State* is coal No. 4. This coal bank, on the property of Hon. Wm. G. Coffin, is of the same quality. It is only two feet thick, or a little more, and gives also an excellent coke much used in the country for burning pottery. It is overlaid by shales and sandstone.

At the Lock on Sugar creek a high bank, about 150 feet, mostly composed of soft shales, exposed a fine section of the first part of the measures. Coal No. 4 is near the top of the bank and of the hill, and is separated from coal No. 3 by alternate layers of clay and soft shales, containing plants. Both coal strata are here thin and of no value for working. It is generally the case where the shales take a great development and alternate with layers of fire-clay. At this section there is also no sandstone and no limestone. The sandstone is higher, and large pieces of limestone are found in the creek which runs at the base of the bluff.

Along Sugar creek to the eastward the disposition of the measures is the same for a few miles. The Mahoning sandstone is generally cut in high perpendicular banks, and overlaid by shales of coal No. 4. At the mouth of Roaring creek this sandstone is thirty-five feet high, and the coal at its base is three feet thick, separated in two members by a parting of a few inches of shales, containing species of fossil plants characteristic of this horizon. At another place, near by, the coal is separated from the sandstone by its soft, gray, micaceous and fossiliferous shales. On the property of Dr. Dare, on Roaring creek, coal No. 4 is four feet thick, and covered by six feet of the same shales as the former. Sometimes this coal, as near Rockville, is broken in two or three thin strata, which ascend and are lost in the hard sandstone. Sometimes, also, the Mahoning sandstone entirely disappears, and is replaced by a bank of shales, and sometimes, also, even when it is hard and well developed, the shales are found under it without any trace of coal. Sections like those which have been cut at and below the horizon of the Mahoning sandstone, and its accompanying coal strata, are of the greatest interest to the Geologist. Thus, along Roaring creek and Sugar creek, every turn of the creeks, which are running in numerous circuits, exposes a new appearance, a new change in the measures. Coal No. 4, for example, is seen passing from a homogenous bank of coal three feet thick to four thin beds of coal, separated by partings of shales varying from a few inches to three and four feet thick.

Coal No. 3 has generally its horizon lower than the water level o the creeks. It is nevertheless exposed sometimes, as it is remarked above. Near the mouth of Roaring creek it is separated from coal No. 4 by forty feet of shales, containing a bed of rich carbonate of iron ore. This ore is abundant enough to supply a furnace; but it is very hard and compact, and would probably be too refractory in the fire. The shales separating both coal strata have the fossil plants of coal No. 3, and the coal itself is seen just at the level of the creek, where its upper part, the only portion exposed, is a fine Cannel coal. I could not ascertain its thickness.

Some more details about the coal of Parke county are found in the first report of Dr. D. Dale Owen, pages 33 and 34.

It results from the observations reported above, that Parke county has a great abundance of coal, its geological horizon being mostly between No. 1 A and No. 5 coal, and attaining perhaps as high as No. 9. On the eastern side of the county, on Raccoon creek, &c., coal No. 1 B and 1 C are found, at least most probably exposed, at some places. Coal No. 4 and No. 3 occupy the middle part of the county, and are found there finely developed and of the best quality. Near the Wabash river and along its alluvial deposits coal No. 5 crops out, or is found near the surface. Coal No. 9 occupies probably the hills along the western boundary line of the county.

The Mahoning sandstone is still exposed along Sugar creek, on the road to Covington or to Lodi. It is here thirty-feet thick and coal No. 4 is seen cropping out at its base.

FOUNTAIN COUNTY.

One of the most interesting sections of the Coal Measures of Indiana, and also one of the most difficult to explain, is exposed on a bank cut by the Wabash river near the mouth of Coal creek. This section has already been reported by Dr. D. Dale Owen, (first report, page 32,) and somewhat differs from my own. But we have seen already for the section at Grayville, what numerous and repeated variations are found along banks of the Coal Measures, exposed on a certain extent, and how difficult it is to compare two sections, even when they have been made at the same place. My own section, made with a pocket level at a place where the bank is accessible, is:

	FEET. INCHES.
1. Drift overlying sandstone, (covered)	10
2. Shaly coal	2
3. Fire-clay and yellow soft shales	8
4. Shaly coal	2
5. Black shales and fire-clay	3
6. Soft shales, with plants and sandstone	8

		FEET.	INCHES.
7.	Coal, shaly, No. 3?	2	6
8.	Hard fire-clay, with stigmaria		6
9.	Black shales and some coal		6
10.	Limestone overlying fire-clay		4
11.	Limestone		2
12.	Coal, (main)		4
13.	Fire-clay		3
14.	Yellow shaly sandstone to level of the river		8
		68	6

Dr. Owen's section measures eighty-four feet. But its lowest stratum, the yellow sandstone, descends to low water and is fifteen feet thick. The difference is accordingly reduced to eight and a half feet. A small difference indeed for a bank like this, which truly varies in composition and thickness at every foot of its horizontal, exposed surface.

I was first disposed to admit the extraordinary multiplication of coal strata in this section as one of those divisions to which coal No. 9, No. 10 and No. 11 are sometimes subjected, when the intermediate measures are wanting, and they become thus blended in one thick, irregular bed, separated in members of various thickness by partings of shales and of limestone. But on comparing the upper part of the bluff with some of the divisions of coal No. 4, at Sugar creek, and especially by finding under sandstone No. 6 (of the section) some fossil plants of species generally connected with coal No. 4, I now believe that we have at this place a union and sub-division of coal No. 4 and No. 3, and even perhaps of No. 5 above. At Cloverland coal No. 5 is divided in two, or, in some places, three strata, and supposing that the Mahoning sandstone has been here greatly reduced, the section would be easily explained. That there is no great anomaly in the reduction of the Mahoning is proved by what is seen along Sugar creek, where this sandstone is sometimes replaced by shales, or is entirely absent. In a boring for salt, opened from under the main coal, or from the top of the sandstone exposed above the mean water level of the Wabash river, Mr. Thomas, the proprietor of the coal bank, reached the conglomerate salt at 209 feet. (See section in the first report of Dr. D. Dale Owen, page 45.) This would give us about 300 feet of measures from the top of the Millstone Grit to the top of my section of Coal creek. In admitting the ordinary reduction of the Coal Measures towards their borders, we would have the average distance from the top of the Millstone Grit to the Mahoning sandstone. For these reasons I consider the strata Nos. 2, 3, 4 and 5 of the above section as belonging to coal No. 5; the No. 6 of the section as representing the Mahoning sandstone and the shales of coal No. 4; the strata No. 7, 8 and 9 as belonging to coal No. 4, and Nos. 10 and 11 of the section as the shales and the limestone of coal No. 3, which underlies them. In the boring reported by Dr. Owen a sandstone, seventy feet thick, is marked as overlying a bed of coal nine to ten feet in thickness. The last is probably formed by an alternation of the very black, bituminous, somewhat cannel shales of No. 1 C, united with coal No. 1 B, and the 130 feet shales and sandstone to the first salt, represent the base of the Coal Measures above the Millstone Grit. These conclusions appear to me satisfactory enough; nevertheless they can not be considered as perfectly reliable. For it may be, as it has been stated before, that the strata of the bluff of Coal creek show the geological horizon of the part of the measures containing coal No. 9, No. 10 and No. 11. But, if it was so, we should be called to explain the difference of 400 feet at least, which would be wanting in the measures between the top of the Millstone Grit and the base of the Anvil Rock sandstone. And to do it, it would be necessary to suppose: that according to what has happened at many places, especially along the margins of the Coal Measures, whole strata are entirely wanting, either from deundation or from non-formation. In Illinois, for example, near Colchester, coal No. 3 immediately overlies the conglomerate formation, and at Lasalle the highest strata of the Coal Measures almost immediately overlie a formation of the middle Silurian. But at Coal creek the distance to the border of the Coal Measures is still too great to permit the supposition of such disposition of the entire strata. Moreover, the section of the borings through the lower part of the Coal Measures and of the sub-carboniferous formations, show all the principal members in their normal position.

The first salt by boring, or the soft brine, as it is generally called, is found within the Millstone Grit, at various depths from its upper surface. In Kentucky, on the Little Sandy river, near Grayson, it is found at nine feet deep. This salt water, as it is generally called, is too soft to give a remunerative result by evaporation. The average distance from the top of the Millstone Grit to the salt of the sub-carboniferous sandstone is 500 feet. I have measured it in many places in Virginia, along the great Kanawha river, in Pennsylvania, in Ohio, in Kentucky, even in Arkansas. By mentioning the result of these observations to Mr. Thomas, who is a practical miner especially well acquainted with the researches for salt, he confirmed these views by his own experience. Thus a great deal of useless expense and disappointment in the searches for salt would be spared by a preliminary and geological examination of the country, and the exact determination of the geological horizon. From this determination one could know within about fifty feet the depth at which the soft and the hard brine could be reached.

Except the borings of Mr. Thomas, I do not know of any other that have been made along the Wabash river, though it is evident that strong brine could be found all along the Wabash river, at a depth corresponding with the geological horizon of each county.

A few other exposures of coal in Fountain county are mentioned in Dr. D. Dale Owen's report, pages 31, 32 and 33. The whole section of the boring at Mr. Thomas' is mentioned in the same volume, pages 45 to 49.

WARREN COUNTY.

I have examined the Coal Measures of this county only at Williamsport and around that place. Just opposite the railroad depot of Williamsport, the Millstone Grit is exposed on a perpendicular bank at the falls of the creek. The bank is quarried and gives an excellent building stone. It is seventy to eighty feet thick, and underlaid by micaceous shales. These shales, at least where I have seen them exposed, do not contain any coal.

The town of Williamsport is built just at the base of a high bank of soft black shales, intermixed with carbonate of iron. They are marked with remains of fossil plants, and have some traces of coal. Above this bank of shales (twenty feet thick) there is a somewhat conglomeratic and ferruginous sandstone, that is referable to the Millstone Grit, at least to an upper member of this formation. At Williamsport I could get only scant information about the coal strata of the county, and found nobody to direct me to some of its outcrops. From the direction of the limiting line of the coal basin of Indiana, which passed from Independence northward to Pine Village, it is pretty certain that two beds of coal, which are said to be exposed on Big Pine creek, the one five miles, the other twelve miles north-west of Williamsport, belong to the sub-conglomerate coal. These beds are two feet thick. Of course no other coal can be found by boring below this one, and the place of coal Nos. 1 A and 1 B can be looked for only in the hills or above the general level, along the western boundary line of the county.

VERMILLION COUNTY.

This county has the same coal as that of Fountain county, at the mouth of Coal creek, but at a lower level. Thus, on Vermillion river, the Hughs' bank (reported by Dr. D. Dale Owen, vol. 1, page 35,) is exactly the upper part of the high bank exposed at Mr. Thomas' on Coal creek. It is the part referable to coal No. 5, and it is here near the level of the creek. A remark of D. D. Owen about the general geological features of Vermillion county confirms my opinion about the place of this coal. He says, page 37, that he has not seen sandstone in this county suitable for the furnace hearth-stones; but that probably some can be found, if not in Vermillion, at least in Parke and Fountain counties. The Mahoning sandstone is often hard and compact enough to give good building and hearth stones. From the direction of the dip this sandstone passes under the Wabash river in entering Vermillion county, and the upper part of the measures, in their whole thickness of 400 feet, has no sandstone hard enough to be used for building purpose.

In the neighborhood of Newport, on a branch of Bennett's creek, a coal bank is exposed and somewhat worked. It is overlaid by about five feet of black, bituminous, fossiliferous shales, which have the characters of those of coal No. 9. As this bed is pretty high up in the hills it is probably the equivalent to coal No. 9. It is separated by a bank of twenty feet of black, and sometimes yellow, ferruginous shales from another coal bed, one foot thick. The shales have all the lithological characters of coal No. 8, but do not have any fossils. Thus, in Vermillion county, near its western boundary line, and where the hills are somewhat high, coal No. 9, No. 8, and even No. 11 can be found. The first one is generally of good workable thickness and of good quality. Along the Wabash river and near the base of the hills, coal No. 5 has its geological horizon, and may be found at many places.

SULLIVAN COUNTY.

Near Farmersburg a coal bank is owned by Messrs. Elliot and Sharp, and worked by a shaft of about fifty feet, that has the following section:

	FEET.	INCHES.
Drift and soil	10	
Soft micaceous shales	8	
Sandstone (hard and compact)	23	
Soft stone	3	
Shales, soft and soapstone, with plants	3	
Coal	3	6

The soft shales overlying the coal have the fossil plants characteristic of coal No. 4. I thus ascertain the geological horizon of this coal, and of the overlying sandstone, the Mahoning. The coal has the average quality of coal No. 4. It is compact, has much carbon, is free of sulphuret, and would make an excellent coke. In shafting twenty to thirty feet lower, coal No. 3 could be reached and both banks worked together.

Three miles east of Currysville, on Busseron creek, a bank of coal four to five feet thick is exposed on the property of Mr. Ladson. It is overlaid by two inches of black, soft, brittle shales, a true brash coal, and then by four to five feet of soft gray shales or soapstone, where I could not find any fossil plants. Over this there is a bed of coarse, shaly sandstone, passing upward to a hard sandstone. Its thickness could not be ascertained. This coal is the equivalent to the former and the section is nearly the same.

I did not examine any other coal in Sullivan county; but from information received, it appears that three miles east of Farmersburg, at Mr. Duffey's, a coal bank, five feet thick, is exposed and worked. On the farm of Busseron creek, on the farm of Mr. Thos. Creager, section 13, township 7 north, range 9 west, a coal of unknown thickness was reached at fifteen feet below the surface, by digging a well. Judging from the general direction of the dip of the strata, and from the limestone that is exposed in the creek above this coal, it is referable to No. 3.

Some other coal strata of this State are mentioned in the first report of D. D. Owen, page 39.

The position of Sullivan county indicates coal No. 4 all along the line of the railroad, and on both sides of it. Coal No. 3 and No. 2 on its eastern parts, where coal No. 1 B may be reached by borings at 100 feet, and coal No. 5 and No. 6 on the Wabash valley.

KNOX COUNTY.

The middle and western part of this county is occupied by the Mahoning sandstone, forming high banks along the Wabash valley, north of Vincennes and near the Ohio and Mississippi railroad, especially near Wheatland. Coal is seen at different places underlying this sandstone. About three miles east of Wheatland it is worked somewhat high in the hills, four to five feet thick, and gives a fine coal of excellent quality, free from sulphuret and with the general characters of this coal. It is there overlaid by a great bank of hard sandstone, from which it is separated by only four to six inches of soft brittle shales, full of fossil plants. The sandstone is quarried for constructions. It is compact, fine grained, hard and resisting to atmospheric influence.

After passing the meridian of Vincennes to the eastward, and on the other side of the Wabash river, coal No. 4 is only attainable by shafts. On the west fork of White river coal No. 2 and No. 3 may be found exposed along the eastern boundary line of this county, and coal No. 1 can be reached by shafts at about 130 feet from the low water level.

GIBSON COUNTY.

The coal exposed in Gibson county, mostly along White river, could not be examined on account of high water. It is the only county of the State where I have not examined an out-crop of coal. From what I have seen of the distribution of the strata in Pike county, the geological horizon of Gibson county is mostly between coal No. 4 and No. 9, an unfavorable place to see coal out-croping at the surface, while the whole thickness between these coal beds is mostly a series of strata barren of coal.

Coal No. 4 may be still seen out-cropping near the eastern boundary line of the county, as the Mahoning sandstone, according to Dr. D. Dale Owen's statement, is exposed on Patoka river, three miles above Columbia, where it has been quarried for building bridges along the railroad track. The same coal, and also No. 3, may be reached on both sides of the railroad by borings from twenty to one hundred feet deep.

CONCLUSIONS.

I sincerely regret that want of time prevented the examination of a greater number of exposed coal banks, and that I am unable to give to each proprietor of coal satisfactory information about the geological position and the value of his coal lands. My general remarks, nevertheless, will suffice to direct the researches for coal, to appreciate the riches of the combustible mineral of the coal measures of Indiana, and to clearly demonstrate the value of a detailed survey and the advantages that would result to every proprietor of coal lands of an exact acquaintance with the useful minerals which they may contain at various geological horizons.

Until such a survey can be made in Indiana, or rather in prosecuting it, the determination of the strata along the Wabash river, along White river and its principal branches, and along the lines of the railroad, would be for the State of the greatest importance. It would certainly excite an enterprising spirit, bring money to such places where speculation promises to be profitably rewarded, and in any case greatly increase the price of the land.

REPORT OF PROF. LESLEY

PHILADELPHIA, December, 1860.

Richard Owen, M. D.:

DEAR SIR:—As per instructions received from your brother, Dr. David Dale Owen, I made, in the month of October last, a topographical and geographical survey of that portion of the Indiana coal field lying around the town of Cannelton, in fractional township 7 south, range 3 west, Perry county.

The plan pursued was the same as that in the Fourche Cove, Arkansas, examination, and the survey so made presents the following results:

1st. That the cost of extending such a series of examinations over the whole State of Indiana I estimate at $150 per township—field and office work included.

2d. That 477 feet of Coal Measures and Millstone Grit formation are exposed in the district examined, and occur in the following order and thickness, from the top of the highest ground in the township down to the top of the sub-carboniferous limestone, which last shows itself in the bed of Deer creek, on the east edge of the township:

		FEET.	INCHES.
477	Thin bedded sandstone, shaly towards the upper part	77	
	Band of highly ferruginous sandstone	4	
	The so-called "Top Rock," being a thick-bedded, homogenous, fine grained, cream-colored sandstone, extensively quarried for building purposes. It is easily worked, but hardens by exposure	70	
325	Top coal vein	1	6
	Gray shales containing thin bands of nodular iron ore	48	
273	Main Cannelton coal vein	4	
	Fire-clay	5	

		FEET.	INCHES.
	Shales and schistose sandstone, containing a heavy band of Kidney iron ore 35 feet below the main coal vein...	53	
214	Lower Cannelton coal vein...	1	1
	Fire-clay...	4	
	Shales	10	
	The so-called "Bottom Rock." Thick bedded sandstone sometimes quarried for buildings and tombstones.......	40	
	Thin bedded sandstone..	35	
125	Coal streak...	00	
	Massive sandstone and conglomerate.............................	70	
	Covered space, probably sandstone................................	55	
0	Top of the sub-carboniferous limestone, at the mouth of Deer creek..	00	

3d. That the iron ores in this district are so thinly disseminated through the shales as not to warrant the erection of a blast furnace, with the exception perhaps of a band of kidney ore thirty-five feet below the main coal, which might be mined by stripping, and the ore then mixed with Missouri ore and thus form a good iron.

4th. That four beds of coal are seen—the first or lowest is but a streak, and lies immediately upon the conglomerate, or 125 feet above the sub-carboniferous limestone.

The second, or "Lower Cannelton vein," lies 90 feet above the first, or 214 feet above the limestone, and has an average thickness of thirteen inches. It is bright and fractures in very small and irregularly shaped pieces. A bed of fine fire-clay, four feet in thickness, underlies it.

The third, or "Main Cannelton coal," lies 60 feet above the last named, or 237 feet above the sub-carboniferous limestone, and is that which has been so extensively worked for a number of years past for steamboats and manufactories. This coal shows a laminated structure and breaks in cubes. A bench of sulphurous coal divides the vein into two bands—the upper averaging twelve inches in thickness, and the lower thirty-six inches. Locally this coal vein thins to a mere streak, as will be hereafter shown. It lies upon a five foot bed of fire-clay, and has for the most part a compact, gray, sandy shale roof, though in some localities this shale has disappeared and the roof is composed of the sandstone of the "Top Rock."

The fourth, or "Upper Cannelton vein," lies 52 feet above the main vein, or 325 feet above the limestone, and averages eighteen inches in thickness. It has the same general characteristics as the second or lower vein.

5th. That the general dip of the strata is, as shown upon the accompanying map, N. 76¼ W., its average fall being 33 feet to the mile.

6th. That this dip is not regular, but in long, low waves, and that these last are crossed at right angles by similar waves. These waves cause the leading peculiarity of this portion of the coal fields, and also have been the cause of much perplexity and pecuniary loss to those who have undertaken to develop the resources of this district, for the main coal vein has always been found to become thin and sometimes even to disappear upon the crests of these waves, thus reducing very much the area of workable coal, and throwing it, so to speak, into pockets which are difficult to strike, without a previous careful geological and topographical survey, the eye of the practical miner, even trained as it may be, not being so certain to detect these disturbances as a careful examination by compass and level.

In the tunnel north of Cannelton this thinning out of the coal vein can be plainly followed. A shaving off of the coal vein would express this better—as will be seen by the accompanying section—the whole four feet of coal, with its sulphur band, not being compressed into a streak only, but just the upper bench of coal disappears, then the sulphur band, and finally the lower bench of coal, thus:

7th. That the strata decrease in thickness westward, even to entire absence, as in the case of the shales overlying the main coal in the tunnel, where these measure forty-six feet in thickness, whilst at the old Fulton banks, two miles to the westward, they have entirely disappeared—the roof of the coal being formed of the so-called "Top Rock" mentioned in the section above; and

8th. That besides these waves there is a fault running along the south side of the valley of Caney Fork, of Deer creek, and in a direction parallel to that of the general dip of the strata. At right angles to this fault and running into it is another, not so long, and showing itself on the east side of the valley of "Hayden Meadow." These faults are occasioned by an upthrow of the strata of the sub-carboniferous limestones, which, along Caney Fork of Deer creek, form the bluffs along that stream, and dip into the hills at an angle of 60° in a S. S. W. direction.

APPENDIX.

TABLE OF ALTITUDES IN INDIANA.

COUNTY.	PLACE.	Height above high tide.
		Feet.
Allen	Fort Wayne, surface of Maumee	720
Benton, Jasper and White	Grand Prairie, average	880
Clarke	Jeffersonville	441
Cass	Logansport	562
Dearborn	Lawrenceburgh, (high water in the Ohio)	482
Decatur	Greensburgh	913
Daviess	Washington	492
Delaware	Muncietown, (White river surface)	919
Dubois	Patoka above mill dam, at Jasper	428
Fayette	Milton	926
Fayette	Connersville	823
Franklin	White Water, at Brookville	598
Floyd	Low water in Ohio, at New Albany	353
Floyd	New Albany, Court House	426
Greene	White river, at Bloomfield	467
Gibson	Princeton, (sill of Court House door)	481½
Grant	Mississinewa, at Marion	784
Hamilton	Noblesville	750
Jackson	Rockford, (below mill-dam)	548
Jefferson	Madison, (high water in the Ohio)	450
Jennings	Vernon	639
Johnson	Edinburgh	655
Knox	Vincennes, (high water in Wabash)	399
Lawrence	Bedford Court House	680
Lake, Porter and Laporte	Surface of Lake Michigan	595
Marion	Indianapolis	698
Martin	Mt. Pleasant	584
Monroe	Bloomington Court House	771
Montgomery	Crawfordsville Court House	744
Madison	White river, at Andersontown	822
Owen	Gosport, on the hill	593
Orange	Paoli, Court House	599
Putnam	Greencastle, Court House	830
Posey	New Harmony	400
Posey	Low water in Wabash, at New Harmony	350
Posey	Mouth of Wabash	297
St. Joseph	Devil Lake	701
St. Joseph	Little Fish Lake	878
Shelby	Blue river, at Shelbyville	740
Tippecanoe	LaFayette, Court House	538
Tippecanoe	Wabash, at LaFayette	492
Vigo	Terre Haute, southern part of town	483
Vigo	Wabash, at Terre Haute	433
Vanderburgh	High water in the Ohio, at Evansville	361
Wabash	Bottom of Canal, at the town of Wabash	650
Wabash	Surface of Wabash, at Wabash	638
Washington	Salem, Court House	733
Average height of land in the State of Indiana		678

EXTRA-LIMITAL ALTITUDES, FOR COMPARISON.

UNITED STATES.

	Above Sea. Feet.
Alabama, Huntsville	600
Arkansas, Hot Springs	718
Arkansas, Fort Smith	460
California, Sierra Nevada	10,000 to 12,000
California, Passes to Sierra Nevada	4,700
Florida, Pensacola	20
Georgia, Savannah	30
Georgia, Augusta	300
Illinois and Wisconsin, Prairie of Illinois and Wisconson	950
Illinois, Kentucky and Missouri, mouth of Ohio	275
Kansas, Fort Scott	1,000
Kansas, Fort Leavenworth	896
Kansas, Fort Riley	1,300?
Kentucky, Louisville	441
Kentucky, Central Kentucky	800
Louisiana, New Orleans	20
Louisiana, Baton Rouge	41
Massachusetts, Cambridge Observatory	71
Maryland, Baltimore	90
Maryland and Virginia, Blue Ridge, (average)	1,800
Missouri, St. Louis	480
Missouri, Jefferson Barracks	472
Mississippi, Vicksburg	350
Minnesota, Fort Snelling	820
New York, West Point, (U. S. M. A.)	167
New York, Niagara Fort	250
New York, Mt. Marcy	5,344
Nebraska, Fort Kearney	2,360
Nebraska, Fort Laramie	4,519
North Carolina, Blue Ridge	2,200
New Mexico, El Paso	3,830
New Mexico, Fort Webster	6,350
New Mexico, Taos	8,000
Ohio, Lake Erie	565
Ohio, Cincinnati, lower part 432 feet, upper part	550
Ohio, Portsmouth	540
Oregon, Rocky Mountains, at 115° longitude, (W.)	8,000
Oregon, Fremont's Peak	13,570
Pennsylvania, Philadelphia, (Penn. Hospital)	30
South Carolina, Charleston	30
Texas and New Mexico, Staked Plains, from	3,000 to 4,000
Tennessee, Nashville*	460
Tennessee, Knoxville	960
Tennessee, Cumberland river, at Nashville	388
Tennessee, Memphis	400
Utah, Humboldt River Valley	6,506
Virginia, Fort Monroe	8
Wisconsin, Milwaukee	593

BRITISH POSSESSIONS.

Lake Winipeg	853

RUSSIAN AMERICA.

Mt. St. Elias, highest point in North America	17,900

*According to some authorities 533 feet above the ocean; probably this is on the summit of University H'l.

WEST INDIES.	Above Sea. Feet.
Island of Cuba, Havanna	50

SOUTH AMERICA.

Chile, Tupangata, (highest point in South America)	22,480

EUROPE.

England, London	50
Scotland, Ben Nevis	4,368
Germany, Berlin	115
France, Paris	222
Switzerland, Geneva	1,280
Switzerland, Mt. Blanc, (highest point in Europe)	15,750
Italy, Rome	170
Russia, St. Petersburgh	20

ASIA.

China, Canton	40
Himalaya Mountains, Kunchinjinga, highest known point in Asia and in the world	28,178

AFRICA.

Algiers	310

COMPARATIVE VIEW OF RECENT AND FOSSIL PLANTS.*

Classification of Plants into Large Divisions.	Sub-divisions applied to Recent Plants.	The same applied to Fossil Plants.	Geological Groups in which they chiefly occur.	Great Geological Formations embracing those groups or strata
No. 1.—Agamia or Thallogens: Plants made up of cellular tissue, or the interlacing of tubular filaments; reproduction, partly cellular, partly by spores or seminoles, apparently not fecundated; no leaves, stems, or proper vessels.	Algæ or Sea Weeds.	Confervites and Fucoides.	Chiefly in Silurian.	
	Fungi or Mushrooms.			
	Lichenes or Tree Mosses.			
	Hepaticæ or Liverworts.			
No. 2.—Cellular Cryptogamia, or Anophytes: Plants consisting still of cellular tissue only; but stem and foliage distinct, or sometimes the two confluent into a foliaceous body (frond). No proper vessels yet, but breathing pores.	Musci or True Mosses.	Muscites.	[Not found until the Tertiary Period.]	
	Equisetaceæ or Horse-tail family of plants.	Asterophillites. { Equisetites. Asterophillites. Sphenophyllum. Annularia. Calamites. Noeggarathia.	Chiefly in high coals.	PALÆOZOIC. [Silurian to Permian, inclusive.]
No. 3.—Vascular Cryptogamia, or Acrogens: Plants growing by the extension of the point (akres) only. A stem, sometimes large and arborescent, containing woody fibre, and proper vessels; leaves, with cortical pores.	Filices, or Ferns or Brakes.	1. Neuropterideæ. (Nerve-leaf ferns.) { Odontopteris. Dictyopteris. Cyclopteris. Neuropteris.	In high and low coals. In the Devonian or old Red Sandstone. In low coals.	
		2. Sphenopterideæ. (Wedge-leaf ferns.) { Sphenopteris. Hymenophyllites. Asplenites. Alecopteris. Callipteris.	In high and low coals.	
		3. Pecopterideæ. (Embroidered ferns.) { Pecopteris. Cremalopteris. Scolopendrites. Stlypteris. Chænophyllites. Cordaites.	{ Chiefly in low coals. In high coals.	
		4. Doubtful. The following are probable stems, &c., of Ferns: Sigillaria, Stigmaria, Lepidodendron (?), Psaronius, Knorria and others allied. Cones of Lepidostrobus, Fruits of Trigonocarpum, &c.	Common to many strata.	
	Lycopodiaceæ or Club-Mosses.	{The Lepidodendron, with its strobili or fruit; the Lepidostrobus is considered by some allied to the Lycopodiaceæ.	Chiefly in low coals.	
	Marsilaceæ.			
	Characeæ.	Fossil Chara, (formerly Gyrogonites.)	[Not found until the period of Miocene Tertiary.]	

No. 4.—Gymnospermous Phanerogamia, or gymnospermous Dicotyledons. Seeds destitute of capsules, but receiving fecundation. Pithwood and bark (Exogens).	Cycadeæ.	Zamites, Cycadites, Pterophyllum, Mantellia, Clathraria (?).	Lias and Oolite. Wealden.
	Coniferæ or Cone-bearing Trees. (Pine family.)	Voltzia, Araucaria, Taxites, (yew); Thuites, (Arbor Vitæ); Abietites (Fir); Pinites (Pine); Walehia.	Mesozoic. (New Red to Chalk, inclusive.
No. 5.—Angiospermous, Monocotyledonous Phanerogamia. Veins of the leaves parallel. Typical composition of whorls, 3. Growth of stem, from within, out (Endogens). No pith, no angular rings.	Glumaceæ. The Gramineæ or Grasses and Plants having a glume or husk.	Poacites.	
	Aglumaceæ. Endogenous Plants without glume, sometimes without floral envelope; sometimes having a perianth representing calix.	Palmacites, Endogenites, (?) Clathraria, (?) Nymphœæ, Antholithes.	No tree Palms in Coal Period; some in Oolite; over 50 species already found in the Tertiary.
No. 6.—Angiospermous, Dicotyledonous Phanerogamia. Venation of leaves crossing like net-work. Typical composition of whorls, 5. Growth or new deposit outside, next the bark. Pithwood, Medullary rays and bank. (Exogens.) Closed ovary.	Apetalous. Monopetalous.	Fossil Poplar (Populus), Willow, Elm, Chestnut, Walnut, Sycamore, Oak, &c.	Some Dicotyledons in Wealden and Chalk Periods, but chiefly in Tertiary.
	Polypetalous.	Fossil Maple, Linden, Buckthorn Vine, Dogwood, &c.	In lignite or brown coal of the Rhine, also in tertiary marl of France and Italy, and the U. S.)	Cainozoic. (Tertiary.)

*Chiefly from the works of Brongniart, Geinitz, Profs. Gray and Lesquereux.

BRONGNIART'S CLASSIFICATION OF FERNS.

As some persons may prefer to have the exact classification of Ferns adopted by Brongniart, it is here subjoined, premising that it is founded upon the distribution of the mid-rib or main tube, transmitting nourishment from the petiole to the apex of the leaf. In French this is termed *nervure*, while the smaller tubes or finer ramifications of the leaf are called, in French, *nervules*, by English botanists, veins:

I. *Nerves pinnated, veins not reticulated.*
   A.—Mid-rib or nerve simple, bifurcated or pinnated.
      1. Frond simple, veins simple or bifurcated..................................*Teniopteris.*
      2. Pinnules simple or semi-pinnatifid, with equal lobes, veins slightly oblique to the mid-rib or median nerve..................................*Pecopteris.*
      3. Pinnules deeply lobed, lobes crossing and diverging; nerve bifurcating or bipinnate; oblique..................................*Sphenopteris.*
   B.—Veins dichotomous, very oblique on the median nerve.
      4. Frond simple..................................*Glossopteris.*
      5. Pinnules adhering to the base of the rachis; veins growing from this rachis; no median nerve..................................*Odontopteris.*
      6. Pinnules not adhering to the rachis:
         (a) Pinnules entire, symmetrical..................................*Neuropteris.*
         (b) Pinnules entire or lobed, very inequilateral, principal nerve almost marginal..................................*Lozopteris.*
         (c) Pinnules flabelliform (fanshaped) lobed..................................*Leptopteris.*
         (d) Pinnules palmated, with pinnated nerves on each lobe........*Cheiropteris.*

II. *Veins flabelliform, no principal nerve.*
   A.—Veins pedunculated..................................*Cyclopteris.*
   B.—Fasciculated veins, radiating dichotomously..................................*Hymenopteris.*
   C.—Frond deeply lobed; lobes one-nerved..................................*Schizopteris.*

III. *Nerves anastomosing.*
   A.—Secondary nerves all equal and reticulated; no free nerve..............*Lonchopteris.*
   B.—Principal nerves forming a square grating; veins reticulated, none free..................................*Clathropteris.*
   C.—Nerves unequal, areolar, a portion terminating freely, in these areoles or inter-spaces..................................*Phlebopteris.*

GENERAL SYNOPSIS OF THE ANIMAL KINGDOM, FROM MILNE EDWARDS.

Departments.	Sub-Departments.	Structure upon which the Sub-Division is Founded.	Classes.
I. Osteozoa or Vertebrates. An internal skeleton; a cerebrospinal nervous system.—The organ of animal life (or that performing the functions of that relation) symmetrical with reference to a direct median line.	Allantoidean Vertebrates. (Respiration, pulmonary in youth; never branchial.)	Organs of lactation. Warm blood; circulation complete, and heart with *four* cavities; respiration pulmonary; simple lobes of the cerebellum united by an annular protuberance; inferior maxillary articulating directly with the cranium; body usually furnished with hair; animal viviparous.	*Mammals.*
		Circulation complete; heart with *four* cavities. Respiration double; blood warm; body covered with feathers.	*Birds.*
		No organs of lactation. Encephalon deprived of annular protuberance. Lower jaw united to the skull by one or two intermediate bones. Animal oviparous.	*Reptiles.*
	Anallantoidean Vertebrates. (Respiration, branchial in youth; sometimes during life.)	Lungs in the adult; body without scales; metamorphoses in early life; heart with *three* compartments.	*Batrachians.*
		No lungs and no metamorphoses; heart with *two* compartments; body usually covered with scales.	*Fishes.*
II. Entomozoa or Annelidans. No interior skeleton, but usually a tegumentary skeleton, composed of movable rings. No cerebrospinal axis. A central nervous system, composed in general of a series of ganglions, united in pairs, along the median line of the body, so as to constitute a long straight chain. The different symmetrical organs disposed with reference to a direct median line.	Arthrodians or Articulated Animals. (Body provided with jointed organs of locomotion.—Ganglionic system well developed.)	Body composed of a distinct head, thorax and abdomen; furnished with three pair of legs, and with tracheæ, (breathing pores); vascular system almost wanting.	*Insects.*
		Body composed of a head and a series of thoracic abdominal rings; twenty-four or more pair of feet; tracheæ; vascular system little developed.	*Myriapods.*
		Head confounded with the thorax; four pairs of feet; sometimes tracheæ, sometimes pulmonary sacks; vascular system generally well developed.	*Arachnidans.*
		Respiration *aquatic*; effected by gills or by the skin; a well developed vascular system; usually five or six pair of claws or feet.	*Crustaceans.*
		Respiration almost always *aërial*; blood almost always colored; nervous system very distinct, and forming a median ganglionic chain; bristle-like tubercles, usually serve as feet.	*Annelidans.*
	Worms. (Body not provided with jointed locomotive organs. Ganglionic system little developed, or rudimentary.)	Cylindrical body, destitute of locomotive organs, of suckers and of distinct annular division; digestive tube simple, and open at its two extremities; sexes distinct, (diœcious).	*Helminthans.*
		Respiration cutaneous and vague; blood almost always without color; nervous system more or less rudimentary and lateral; never having a median ganglionic chain.	
		Body flattened, covered with vibratile cilliæ; deprived of locomotive appendages; few or no rings; digestion of cavity complex, and communicating externally only by one opening.	*Turbellarans.*
		Body flattened, having lynings, and destitute of organs of locomotion; digestive cavity complex and without anus; unisexual or monœcious.	*Cestoids.*
		Body ringed, and bearing on its anterior part lobes furnished with vibratile cilia; digestive cavity tubular, and open at the two extremities of the body.	*Rotifers.*

GENERAL SYNOPSIS OF THE ANIMAL KINGDOM.—Continued.

Departments.	Sub-Departments.	Structure from which the Sub-Division is Founded.	Classes.
III. *Malacenoa or Mollusks.* Neither interior jointed skeleton, nor exterior, annular skeleton. Body sometimes naked, sometimes covered with a shell. No cerebro-spinal axis. Nervous system composed of ganglions, the reunion of which never constitutes a long chain. The principal symmetrical organs, with reference to a median plan, usually curved.	*Mollusks proper.* Nervous system composed of several ganglions united by medullary cords. Generation oviparous.	Organs of locomotion placed *around the mouth*, and having the form of tentacles or of arms.	*Cephalopoda.*
		Organs of locomotion placed *on each side of the neck*, and having the form of swimming paddles. Commonly a univalve shell; never a bivalve.	*Pteropoda.*
		An organ of locomotion occupying the *inferior surface* of the body and having the form of a foot, or fleshy disk.	*Gasteropoda.*
		A *distinct head*, furnished with different appendages, and having usually eyes. Commonly a univalve shell; never a bivalve.	*Acephala.*
	Molluscoida. Nervous system rudimentary or wanting. Reproduction effected usually by budding, as well as by deposition of eggs.	No distinct head; a bivalve shell. Respiration operating by means of *interior* gills, (branchiæ), or sack lining the mouth. No protractile tentacles around the mouth. A vascular system and heart.	*Tunicates.*
		Respiration operating by means of *exterior* gills, which constitute around the mouth a crown of ciliated and protractile tentacles. Neither vascular system nor heart.	*Bryozoa.*
IV. *Zoophytes.* Usually no jointed skeleton, interiorly or exteriorly. Nervous system rudimentary or wanting. The different organs disposed in a manner more or less radiated with reference to an axis or central point, either in the young or adult animal.	*Radiates.* Body offering a distinct radiating structure, either in its whole or in its principal parts. Almost always prehensile appendages, such as tentacles, disposed in a circle round the mouth.	Animals formed for *crawling*. The surface of the body furnished, ordinarily, with small tentacles, terminated by suckers. In general, an anal opening opposed to a mouth. Teguments usually very hard, and often armed with spines.	*Echinoderms.*
		Animals formed for *swimming*. The body usually soft and apparently gelatinous. Arms resembling or contractile sack. Tentacles very soft and apparently gelatinous. Placed by pores, or by the mouth itself.	*Acalepha.*
		Sedentary animals, living almost always fixed to the ground, and never having special organs for locomotion. Digestive cavity offering a single orifice. Individuals usually aggregated, and covered with a fleshy or horny crust.	*Polyps.*
	Sarcodares. Body offering a spherical, rather than a radiated structure, and frequently becoming altered in form by age. Almost always prehensile appendages.	General form approaching that of a spheroid in adult age, as well as while young. Usually vibratile cilia or far-spread appendages, serving for swimming. Body perforated by several interior cavities, which perform the functions of a stomach.	*Infusoria (proper.)*
		Form generally spheroidal only in youth; becoming afterwards irregular and indeterminable; no indications of sensibility, nor movements of locomotion in the adult state. The body hollowed by canals, and sustained by spicules of horny, calcareous or silicious nature.	*Sponges.*

SYNOPSIS OF THE CLASS POLYPI.*

Radiated animals without locomotive organs, with one or more circles of contractile tentacles around the mouth, having a central visceral cavity, presenting but one opening for the reception of food, and discharge of excretions; containing also the reproductive organs when they exist. Reproduction fissiparous, also by gemmæ (buds) and ovules (eggs).

I. Sub-Class Corallaria, (Actinoidea of Dana.)

The Polypary, where it exists, is usually calcareous; may be tubular, cyathoid, discoid, or basilar, but never has tubular, horny stems; gastric cavity surrounded by membranous vertical lamellæ.

II. Sub-Class Hydraria.

Simple cylindrical bodies, *without* polypary.

I. Order Zoantharia.

Polyps furnished with conical, tubular, simple or arborescent tentacles, not bipinnated, and with numerous membranous lamellæ.

Sec. 1. *Zoantharia Malacodermata.*

Derma tissue fleshy, not a true calcareous polypary.

1. Fam. Actinidæ.
2. Fam. Cerianthidæ.
3. Fam. Minyadæ.

Sec. 2. *Zoantharia Aporosa.*

Dermal, sclerotic, calcareous polypary; septal system well developed, imperforate, derived from six primitive rays. No diaphragms, (floors.) The most star-shaped of all corals.

	Genera.
1. Fam. Turbinolidæ.	Turbinolia. Flabellum, &c.
2. Fam. Oculinidæ.	
3. Fam. Astreidæ.	Caryophyllia. Meandrina. Stylina. Astrea, &c.
4. Fam. Fungidæ.	Fungia. Stephanoseris. Agaricia, &c.

Sec. 3. *Zoantharia Perforata.*

Lamellæ of the polypary not imperforate as in the previous section, but porous or reticulated; without diaphragm.

1. Fam. Madreporidæ.	Madrepora. Turbinaria, &c.
2. Fam. Poiritidæ.	Porites.

Sec. 4. *Zoantharia Tabulata.*

Polypary composed of a well developed mural system, having the visceral chambers entirely divided into stories or compartments by a series of complete diaphragms or trans-

II. Order Alcyonaria.

Polypary having eight bipinnated tentacles, or eight perigastric lamellæ, containing the reproductive organs. Never divided into radiating longitudinal chambers.

1. Fam. Alcyonidæ.	Genera. Alcyonium. Tubipora, &c.
2. Fam. Gorgonidæ.	Gorgonia. Isis. Corallium, &c.

3. Fam. Pennatulidæ.

verse floors. Septal apparatus rudimentary, or almost wanting; never crucial.

1. Fam. Milleporidæ.	Heliolites. Plasmopora. Michelinia. Favosites. Alveolites.
2. Fam. Favositidæ.	Chaetetes. Constallaria. Halysites. Syringopora. Pocillypora, &c.

3. Fam. Seriatoporidæ. Seriatopora, &c.
4. Fam. Thecidæ. Columnaria. Thecia.

Sec. 5. *Zoantharia Tubulosa.*

Walls of the simple or compound polypary, not perforated; visceral cavity presenting neither columella, floors, nor septa, only costal striæ, not projecting from the interior of walls.

Fam. Auloporidæ.	Aulopora. Pyrgia.

Sec. 6. *Zoantharia Rugosa.*

Simple or compound polypary, deriving its septal system from *four*, not

III. Order Podactynaria.

Polyps having the gastric cavity surrounded by 4 membranous vertical septa, surmounted by four pairs of intestiniform reproductive organs. Tentacles discoid, pedunculated, not tubular, mouth proboscidiform.

Fam. Lucernaridæ.

Fam. Hydridæ.

six, primitive elements, and having the visceral chambers usually provided with floors or with vascular tissue. Reproduction fissiparous, not by gemmation.

1. Fam. Stauridæ.	Stauria, &c.
2. Fam. Cyathoxonidæ.	Cyathoxonia.
3. F. Cyathophyllidæ.	Zaphrentis. Amplexus. Cyathophyllum. Streptelasma. Acervularia. Strombodes. Lithostrotion. Axophyllum. Lonsdalia.
4. Fam. Cystiphyllidæ.	Cystiphyllum.

Sec. 7. *Zoantharia Cauliculata.*

Polyps supported on a sclerobase; polypary resembling that of Isis, Gorgonia, &c., among Alcyonaria; but distinguished by a spiny or smooth surface, while alcyonaria is marked by striæ.

Sec. 8. *Zoantharia Incertæ Sedis.*

A few genera, the exact affinities of which are somewhat undetermined, are arranged under this head.

---

*From the monograph of Milne Edwards and Jules Haime on fossil polyparies, the coral framework of the polyp.

WOODWARD'S CLASSIFICATION OF THE MOLLUSCA.

Department.	Class.	Order.	Section.	Family.	Genera.
DEPARTMENT MOLLUSKS.	Class I. *Cephalopoda.*	Order I. Dibranchiata......	Sec. A. Octopoda....... Sec. B. Decapoda.......	I. Argonautidæ. II. Octopodæ. III. Teuthidæ. IV. Belemnitidæ. V. Sepiadæ. VI. Spirulidæ.	
		Order II. Tetrabranchiata...		I. Nautilidæ. II. Orthoceratidæ. III. Ammonitidæ.	Nautilus, Lituites, Trochoceras, Clymenia. Orthoceras, Gomphoceras, Oncoceras, Phragmoceras, Cyrtoceras, Ascoceras, Gyroceras. Goniatites, Bactrites, Ceratites, Ammonites, Crioceras, Turrilites, Hamites, Ptychoceras, Baculites.
	Class II. *Gasteropoda.*	Order I. Prosobranchiata...	Sec. A. Siphonostomata. Sec. B. Holostomata.		
		Order II. Pulmonifera......	Sec. A. Inoperculata. Sec. B. Operculata.		
		Order III. Opisthobranchiata.	Sec. A. Tectibranchiata. Sec. B. Nudibranchiata.		
	Class III. *Pteropoda...*	....	A. Thecosomata. B. Gymnosomata.		
	Class IV. *Brachiopoda.*			I. Terebratulidæ. II. Spiriferidæ. III. Rhynchonellidæ. IV. Orthidæ. V. Productidæ. VI. Craniadæ. VII. Discinidæ. VIII. Lingulidæ.	Terebratula, Terebratella, Argiope, Thecidium, Stringocephalus. Spirifera, Athyris, Retzia, Uncites. Rhynchonella, Camarophoria, Pentamerus, Atrypa. Orthis, Strophomena, Leptæna, Koninckia, Davidsonia, Calceola. Producta, Aulosteges, Strophalosia, Chonetes. Crania. Discina, Siphonotreta. Lingula, Obolus.
	Class V. *Conchifera...*		Asiphonida...... Sec. A.	I. Ostreidæ. II. Aviculidæ. III. Mytilidæ. IV. Arcadæ. V. Trigoniadæ. VI. Unionidæ.	
			Siphonida Integropallialia	VII. Chamidæ. VIII. Hippuridæ. IX. Tridacnidæ. X. Cardiadæ. XI. Lucinidæ. XII. Cycladidæ. XIII. Cyprinidæ. XIV. Veneridæ.	
			Sec. B. Siphonida Sinupallialia....	XV. Mactridæ. XVI. Tellinidæ. XVII. Solenidæ. XVIII. Myacidæ. XIX. Anatinidæ. XX. Gastrochœnidæ. XXI. Pholadidæ.	
	Class VI. *Tunicata......*			I. Ascidiadæ. II. Clavellinidæ. III. Botryllidæ. IV. Pyromidæ. V. Salpidæ.	

SUGGESTIVE TABLE OF GEOLOGICAL DIVISIONS AND SUB-DIVISIONS OF TIME AND VERTICAL SPACE.

Decamyriocentenaries.	Myriocentenaries.	Kilocentenaries.	Hectocentenaries.	Decacentenaries.	Centenaries.
Hundreds of thousands of centuries.	Tens of thousands of centuries.	Thousands of centuries.	Hundreds of centuries.	Tens of centuries.	Centuries or hundreds of years.
Period.	Age.	System or Era.	Formation.	Group.	Member.
From tens to hundreds of thousands of feet.	From thousands to tens of thousands of feet.	From hundreds to thousands of feet.	From tens to hundreds of feet.	From several feet to tens of feet.	From tenths of feet to several feet.

Palæozoic, or Primary.	Quaternary. (From 500 to 1,000 feet.)	Newer Quaternary.	The result of { Growth, Accretes. Filtration, Concretes.	{ Peat bogs, jungles, &c. Coral reefs, shell-beds, &c. Shell marl, bone breccias, &c. Bog iron ore. Travertine, stalactites and stalagmites. Tufaceous and clay marls. Siliceous sinter. Saline, sulphurous, and bituminous deposits. Lacustrine silt, clay and marl. Fluviatile valley deposits and successive terraces. Loess of Rhine. Delta and estuary deposits. Submarine forests and accumulations.	
		Older Quaternary.	The result of { Washing, Subsidents. Extensive transportation, Erratics.	Newer drift of Sweden. Ossiferous caves. Moraines. [drift. Light and dark and pebbly clays of Shingle beaches, sand and gravel. Glacial bowlders of northern drift. Newer, (*700–1,000 feet). Older.	
Mesozoic, or Cainozoic, or Secondary. Tertiary. (About 7,000 feet (From 4,000 to or about one mile 5,990 feet.) and a third.)	Tertiary.	Pliocene (*3,000 feet). Miocene (*730 feet). Eocene (*1,800 feet).	Upper. Middle (*170 feet). Lower (*1,000 feet or more). Upper (*600 feet). Maestricht beds. Upper White Chalk. Lower White Chalk.	
		Cretaceous (†1,100 ft.)	Upper cretaceous. Lower cretaceous. Lower green sand.	Upper Green Sand (*100 feet.) Gault (*100 feet). Sand with limestone and schist (*70 [n.). Sand with green matter (*100 feet.) Kentish Rag (*60–90 feet).	

Agrees or Stratified Rocks, Comprising the entire *Sedimentary Deposits.*

Assuming the thickness of the Azoic or non-fossiliferous rocks at about thirteen miles, and the total Zoic or fossiliferous ages being, as seen in this table, nearly the same, we have an aggregate of about twenty-six miles, (or nearly the distance between the Equatorial and Polar diameters;) for the entire thickness of the Stratified Rocks, resting on an equal or probably greater thickness of Igneous or Unstratified Rocks.

SUGGESTIVE TABLE OF GEOLOGICAL DIVISIONS AND SUB-DIVISIONS OF TIME AND VERTICAL SPACE.—CONTINUED.

Decamyriocentenaries.—Continued.	Myriocentenaries.—Continued.	Kilocentenaries.—Continued.	Hectocentenaries.—Continued.	Decacentenaries.—Continued.	Centenaries.—Continued.
Mesozoic, or Secondary. (About 7,000 feet, or about one mile and a third.)		Wealden (†900–†1,500 ft.)	Weald clay (†140–280 feet.) Hastings sand (†400–1,000 ft.)	Purbeck beds. Portland beds. Kimmeridge clay. Coral Rag. Oxford clay. Great or Bath Oolite. Inferior Oolite.	
		Jurassic or Oolitic (†1,250 feet.)	Upper Oolite (†400 feet.) Middle Oolite (†450 feet.) Lower Oolite (†400 feet.)		
		Liassic (†1,000 feet.)	Upper Lias. Lower Lias.		
		Triassic (2,500 feet.)	Upper Trias or Keuper, (†100 feet.) Lower Trias or Muschelkalk, (100 to 500 feet.) Lower Trias or Bunter Sandstein, (†1,000 feet.)		
Palæozoic, or Primary Fossiliferous. (From about 25,000 feet to about 35,000 feet, or nearly 5 to over 10 miles.)		Permian, or Lower New Red or Magnesian Limestone (900 to 1,000 feet.)	Upper Permian. Lower Permian.		
		Carboniferous, (1,750 to 6,500 feet.)	Coal measures, (†1,000 to †3,000 feet.)	High or barren coal group.	Sandstone, shales and thin limestone. Nos. 12, 13 and 14, coals of Ky. report. Anvil Rock Sandstone. No. 11 coal, a light fire-clay. Shales and Sandstone. No. 9 coal, and dark fire-clay. Marls and shales and limestone. Nos. 6, 7 and 8 coal. Thick sandstone. Limestone and shales. No. 5 coal. Mahoning Sandstone. No. 4 coal. Limestone and shales. Nos. 3, 2 and 1 coal.
				Middle coal group.	
				Low coal group.	
			Millstone grit, (†300 to †1,500 feet.)	Pebbly Sandstone conglomerate.	Whetstone grit. Lepidodendron beds. Thin coals, sometimes workable. Gritstone grit. Thin limestone.
				Sandstone grit.	

AGGREGATE OR STRATIFIED ROCKS, Comprising the entire *Sedimentary Deposits.*

Assuming the thickness of the Azoic or non-fossiliferous rocks at about thirteen miles, and the total Zoic or fossiliferous ages being, as seen in this table, nearly the same, we have an aggregate of about twenty-six miles, (or nearly the difference between the Equatorial and Polar diameters,) for the entire thickness of the Stratified Rocks, resting on an equal or probably much greater thickness of Igneous or Unstratified Rocks.

		Shaly Sandstone.
		Upper Encrinital and Bryozoic, often Oolitic limestone.
		Schist and red clay beds.
		Middle Encrinital and Lithostrotion limestone, also sometimes Oolitic.
	Sub-carboniferous or mountain limestone, (350 to 450 ft.) cavernous.	Shales.
Lower carboniferous formation, *400 to *2,000 feet.		Magnesian limestone.
		Lower Encrinital and brachiopodous limestone.
		Ironstone passing to shaly sandstone.
	Sub-carboniferous sandstone (300 to 500 feet).	Shales, aluminous and silicious.
		Waverly sandstone.
Upper Devonian.	Catskill group of N.Y. (13,000 ft.)	
	Chemung group of N.Y. (1500 ft.)	Tully limestone, (*20 feet.)
	Portage group of N.Y. (1000 ft.)	Marcellus shale, (*50 feet.)
	Genesee slate, N.Y. (250 ft.)	Corniferous limestone, (*14 feet.)
Lower Devonian.	Hamilton group of N.Y. (11,000 ft.)	Onondago limestone, (*14 feet.)
	Upper Helderberg limestone.	Schoharie grit.
	Cauda-galli grit.	Upper Pentamerous limestone.
	Oriskany sandstone, (?700 ft.)	Encrinital limestone.
Upper Silurian or Ludlow and Wenlock, *4,000 feet.	Lower Helderberg limestones.	Delthyris shaly limestone, (*500 feet.)
		Pentamerus galeatus limestone, *50 ft.
	Tentaculite limestone (1100 ft.).	
	Onondaga salt group, (1000–1,000 ft.)	
	Niagara group, (*264 ft.)	
Middle or Caradoc, *2,000 ft.	Clinton group. (50 ft.)	
	Medina sandstone, (350 ft.)	
	Oneida conglomerate, (500 ft.)	
	Gray sandstone.	
	Hudson River group, (?1700 ft.)	
Lower Silurian or Llandeilo, ?3,500 to *20,000 feet.	Utica slate, (1100 ft.)	
	Trenton limestone, (1400 ft.)	
	Black River limestone.	
	Bird's-eye limestone, (1101 ft.)	
	Chazy limestone.	
	Calciferous sandstone, (1300 ft.)	
Upper { Arenig slates, *7,000 ft. Tremadoc slates, *1,000 feet. Lingula Flagz, of North Wales, *1,500–2,000 ft.	Potsdam sandstone, (*300 ft.)	
	Wisconsin and Minnesota sandstones and Magnesian limestones.	
Lower { Barlic grits, *500 ft. Llanberis slates, *1,000 feet.	Primordial Zone of Bohemia (*1,200 feet.	
	Bangor group.—Wicklow group, Ireland.	

Devonian or Old Red sandstone, *1200 to *10,000 feet, ?7,000 feet.

Silurian, ?7,500 to *25,000 feet.

Cambrian, ?10,000 to ?25,000 feet.

Azoic, or Metamorphic. Thickness unknown, probably greater than that of the entire series of ages, say therefore 70,000 ft., or about 13 miles.

Clay slates.
Quartzites.
Hornblende Schists.
Mica Schists.
Talcose and Chloritic Slates.
Hypogene limestone.
Gneiss.

NOTE.—The numbers marked with a * are given on the authority of Sir C. Lyell, in his "Elements," edition of 1856; those marked with a † on the authority of Mr. Trimmer, 1843; ‖ Prof. Sedgwick; ‡ Hitchcock's "Geology of the Globe;" § Prof. Hall.

ANALYSIS OF THE LOUISVILLE AND JEFFERSONVILLE HYDRAULIC LIMESTONE.*

BY DR. PETER, SEE PAGE 220, SECOND VOLUME, KENTUCKY REPORT.

Hydraulic Limestone, (unburnt,) "*from the Falls of the Ohio River, at Louisville, Jefferson County, Kentucky.*"

A greenish-gray, dull, fine granular limestone, adheres slightly to the tongue; powder light gray. Composition dried at 212° F.

Carbonate of Lime,	50.43	28.29 Lime.	
Carbonate of Magnesia	18.67	8.89 Magnesia.	
Alumina and oxides of Iron and Manganese	2.93		
Phosphoric acid	.06		
Sulphuric acid	1.58		
Potash	.32	Silica	22.58
Soda	.13	Alumina, colored with oxide of iron	2.88
Silica and insoluble silicates	25.78	Lime, Magnesia and Loss	.32
Loss	.10		25.78
	100.00		

The air-dried rock lost .70 per cent. of moisture at 212° F.

ANALYSIS OF A HYDRAULIC LIMESTONE FROM PENDLETON, MADISON CO., IND.
BY R. OWEN.

Moisture expelled from one gramme at 300° F	0.0085
Insoluble silicates	0.3400
Peroxide of iron, and alumina	0.0020
Carbonate of lime	0.1340
Carbonate of Magnesia	0.2409
Alkalies, loss, &c., undetermined	0.2746
	1.0000

ANALYSIS OF A HYDRAULIC LIMESTONE FROM LOGANSPORT, CASS CO., INDIANA.
BY R. OWEN.

Moisture expelled from one gramme at 300° F	0.0030
Insoluble silicates	0.0215
Peroxide of iron, and alumina	0.0360
Carbonate of lime	0.1840
Carbonate of Magnesia	0.2750
Alkalies, loss, &c., undetermined	0.4805
	1.0000

*This limestone is exactly of the same character on the two sides of the river; see page 107 of this report.

DESCRIPTION OF FOSSILS.*

No. 1. *Syphonia digitata.*—Formerly named and described by Dr. D. D. Owen, in the Kentucky Geological Report, but never figured. The cut represents only a part of the body of this amorphozoon, and is one third the natural size. Usually the body is spheroidal with seven or nine points of attachment, three of which, broken, are represented in the cut. The central elevation appears to have been tubular, and the substance made up of cellular tissue, containing probably calcareous spicules, constituting altogether a type of the simplest form of animal life, similar to other spongy Sarcodans, such as our present sponge of commerce. This fossil occurs abundantly in the Lower Silurian, near Frankfort, Ky., and will probably be found in portions of Indiana with a similar geological horizon. Although extra limital, this interesting specimen of the first dawn of animal existence was thought worthy of being lignographed.

No. 2. *Halysites sexto-catenatus.*—This is perhaps only a variety of the H. catenularia or H. escharoides; but seems to differ from the other chain corals in having each individual polypary more uniformly enclosed by *six-sided* vertical sutures than in any other species; this suggested the specific name. This may, however, not be a sufficiently constant or distinctive form of the mural system, to constitute a new species, as we know that a mass of spherical semi-solid bodies exposed to considerable lateral pressure is finally disposed to assume the hexagonal form. The diaphragms or floors are 1-16 of an inch apart. This fossil is from the Upper Silurian of Huntington county, Indiana.

No. 3. *Bucania euomphaloides.*—This closely resembles some species of the genus Bucania employed by Prof. Hall to describe certain gasteropod mollusks similar to the Bellerophon. The specimen is not quite perfect enough to determine all its characters, but its deep umbilication suggested a specific name showing its resemblance to the Euomphalus.

No. 4. *Gyroceras rhombolinearis.*—Unfortunately the exact position of the siphuncles can not be determined from this specimen, but the external markings of ridges produced by the septal apparatus, and crossed by an occasional increase or thickening of the shell into ridges have produced diamond shaped figures so mathematically regular as sometimes to constitute a perfect rhombus, at others a rhomboid, hence the name.

No. 5 Represents *Columnaria inequalis*, (Hall,) a fine specimen, the columns radiating with a regular and rapid divergence. The specimen was found at Peru, Miami county, Indiana.

---

* It was designed to furnish many more wood cuts and to give a full description of fossils, as well as a methodical list of all the fossils hitherto found in the State; but the want of time now prevents this. As some apology for the meagreness of the descriptions actually furnished perhaps it may be admissible to remark that when finally called upon to finish this part of my duty, I was at a distance from specimens, figures, and works essential for comparison, and was occupied with important military duties, which claimed priority over a task that would otherwise have been a labor of love.   R. OWEN.

CAMP MORTON, March 8, 1862.

SILURIAN FOSSILS.

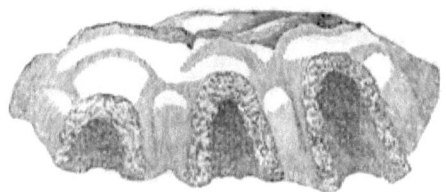

No. 1.

No. 2.

No. 3.

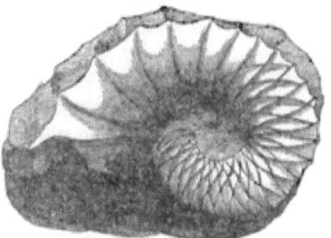

No. 4.

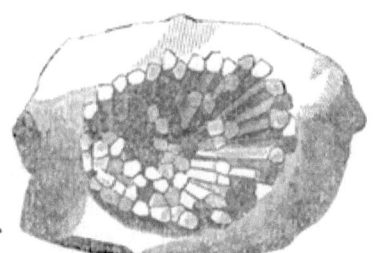

No. 5.

No. 6. *Ceriopora lyra*, formerly described by Owen and Norwood from specimens found in Kentucky, was given to the engraver for comparison. It was designed that he should figure a fine species found four miles west of Fredonia, and which, from the constant, triangular form of the ramifications which support the net work is named Ceriopora tricarinata, (R. Owen.) The lacunose apertures are small and sub-round, the spaces between the apertures are usually double the diameter of the pits.

No. 7. *Lithostrotion Canadense* is from the neighborhood of Lost River, Orange county, Indiana, and is figured to show that in the same aggregate polypary or coral-community the individual polyparies (polyparites) are sometimes round, sometimes hexagonal, depending probably upon external circumstances of lateral pressure, &c., while yet plastic.

No. 8. The engraver here misunderstood the instructions and gave prominence to the crinoid already beautifully figured in Prof. Hall's Iowa Report, showing only in a subordinate manner the gasteropod, which seems to have formed its chief food, and which it was the intention to describe. So invariably is it found partly swalled by the crinoid that I selected for it the name *Pileopsis pabulocrinus*, (R. Owen) to indicate that this Pileopsis furnished food for Crinoids. Some of the detached specimens are quite large, measuring over two inches across the base, and nearly three inches from the apex, following the convexity of the curve to the base. Those partly swallowed are, however, not over half that size. In the neighborhood of Crawfordsville, Montgomery county, where this Capulus or Pileopsis is found, a very large per centage of the Crinoids from the same locality present the appearance of a protuberant Pileopsis, more or less drawn into the mouth of the Croinoid. The locality of these fossils is not far from the junction of the sub-carboniferous limestone and sandstone.

No. 9. *Conularia Crawfordsvillensis*, (R. Owen).—Associated with the above fossils described in No. 8 are found somewhat abundantly, pteropods of the genus Conularia. The specimen furnished the engraver is somewhat crushed, but the markings are more curved than in the lignograph and are half as numerous again as represented. The sub-carboniferous limestone of Crawfordsville is remarkable rich is fossils, and may develop other species of Conularia besides the one which we have specifically denominated from its locality.

Nos. 10 and 11 represent a fucoid found in the sandstones between the upper and middle pentremital limestones, about 150 feet below the whetstone quarry of Mr. T. Powel, in Orange county. From its star-like regularity the name of fucoides asteroides is proposed.

CARBONIFEROUS FOSSILS.

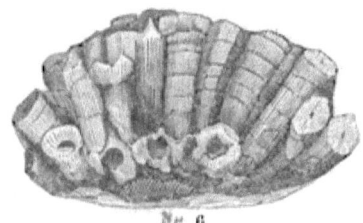

No. 6.

No. 7.

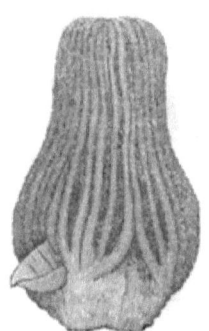

No. 8.

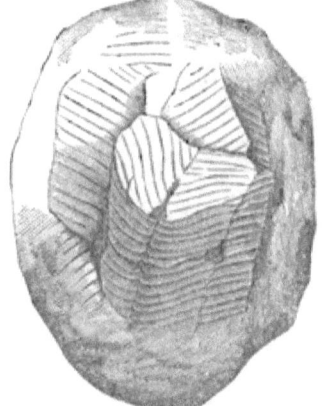

No. 9.

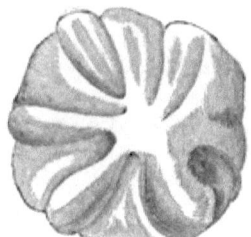

No. 10.

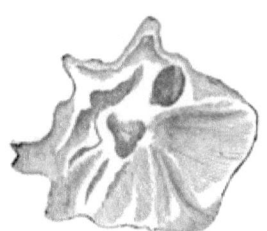

11.

ERRATA.

On page 14, in the Mesozoic part of the table, for "carboniferous" read "cretaceous;" for "palæozotic" read "palæozoic."

On page 20, on 16th line from the the top, for "Shropsbine" read "Shropshire."

On page 21, on 9th line from bottom, for "United Seates" read "United States."

On page 27, on 13th line from top, for "opportunies" read "opportunities."

On page 39, on 18th line from top, for "Protarœa" read "Protarœa."

On page 40, on 16th line from top, for "vertebrate" read "vertebrale."

On page 40, on 25th line from top, for "Cy-therina" read "Cytherina."

On page 44, on 2d line from bottom, for "populu" read "populus."

On page 45, on last line, a foot note, for "Poe" read "Poa."

On page 47, at two places in section, for "Domolite" read "Dolomite."

On page 47, in section, for "silicolus" read "silicious."

On page 48, on 7th line from top, for "inconvenienceon" read "inconvenience, on;" on 4th line, "impevious" read "impervious."

On page 60, on 10th line from top, for "limestones" read "limestone;" 7th line from bottom, for "exists" read "exist."

On page 62, after "Halysites sexto-catenatus," read (R. Owen.)

On page 62, on 12th line from bottom, for "inæqualis" read "inequalis."

On page 63, on 20th line from top, for "Bumastus" read "Bumastis."

On page 63, on 3d line from top, for "Peteramorus" read "Pentamerus."

On page 64, last line, for "their growth by" read "by their growth."

On page 66, near bottom, for "cheet" read "chert."

On page 67, on 3d line, for "zinc blede" read "zinc blende."

On page 68, on 7th line from bottom, for "anticlinal" read "anticlinal."

On page 74, on 6th line from top, for "inlicates" read "silicates."

On page 81, on 6th line from bottom, for "Acelapias" read "Asclepias."

On page 92, on 7th line from top, for "problematium," read "problematicum."

On page 105, on 10th line from top, for "rotalorius" read "rotatorius."

On page 106, last line, for "Amplexus yandelli," read "A. Yandelli."

On page 125, on 16th line from bottom, for "limestones" read "limestone.",

On page 126, on 10th line from top, for "Lithostration" read "Lithostrotion."

On page 126, on 8th line from bottom, for "Archimeidpora" read "Archimedipora."

On page 127, in section, for "Bluer" read "Blue," for "Lithostroion" read "Lithostrotion," and after "locally lithographic" omit the period.

On page 131, last line, after "Conularia Crawfordsvillensis" add (R. Owen.)

On page 132, first line, after "Pileopsis pabulocrinus,' add (R. Owen.)

On page 140, in section, for "penremital" read "pentremital," for "eucrinital" read "encrinital."

On page 144, on 13th line from bottom, for "stigmarciæ" read "stigmariæ."
On page 146, on 4th line from top, for "pentrinital" read "pentremital."
On page 148, on 14th line from top, for "experince" read "experience."
On page 150, on 21st line from top, for "deseription" read "description."
On page 153, on 14th line from top, for "denominate" read "denominated."
On page 158, on 16th line from bottom, for "Lousiville" read "Louisville."
On page 162, on 16th line from bottom, for "Fusulima" read "Fusulina."
On page 164, on 18th line from bottom, for "trunk" read "trucks."
On page 169, on 19th line from bottom, for "cooking" read "coking."
On page 170, on 6th line from top, for "cooking" read "coking."
On page 170, on 9th line from bottom, for "drying" read "dying."
On page 174, on 8th line from top, for "points" read "paints."
On page 177, on 11th line from top, for "sold" read "solid."
On page 179, on 14th line from top, for "draing" read "draining."
On page 181, in three places, for "Potoka" read "Patoka."
On page 183, on 5th line from top, for "one" read "ore," on 15th line, for "lesqui" read "sesqui."
On page 185, on 10th line from top, for "lithographical" read "topographical."
On page 185, on 19th line from top, for "shows" read "show."
On page 190, on 10th line from bottom, after "land" insert "and."
On page 191, in section, for "Owen" read "Owen's."
On page 191, on 13th line from bottom, for "deundation" read "denudation."
On page 193, on 11th line from top, after "well" omit the word "as."
On page 194, on 10th line from top, for "gloosed" read "glossed."
On page 199, on 14th line from top, for "distinguished" read "distinguish."
On page 201, on 13th line from top, for "Gosport" read "Gossert."
On page 201, on 13th line from bottom, for "nervures" read "nervure."
On page 205, on 6th line from top, after "Valparaiso" insert "we find."
On page 206, on 9th line from bottom, for "Orange" read "Osage."
On page 212, on 17th line from top, for "grapes" read "gasses."
On page 216, on 9th line from top, after "Rychænus" omit the comma.
On page 219, on 2d line from top, for "of" read "is."
On page 219, on 13th line from top, for "distant" read "distance."
On page 229, on 20th line from top, for "cat-flag" read "cat-tail flag."
On page 232, in foot note, for "rises" read "raise," and after "filters" insert "it."
On page 239, on 18th line from top, for "board" read "bord."
On page 240, in foot note, for "usefully" read "useful."
On page 247, last line, for "vegetable which" read "vegetable, they."
On page 252, last line, for "productions" read "productiveness."
On page 256, on 6th line from bottom, for "amount" read "amounts."
On page 261, last line, for "entrusted" read "extracted," and for "digestion" read "digesting."
On page 264, on 4th line from top, for "returning" read "retained."
On page 282, on 10th line from bottom, for "called in" read "called on."
On page 286, on 14th line from bottom, for "pecolating" read "percolating."
On page 287, 1st line, for "returing" read "returning."
On page 296, near top of section, for "coal No. B" read "coal No. 18."
On page 311, on 14th line from bottom, for "coal No. 113" read "coal No. 1 B."
On page 323, 6th line from top, for "recalled" read "referred."